MACHINE VISION – APPLICATIONS AND SYSTEMS

Edited by **Fabio Solari, Manuela Chessa** and **Silvio P. Sabatini**

Machine Vision – Applications and Systems
Edited by Fabio Solari, Manuela Chessa and Silvio P. Sabatini

Published by InTech
Janeza Trdine 9, 51000 Rijeka, Croatia

Publishing Process Manager Martina Blecic
Technical Editor Teodora Smiljanic
Cover Designer InTech Design Team

Additional hard copies can be obtained from orders@intechopen.com

Machine Vision – Applications and Systems, Edited by Fabio Solari, Manuela Chessa and Silvio P. Sabatini
p. cm.
ISBN 978-953-51-0373-8

Meet the editors

Fabio Solari received the Laurea degree in Electronic Engineering in 1995 and the Ph.D. in Computer Science in 1999 at the University of Genoa, Italy. Since 2005 he has been appointed as Assistant Professor of Computer Science at the University of Genoa. His research activity concerns the study of visual perception with the aim to develop computational models of cortical vision processing, to devise novel bio-inspired computer vision algorithms, and to design virtual reality environments for ecological visual stimulations. In particular, his research interests are related to neuromorphic architectures of visual cortex, space-variant visual processing, motion and depth estimation and interpretation, active vision systems, and augmented reality systems for the visuo-motor coordination in the peripersonal space.

Manuela Chessa is a Postodoctoral Research scientist at the University of Genoa, Italy. She received her MSc in Bioengineering from the University of Genoa in 2005, and the Ph.D. in Bioengineering from University of Genoa in 2009. She has been working in the PSPC Lab since 2005, and her research interests are focused on the study of biological and artificial vision systems, on the development of bioinspired models for the estimation of optic flow and disparity, on the study of the interplay existing between vision and motion control in the peripersonal space, and on the development of virtual and augmented reality system for the study of the perception of tridimensionality.

Silvio P. Sabatini received the Laurea Degree in Electronics Engineering and the Ph.D. in Computer Science from the University of Genoa in 1992 and 1996. He is currently Associate Professor of Bioengineering at the Department of Informatics, Bioengineering, Robotics and Systems of the University of Genoa. In 1995 he promoted the creation of the "Physical Structure of Perception and Computation" (PSPC) Lab to develop models that capture the "physicalist" nature of the information processing occurring in the visual cortex, to understand the signal processing strategies adopted by the brain, and to build novel algorithms and architectures for artificial perception machines. His research interests relate to visual coding and multidimensional signal representation, early-cognitive models for visually-guided behavior, and robot vision.

Contents

Preface

Vision plays a fundamental role for living beings by allowing them to interact with the environment in an effective and efficient way. The Machine Vision goal is to endow computing devices, and more generally artificial systems, with visual capabilities in order to cope with not a priori predetermined situations. To this end, we have to take into account the computing constraints of the hosting architectures and the specifications of the tasks to be accomplished. These elements lead a continuous adaptation and optimization of the usual visual processing techniques, such as ones developed in Computer Vision and Image Processing. Nevertheless, the fast development of off-the-shelf processors and computing devices made available to the public a large and low-cost computational power. By exploiting this contingency, the Vision Research community is now ready to develop real-time vision systems designed to analyze the richness of the visual signal online with the evolution of complex real-world situations at an affordable cost. Thus the application field of Machine Vision is not more limited to the industrial environments, where the situations are simplified and well known and the tasks are very specific, but nowadays it can efficiently support system solutions of everyday life problems.

This book will focus on both the engineering and technological aspects related to visual processing.

The first four chapters describe solutions related to the recovery of depth information in order to solve video surveillance problems and an helicopter landing task (Chp.1 and Chp. 2, respectively), and to propose a high speed calculation of depth maps from stereo images based on FPGAs (Chp. 3) and a Time-of-Flight sensor as an alternative to stereo video camera (Chp. 4). The next three chapters address typical industrial situations: an approach for robust rotation angle estimation for textures alignment is described in Chp. 5, the characterization of the surface finish of machined parts through Hough transform is addressed in Chp. 6 and through structured light patterns in Chp. 7. A new algorithm based on ensemble of trees for object localization and 3D pose estimation that works for piled parts is proposed in Chp. 8. The detection of geometric shapes like ellipses from real images and a theoretical framework for characterization and identification of a discontinuity are addressed in Chp. 9 and

Chp.10, respectively. The automated visual inspection improvement due to reflectance measuring and modeling in the context of image analysis and synthesis is presented in Chp. 11. The last chapter addresses an analysis of different computing paradigms and platforms oriented to image processing

Fabio Solari, Manuela Chessa and Silvio P. Sabatini
University of Genoa
Italy

1

Bio-Inspired Active Vision Paradigms in Surveillance Applications

Mauricio Vanegas, Manuela Chessa, Fabio Solari and Silvio Sabatini
The Physical Structure of Perception and Computation - Group, University of Genoa
Italy

1. Introduction

Visual perception was described by Marr (1982) as the processing of visual stimuli through three hierarchical levels of computation. In the first level or *low-level* vision it is performed the extraction of fundamental components of the observed scene such as edges, corners, flow vectors and binocular disparity. In the second level or *medium-level* vision it is performed the recognition of objects (e.g. model matching and tracking). Finally, in the third level or *high-level* vision it is performed the interpretation of the scene. A complementary view is presented in (Ratha & Jain, 1999; Weems, 1991); by contrast, the processing of visual stimuli is analysed under the perspective developed by Marr (1982) but emphasising how much data is being processed and what is the complexity of the operators used at each level. Hence, the low-level vision is characterised by large amount of data, small neighbourhood data access, and simple operators; the medium-level vision is characterised by small neighbourhood data access, reduced amount of data, and complex operators; and the high-level vision is defined by non-local data access, small amount of data, and complex relational algorithms. Bearing in mind the different processing levels and their specific characteristics, it is plausible to describe a computer vision system as a modular framework in which the low-level vision processes can be implemented by using parallel processing engines like GPUs and FPGAs to exploit the data locality and the simple algorithmic operations of the models; and the medium and high-level vision processes can be implemented by using CPUs in order to take full advantage of the straightforward fashion of programming these kind of devices.

The low-level vision tasks are probably the most studied in computer vision and they are still an open research area for a great variety of well defined problems. In particular, the estimation of optic flow and of binocular disparity have earned special attention because of their applicability in segmentation and tracking. On the one hand, the stereo information has been proposed as a useful cue to overcome some of the issues inherent to robust pedestrian detection (Zhao & Thorpe, 2000), to segment the foreground from background layers (Kolmogorov et al., 2005), and to perform tracking (Harville, 2004). On the other hand, the optic flow is commonly used as a robust feature in motion-based segmentation and tracking (Andrade et al., 2006; Yilmaz et al., 2006).

This chapter aims to describe a biological inspired video processing system for being used in video surveillance applications; the degree of similarity between the proposed framework and the human visual system allows us to take full advantage of both optic flow and disparity estimations not only for tracking and fixation in depth but also for scene segmentation. The most relevant aspect in the proposed framework is its hardware and software modularity. The proposed system integrates three cameras (see Fig. 1); two active cameras with variable-focal-length lenses (binocular system) and a third fixed camera with a wide-angle lens. This system has been designed to be compatible with the well-known iCub robot interface[1]. The cameras movement control, as well as the zoom and iris control run on an embedded computer PC/104. The optic flow and the disparity algorithms run on a desktop computer equipped with a processor *Intel Core 2 Quad @ 2.40GHz* and a memory RAM of about 8 GB. All system components, namely the desktop computer, the embedded computer PC/104, and the cameras, are connected in a gigabit Ethernet network through which they can interact as a distributed system.

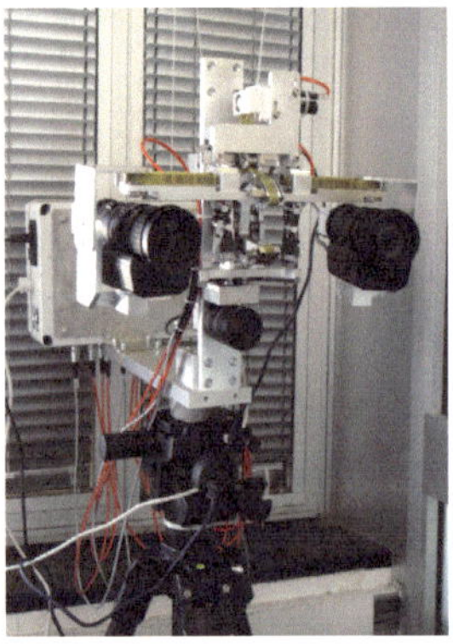

Fig. 1. Trinocular robotic head with 5 degrees of freedom, namely a common tilt movement, and independent zoom-pan movements for left and right cameras, respectively.

The general features of the moving platform are compiled in Table 1. Likewise, the optic features of the cameras are collected in Table 2. Lastly, it is important to mention that the binocular system has a baseline of 30 cm.

Features	**Pan Movement**	**Tilt Movement**
Limits:	$\pm 30^\circ$ (Software limit)	$\pm 60^\circ$ (Software limit)
Acceleration:	$5100^\circ/sec^2$	$2100^\circ/sec^2$
Max. Speed:	$330^\circ/sec$	$73^\circ/sec$
Resolution:	0.03°	0.007°
Optical Encoder:	512 pulses/revolution	512 pulses/revolution
Motor Voltage:	12 V	12 V
Gear Ratio:	1:80	1:80
Motor Torque:	0.59 Nm	0.59 Nm

Table 1. General features of the moving platform.

Most of the video surveillance systems are networks of cameras for a proper coverage of wide areas. These networks use both fixed or active cameras, or even a combination of both, placed

[1] The iCub is the humanoid robot developed as part of the EU project RobotCub and subsequently adopted by more than 20 laboratories worldwide (see `http://www.icub.org/`).

Features	Active Cameras	Fixed Camera
Resolution:	11392 x 1040 pixels	1624 x 1236 pixels
Sensor Area:	6.4 x 4.8 mm	7.1 x 5.4 mm
Pixel Size:	4.65 x 4.65 μm	4.4 x 4.4 μm
Focal Length:	7.3 $\sim$ 117 mm, FOV 47° $\sim$ 3°	4.8 mm, FOV 73°

Table 2. Optic features of the cameras.

at not predetermined positions to strategically cover a wide area; the term *active* specifies the camera's ability of changing both the angular position and the field of view. The type of cameras used in the network has inspired different calibration processes to find automatically both the intrinsic and extrinsic camera parameters. In this regard, Lee et al. (2000) proposed a method to estimate the 3D positions and orientations of fixed cameras, and the ground plane in a global reference frame which lets the multiple cameras views to be aligned into a single planar coordinate frame; this method assume approximate values for intrinsic cameras parameters and it is based on overlapped cameras views; however, others calibration methods have been proposed for non-overlapped cameras views (i.e. Kumar et al., 2008). In the case of active cameras, Tsai (1987) has developed a method for estimating both the matrices of rotation and translation in the Cartesian reference frame, and the intrinsic parameters of the cameras. In addition to the calibration methods, the current surveillance systems must deal with the segmentation and identification of complex scenes in order to characterise them and thus to obtain a classification which let the system to recognise unusual behaviours into the scene. In this regard, a large variety of algorithms have been developed to detect changes in scene; for example the application of a threshold to the absolute difference between pixel intensities of two consecutive frames can lead to the identification of moving objects, some methods for the threshold selection are described in (Kapur et al., 1985; Otsu, 1979; Ridler & Calvar, 1978). Other examples are the adaptive background subtraction to detect moving foreground objects (Stauffer & Grimson, 1999; 2000) and the estimation of optic flow (Barron et al., 1994). Our proposal differs the most of the current surveillance systems in at least three aspects: (1) the use of a single camera with a wide-angle lens to cover vast areas and a binocular system for tracking areas of interest at different fields of view (the wide-angle camera is used as the reference frame), (2) the estimation of both optic flow and binocular disparity for segmenting the images; this system feature can provide useful information for disambiguating occlusions in dynamic scenarios, and (3) the use of a bio-inspired fixation strategy which lets the system to fixate areas of interest, accurately.

In order to explain the system behaviour, two different perspectives were described. On the one hand, we present the system as a bio-inspired mathematical model of the primary visual cortex (see section 2); from this viewpoint, we developed a low-level vision architecture for estimating optic flow and binocular disparity. On the other hand, we describe the geometry of the cameras position in order to derive the equations that govern the movement of the cameras (see section 3). Once the system is completely described, we define an angular-position control capable of changing the viewpoint of the binocular system by using disparity measures in section 4. An interesting case study is described in section 5 where both disparity and optic flow are used to segment images. Finally, in section 6, we present and discuss the system's performance results.

2. The system: a low-level vision approach

The visual cortex is the largest, and probably the most studied part of the human brain. The visual cortex is responsible for the processing of visual stimuli impinging on the retinas. As a matter of fact, the first stage of processing takes place in the lateral geniculate nucleus (LGN) and then the neurons of the LGN relay the visual information to the primary visual cortex (V1). Then, the visual information flow hierarchically to areas V2, V3, V4 and V5/MT where visual perception gradually takes place.

The experiments carried out by Hubel & Wiesel (1968) proved that the primary visual cortex (V1) consists of cells responsive to different kinds of spatiotemporal features of the visual information. The apparent complexity with which the brain extracts the spatiotemporal features has been clearly explained by Adelson & Bergen (1991). The light filling a region of space contains information about the objects in that space; in this regard, they proposed the *plenoptic function* to describe mathematically the pattern of light rays collected by a vision system. By definition, the plenoptic function describes the state of luminous environment, thus the task of the visual system is to extract structural elements from it.

Structural elements of the plenoptic function can be described as oriented patterns in the plenoptic space, and the primary cortex can be interpreted as a set of local, Fourier or Gabor operators used to characterise the plenoptic function in the spatiotemporal and frequency domains.

2.1 Neuromorphic paradigms for visual processing

Mathematically speaking, the extraction of the most important aspects of the plenoptic function can emulate perfectly the neuronal processing of the primary visual cortex (V1). More precisely, qualities or elements of the visual input can be estimated by applying a set of low order directional derivatives at the sample points; the so obtained measures represent the amount of a particular type of local structure. To effectively characterise a function within a neighbourhood, it is necessary to work with the local average derivative or, in an equivalent form, with the oriented linear filters in the function hyperplanes. Consequently, the neurons in V1 can be interpreted as a set of oriented linear filters whose outputs can be combined to obtain more complex feature detectors or, what is the same, more complex receptive fields. The combination of linear filters allow us to measure the magnitude of local changes within a specific region, without specifying the exact location or spatial structure. The receptive fields of complex neurons have been modelled as the sum of the squared responses of two linear receptive fields that differ just in phase for 90° (Adelson & Bergen, 1985); as a result, the receptive fields of complex cells provide *local energy measures*.

2.2 Neural Architecture to estimate optic flow and binocular disparity

The combination of receptive fields oriented in space-time can be used to compute local energy measures for optic flow (Adelson & Bergen, 1985). Analogously, by combining the outputs of spatial receptive fields it is possible to compute local energy measures for binocular disparity (Fleet et al., 1996; Ohzawa et al., 1990). On this ground, it has been recently proposed a neural architecture for the computation of horizontal and vertical disparities and optic flow (Chessa, Sabatini & Solari, 2009). Structurally, the architecture comprises four processing stages (see

Fig. 2): the distributed coding of the features by means of oriented filters that resemble the filtering process in area V1; the decoding process of the filter responses; the estimation of the local energy for both optic flow and binocular disparity; and the coarse-to-fine refinement.

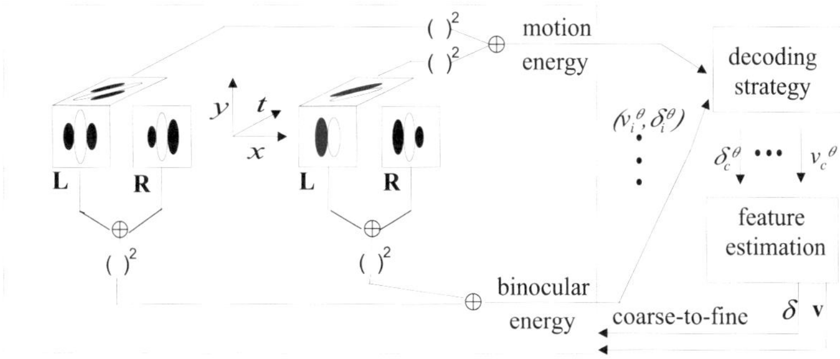

Fig. 2. The neural architecture for the computation of disparity and optic flow.

The neuronal population is composed of a set of 3D Gabor filters which are capable of uniformly covering the different spatial orientations, and of optimally sampling the spatiotemporal domain (Daugman, 1985). The linear derivative-like computation concept of the Gabor filters let the filters to have the form $h(\mathbf{x}, t) = g(\mathbf{x})f(t)$. Both spatial and temporal terms in the right term are comprised of one harmonic function and one Gaussian function. This can be easily deduced from the impulse response of the Gabor filter.

The mathematical expression of the spatial term of a 3D Gabor filter rotated by an angle θ with respect to the horizontal axis is:

$$g(x, y; \psi, \theta) = e^{\left(-\frac{x_\theta^2}{2\sigma_x^2} - \frac{y_\theta^2}{2\sigma_y^2}\right)} e^{j(\omega_0 x_\theta + \psi)}, \tag{1}$$

where $\theta \in [0, 2\pi)$ represents the spatial orientation; ω_0 and ψ are the frequency and phase of the sinusoidal modulation, respectively; the values σ_x and σ_y determine the spatial area of the filter; and (x_θ, y_θ) are the rotated spatial coordinates.

The algorithm to estimate the binocular disparity is based on a *phase-shift* model; one of the variations of this model suggests that disparity is coded by phase shifts between receptive fields of the left and right eyes whose centres are in the same retinal position (Ohzawa et al., 1990). Let the left and right receptive fields be $g^L(\mathbf{x})$ and $g^R(\mathbf{x})$, respectively; the binocular phase shift is defined by $\Delta\psi = \psi^L - \psi^R$. Each spatial orientation has a set of k receptive fields with different binocular phase shifts in order to be sensitive to different disparities ($\delta^\theta = \Delta\psi/\omega_0$); the phase shifts are uniformly distributed between $-\pi$ and π. Therefore, the left and right receptive fields are applied to a binocular image pair $I^L(\mathbf{x})$ and $I^R(\mathbf{x})$ according to the following equation:

$$Q(\mathbf{x_0}; \delta^\theta) = \int_{-\infty}^{\infty} g^L(\mathbf{x_0} - \mathbf{x}) I^L(\mathbf{x}) d\mathbf{x} + \int_{-\infty}^{\infty} g^R(\mathbf{x_0} - \mathbf{x}) I^R(\mathbf{x}) d\mathbf{x}, \tag{2}$$

so, the spatial array of binocular energy measures can be expressed as:

$$E(\mathbf{x}; \delta^\theta) = |Q(\mathbf{x}; \delta^\theta)|^2 = |Q^L(\mathbf{x}; \delta^\theta) + e^{-j\Delta\psi} Q^R(\mathbf{x}; \delta^\theta)|^2. \tag{3}$$

Likewise, the temporal term of a 3D Gabor filter is defined by:

$$f(t;\omega_t) = e^{\left(-\frac{t^2}{2\sigma_t^2}\right)} e^{j\omega_t t} 1(t), \tag{4}$$

where σ_t determines the integration window of the filter in time domain; ω_t is the frequency of the sinusoidal modulation; and $1(t)$ denotes the unit step function. Each receptive field is tuned to a specific velocity v^θ along the direction orthogonal to the spatial orientation θ. The temporal frequency is varied according to $\omega_t = v^\theta \omega_0$. Each spatial orientation has a set of receptive fields sensitive to M tuning velocities; M depends on the size of the area covered by each filter according to the Nyquist criterion.

The set of spatiotemporal receptive fields $h(\mathbf{x},t)$ is applied to an images sequence $I(\mathbf{x},t)$ according to the following equation:

$$Q(\mathbf{x_0},t;v^\theta) = \int_{-\infty}^{\infty}\int_{-\infty}^{\infty} h(\mathbf{x_0}-\mathbf{x},t-\tau)I(\mathbf{x},\tau)d\mathbf{x}d\tau, \tag{5}$$

so, the motion energy $E(\mathbf{x_0},t;v^\theta)$ equals:

$$E(\mathbf{x_0},t;v^\theta) = |Q(\mathbf{x_0},t;v^\theta)|^2 = \left| e^{j\psi(t)} \int_0^t Q(\mathbf{x_0},\tau;v^\theta)e^{-j\omega_t\tau}d\tau \right|^2. \tag{6}$$

where $\psi(t) = \psi + \omega_t t = \psi + \omega_0 v^\theta t$.

So far, we have described the process of encoding both binocular disparity and optic flow by means of a $N \times M \times K$ array of filters uniformly distributed in space domain. Now, it is necessary to extract the component velocity (v_c^θ) and the component disparity (δ_c^θ) from the local energy measures at each spatial orientation. The accuracy in the extraction of these components is strictly correlated with the number of filters used per orientation, such that precise estimations require a large number of filters; as a consequence, it is of primary importance to establish a compromise between the desired accuracy and the number of filters used or, what is the same, a compromise between accuracy and computational cost.

An affordable computational cost can be achieved by using weighted sum methods as the *maximum likelihood* proposed by Pouget et al. (2003). However, the proposed architecture uses the centre of gravity of the population activity since it has shown the best compromise between simplicity, computational cost and reliability of the estimates. Therefore, the component velocity v_c^θ is obtained by pooling cell responses over all orientations:

$$v_c^\theta(\mathbf{x_0},t) = \frac{\sum_{i=1}^{M} v_i^\theta E(\mathbf{x_0},t;v_i^\theta)}{\sum_{i=1}^{M} E(\mathbf{x_0},t;v_i^\theta)}, \tag{7}$$

where v_i^θ represent all the M tuning velocities; and $E(\mathbf{x_0},t;v_i^\theta)$ represent the motion energies at each spatial orientation. The component disparity δ_c^θ can be estimated in a similar way.

Because of the aperture problem a filter can just estimate the features which are orthogonal to the orientation of the filter. So we adopt k different binocular and M different motion receptive fields for each spatial orientation; consequently, a robust estimate for the full velocity $\mathbf{v}$ and for the full disparity $\boldsymbol{\delta}$ is achieved by combining all the estimates v_c^θ and δ_c^θ, respectively (Pauwels & Van Hulle, 2006; Theimer & Mallot, 1994).

Finally, the neural architecture uses a coarse to fine control strategy in order to increase the range of detection in both motion and disparity. The displacement features obtained at coarser levels are expanded and used to warp the images in finer levels in order to achieve a higher displacement resolution.

3. The system: a geometrical description

In the previous section we presented the system from a biological point of view. We have summarised a mathematical model of the behaviour of the primary visual cortex and we have proposed a computational architecture based on linear filters for estimating optic flow and binocular disparity. Now it is necessary to analyse the system from a geometrical point of view in order to link the visual perception to the camera movements, thus letting the system to interact with the environment.

To facilitate the reference to the cameras within this text, we are going to refer the fixed camera as *wide-angle camera*, and the cameras of the binocular system as *active cameras*. The wide-angle camera is used for a wide view of the scene, and it becomes the reference of the system. In vision research, the cyclopean point is considered the most natural centre of a binocular system (Helmholtz, 1925) and it is used to characterise stereopsis in human vision (Hansard & Horaud, 2008; Koenderink & van Doorn, 1976). By doing a similar approximation, the three-camera model uses the wide-angle-camera image as the cyclopean image of the system. In this regard, the problem statement is not trying to construct the cyclopean image from the binocular system, but using the third camera image as a reference coordinate to properly move the active cameras according to potential targets or regions of interest in the wide range scenario.

Each variable-focal-length camera can be seen as a 3DOFs pan-tilt-zoom (PTZ) camera. However, the three-camera system constraints the active cameras to share the tilt movement due to the mechanical design of the binocular framework. One of the purposes of our work is to describe the geometry of the three-camera system in order to properly move the pan-tilt-zoom cameras to fixate any object in the field of view of the wide-angle camera and thus to get both a magnified view of the target object and the depth of the scene.

We used three coordinates systems to describe the relative motion of the active cameras with respect to the wide-angle camera (see Fig. 3). The origin of each coordinate system is supposed in the focal point of each camera and the Z-axes are aligned with the optical axes of the cameras. The pan angles are measured with respect to the planes $X_L = 0$ and $X_R = 0$ respectively; note that pan angles are positive for points to the left of these planes ($X_L > 0$ or $X_R > 0$). The rotation axes for the pan movement are supposed to be parallel. The common tilt angle is measured with respect to the horizontal plane; note that the tilt angle is positive for points above the horizontal plane ($Y_L = Y_R > 0$).

The point $P(X, Y, Z)$ can be written in terms of the coordinate systems shown in Fig. 3 as follows:

$$(X, Y, Z) = (X_L, Y_L, Z_L) - O_L, \tag{8}$$
$$(X, Y, Z) = (X_R, Y_R, Z_R) - O_R, \tag{9}$$

where $O_L = (dx_L, dy_L, dz_L)$ and $O_R = (-dx_R, dy_R, dz_R)$ are the origin of the coordinate system of the left and right cameras with respect to the wide-angle camera coordinate system.

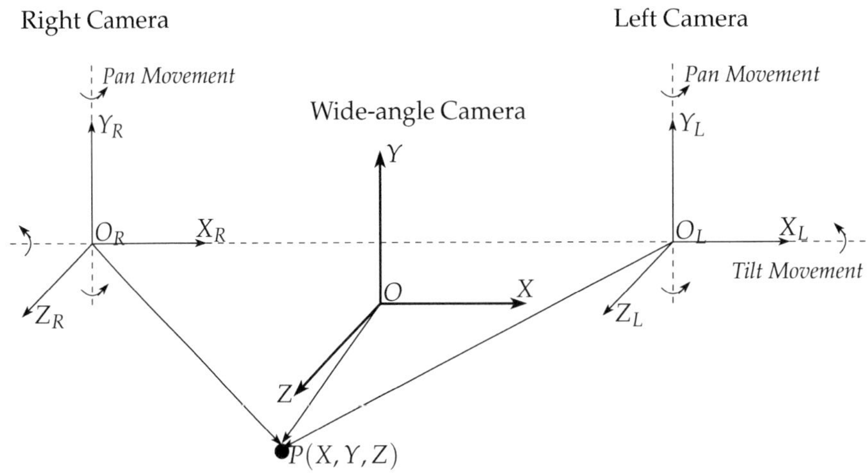

Fig. 3. The coordinate systems of the three cameras in the binocular robotic head.

It is considered f_w as the focal length of the wide-angle camera and f as the focal length of the active cameras. The Equations 8 and 9 can be written in terms of the image coordinate system of the wide-angle camera if these equations are multiplied by factor $\frac{f_w}{Z}$:

$$\frac{f_w}{Z}(X_L, Y_L, Z_L) = (x, y, f_w) + \frac{f_w}{Z}(dx_L, dy_L, dz_L), \tag{10}$$
$$\frac{f_w}{Z}(X_R, Y_R, Z_R) = (x, y, f_w) + \frac{f_w}{Z}(-dx_R, dy_R, dz_R). \tag{11}$$

Now, it is possible to link the image coordinate system of the wide-angle camera to the image coordinate system of the active cameras by multiplying the Equations 10 and 11 by the factors $\frac{f}{Z_L}$ and $\frac{f}{Z_R}$, respectively:

$$\frac{f_w}{Z}(x_L, y_L, f) = \frac{f}{Z_L}(x, y, f_w) + \frac{f_w f}{Z_L Z}(dx_L, dy_L, dz_L), \tag{12}$$
$$\frac{f_w}{Z}(x_R, y_R, f) = \frac{f}{Z_R}(x, y, f_w) + \frac{f_w f}{Z_R Z}(-dx_R, dy_R, dz_R). \tag{13}$$

Assuming that the position of the origin with respect to the Z-axis is small enough compared to the distance of the real object in the scene, it can be done the next approximation $Z \approx Z_L$ and $Z \approx Z_R$. Accordingly, the Equations 12 and 13 can be rewritten to obtain the *wide-to-active camera mapping equations* as follows:

$$(x_L, y_L, f) = \frac{f}{f_w}(x, y, f_w) + \frac{f}{Z}(dx_L, dy_L, 0), \tag{14}$$

$$(x_R, y_R, f) = \frac{f}{f_w}(x, y, f_w) + \frac{f}{Z}(-dx_R, dy_R, 0). \tag{15}$$

These equations describe the position of any point in the field of view of the wide-angle camera into the image coordinate of the active cameras.

So far, we have described the geometry of the cameras system, now the problem is to transform the wide-to-active camera mapping equations to motor stimuli in order to fixate any point in the wide-angle image. The fixation problem can be defined as the computation of the correct angular position of the motors in charge of the pan and tilt movements of the active cameras, to direct the gaze to any point in the wide-angle image. In this sense, the fixation problem is solved when the point $p(x, y)$ in the wide-angle image can be seen in the centres of the left and right camera images.

From the geometry of the trinocular head we can consider $dx_L = dx_R$, and $dy_L = dy_R$. In this way, both pan (θ_L, θ_R) and tilt (θ_y) angles of the active cameras, according to the wide-to-active camera mapping equations, can be written as:

$$\theta_L = \arctan\left(\frac{c}{f_w}x + \frac{c}{Z}dx\right), \qquad \theta_R = \arctan\left(\frac{c}{f_w}x - \frac{c}{Z}dx\right), \tag{16}$$

$$\theta_y = \arctan\left(\frac{c}{f_w}y + \frac{c}{Z}dy\right), \tag{17}$$

where c is the camera conversion factor from pixel to meters; dx, dy are the terms $dx_L = dx_R$ and $dy_L = dy_R$ in pixel units.

Bearing in mind the wide-to-active camera mapping equation, in the following section we will describe the algorithm to move the active cameras to gaze and fixate in depth any object in the field of view of the wide-angle camera.

4. Fixation in depth

Two different eyes movements can be distinguished: version movements rotate the two eyes by an equal magnitude in the same direction, whereas vergence movements rotate the two eyes in opposite direction. The vergence angle, together with version and tilt angles, uniquely describe the fixation point in the 3D space according to the Donders' law (Donders, 1969).

Fixation in depth is the coordinated eye movement to align the two retinal images in the respective foveas. Binocular depth perception has its highest resolution in the well-known Panum area, i.e. a rather small area centred on the point of fixation (Kuon & Rose, 2006). The fixation of a single point in the scene can be achieved, mainly, by vergence eye-movements which are driven by binocular disparity (Rashbass & Westheimer, 1961). It follows that the amount of disparity around the Panum area must be reduced in order to properly align the two retinal images in the respective foveas.

4.1 Defining the *Panum* area

The Panum area is normally set around the centre of uncalibrated images. This particular assumption becomes a problem in systems where the images are captured by using variable-focal-length lenses; consequently, if the centre of the image is not lying on the optical axis, then any change in the field of view will produce a misalignment of the Panum area after a fixation in depth. Lenz & Tsai (1988) were the first in proposing a calibration method to determine the image centre by changing the focal length even though no zoom lenses were available at that time. In a subsequent work (Lavest et al., 1993) have used variable-focal-length lenses for three-dimensional reconstruction and they tested the calibration method proposed by (Lenz & Tsai, 1988).

In a perspective projection geometry the parallel lines, not parallel to the image plane, appear to converge to a unique point as in the case of the two verges of a road which appear to converge in the distance; this point is known as the vanishing point. Lavest et al. (1993) used the properties of the vanishing point to demonstrate that, with a zoom lens, it is possible to estimate the intersection of the optical axis and the image plane, i.e. the image centre.

The Equation 18 is the parametric representation of a set of parallel lines defined by the direction vector $\vec{D} = (D_1, D_2, D_3)$ and parameter $t \in [-\infty, +\infty]$. The vanishing point of these parallel lines can be estimated by using the perspective projection as shown in Equation 19:

$$\begin{aligned} X(t) &= X(0) + D_1 t, \\ Y(t) &= Y(0) + D_2 t, \\ Z(t) &= Z(0) + D_3 t. \end{aligned} \tag{18}$$

$$\begin{aligned} x &= \lim_{t\to\infty} f\frac{X(t)}{Z(t)} = f\frac{D_1}{D_3}, \\ y &= \lim_{t\to\infty} f\frac{Y(t)}{Z(t)} = f\frac{D_2}{D_3}, \\ z &= \lim_{t\to\infty} f\frac{Z(t)}{Z(t)} = f. \end{aligned} \tag{19}$$

The result shown in Equation 19 demonstrates that the line passing through the optical centre of the camera and the projection of the vanishing point of the parallel lines is collinear to the director vector ($\vec{D}$) of these lines as shown below:

$$\begin{bmatrix} x \\ y \\ f \end{bmatrix} = \frac{f}{D_3} \begin{bmatrix} D_1 \\ D_2 \\ D_3 \end{bmatrix}. \tag{20}$$

According to the aforementioned equations and taking into account that, by convention, the centre of the image is the intersection of the optical axis and the image plane; it is possible to conclude that the vanishing point of a set of lines parallel to the optical axis lies in the image centre. The optical zoom can be considered as a virtual movement of the scene throughout the optical axis; in this regard, any point in the scene follows a virtual line parallel to the optical axis. This suggests that, from the tracing of two points across a set of zoomed images, it is possible to define the lines $L1$ and $L2$ (see Fig. 4) which represent the projection of these virtual lines in the image plane. It follows that the intersection of $L1$ and $L2$ corresponds with the image centre.

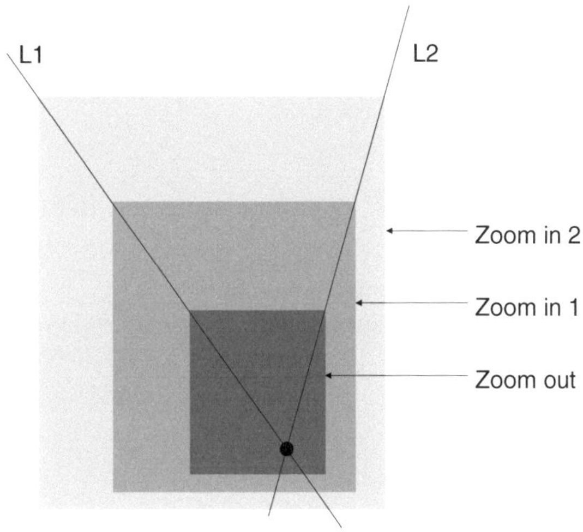

Fig. 4. Geometric determination of the image centre by using zoomed images. The intersection of the lines $L1$ and $L2$, defined by the tracing of two points across the zoomed images, corresponds with the image centre.

Once the equations of lines $L1$ and $L2$ have been estimated, it is possible to compute their intersection. Now, the Panum area is defined as a small neighbourhood around the intersection of these lines and thus it is possible to guarantee the fixation of any object even under changes in the field of view of the active cameras.

4.2 Developing the fixation-in-depth algorithm

Once the Panum area is properly defined, it is possible to develop an iterative angular-position control based on disparity estimations to fixate in depth any point in the field of view of the wide-angle camera. Fig. 5 shows a scheme of the angular-position control of the three-camera system. Any salient feature in the cyclopean image (wide-angle image) provides the point (x, y), in image coordinate, in order to set the version movement. Once the version movement is completed, the disparity estimation module can provide information about the depth of the object in the scene; this information is used to iteratively improve the alignment of the images in the active cameras.

Considering that the angular position of the cameras is known at every moment, it is possible to use the disparity information around the Panum area to approximate the scene depth; this is, a new Z in the wide-to-active camera mapping equations (see Equation 16). If we take the left image as reference, then the disparity information tells us how displaced the right image is; hence, the mean value of these disparities around the Panum area can be used to estimate the angular displacement needed to align the left and right images. As the focal length of the

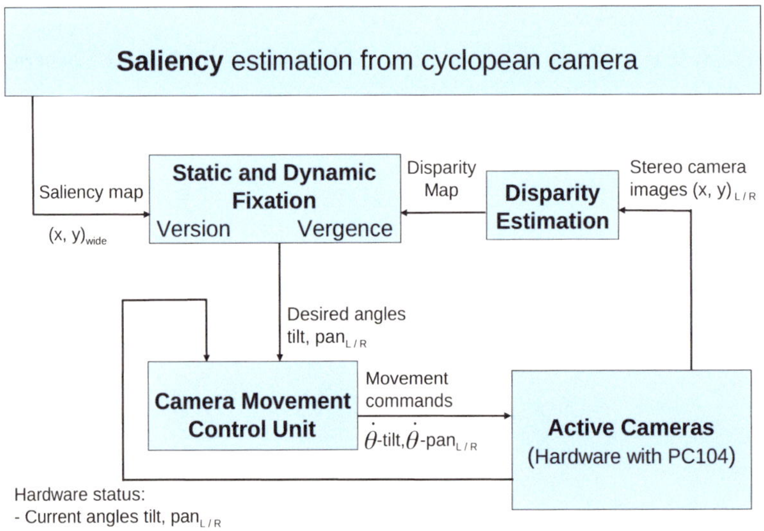

Fig. 5. Angular-position control scheme of the trinocular system.

active cameras can be approximated from the current zoom value, the angular displacement θ can be estimated as follow:

$$\theta = \arctan\left(\frac{cdx}{f}\right). \tag{21}$$

Once the angular displacement is estimated, the new Z parameter is obtained according to Equation 22:

$$Z = \frac{f_w cdx}{f_w \tan(\theta_L + \theta_{verg}) - cx}. \tag{22}$$

The angle θ_{verg} is half of the angular displacement θ according to (Rashbass & Westheimer, 1961). In order to iteratively improve the alignment of the images in the active cameras, the angle θ_{verg} is multiplied by a constant ($q < 1$) in the angular-position control algorithm; this constant defines the velocity of convergence of the iterative algorithm.

5. Benefits of using binocular disparity and optic flow in image segmentation

The image segmentation is an open research area in computer vision. The problem of properly segment an image has been widely studied and several algorithms have been proposed for different practical applications in the last three decades. The perception of what is happening in an image can be thought of as the ability for detecting many classes of patterns and statistically significant arrangements of image elements. Lowe (1984) suggests that human perception is mainly a hierarchical process in which prior knowledge of the world is used to provide higher-level structures, and these ones, in their turn, can be further combined to yield new hierarchical structures; this line of thoughts was followed in (Shi & Malik, 2000). It is worth noting that the low-level visual features like motion and disparity (see Fig. 6) can offer a first description of the world in certain practical application (cf. Harville, 2004; Kolmogorov et al., 2005; Yilmaz et al., 2006; Zhao & Thorpe, 2000). The purpose of this section is to show the benefits of using binocular disparity and optic flow estimates in segmenting surveillance video sequences rather than to make a contribution to the solution of the general problem of image segmentation.

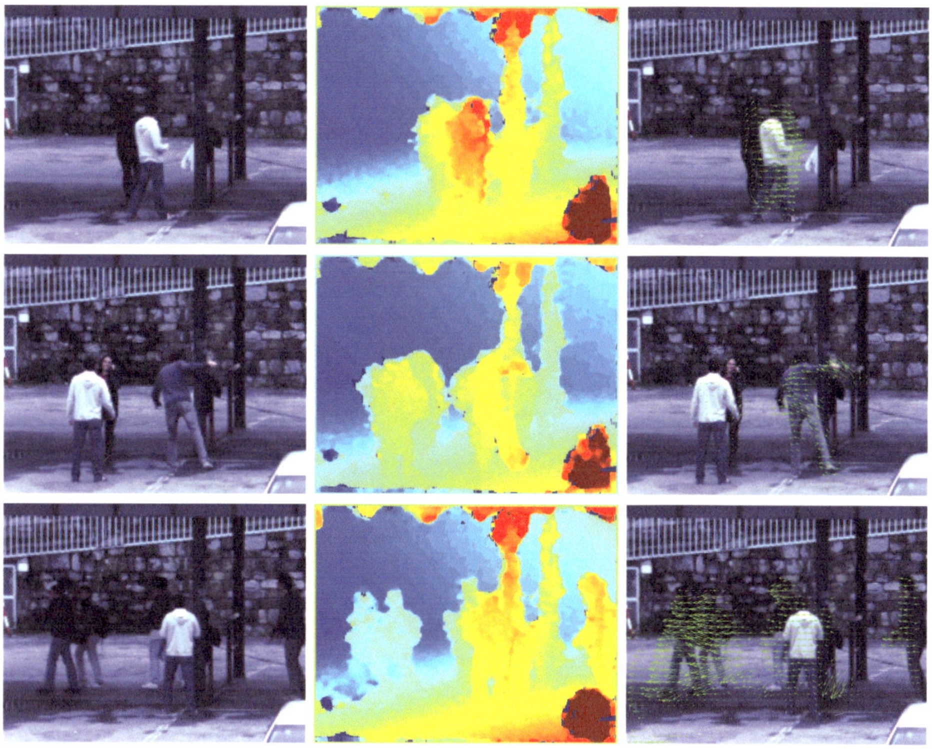

(a) World. (b) Depth map. (c) Motion map.

Fig. 6. Example of how different scenes can be described by using our framework. The low-level visual features refer to both disparity and optic flow estimates.

The following is a case of study in which the proposed system is capable of segmenting all individuals in a scene by using binocular disparity and optic flow. In a first stage of processing, the system fixates in depth the individuals according to the aforementioned algorithm (see section 4); that is, an initial fast movement of the cameras (version) triggered by a saliency in the wide-angle camera, and a subsequent slower movement of the cameras (vergence) guided by the binocular disparity. In a second stage of processing, the system changes the field of view of the active cameras in order to magnified the region of interest. Finally, in the last stage of processing, the system segments the individuals in the scene by using a threshold in the disparity information (around disparity zero or point of fixation) and a threshold in the orientation of the optic flow vectors. The results of applying the above mentioned processing stages are shown in Fig. 7. Good segmentation results can be achieved from the disparity measures by defining a set of thresholds (see Fig. 7b), however, a better data segmentation is obtained by combining the partial segments of binocular disparity and optic flow, respectively; an example is shown in Fig. 7c.

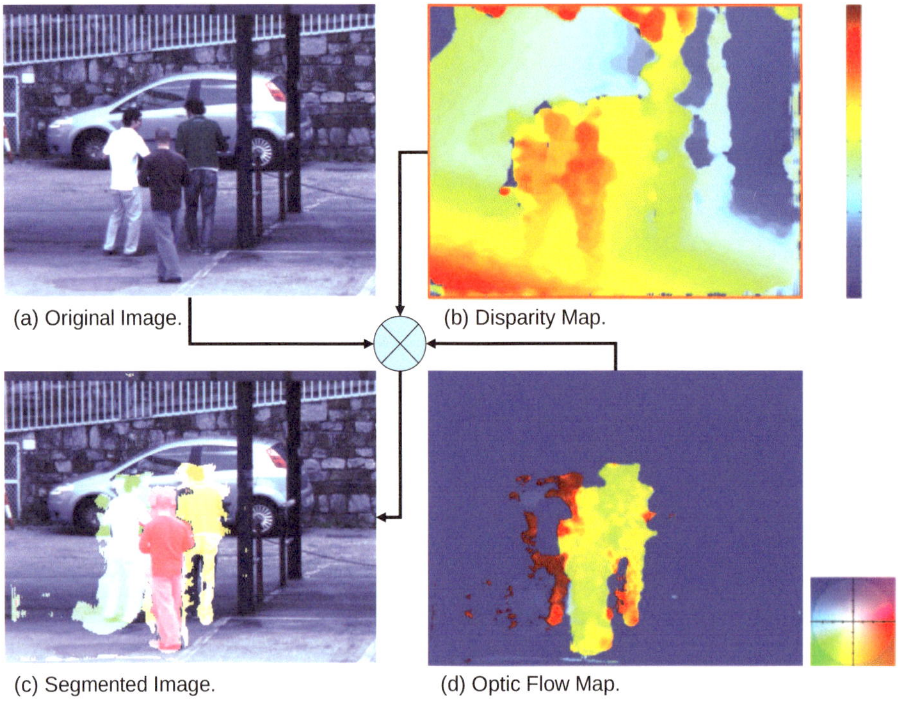

Fig. 7. Case of study: the segmentation of an image by using disparity and optic flow estimates.

The results in segmentation are constrained by the estimates of disparity and optic flow. For this reason, it is necessary to follow segmentation strategies, like the one proposed by Shi & Malik (2000), in order to achieve the appropriate robustness in the data segmentation. In fact, they argue the necessity of combining different features like colour, edge or in general any kind of texture information to create a hierarchical partition of the image, based on graph theory, in which prior knowledge is used to confirm current grouping or to guide further classifications.

6. The system performance

So far, we have presented an active vision system capable of estimating both optic flow and binocular disparity through a biologically inspired strategy, and capable of using these information to change the viewpoint of the cameras in an open, uncontrolled environment. This capability lets the system interact with the environment to perform video surveillance tasks. The purpose of this work was to introduce a novel system architecture for an active vision system rather than to present a framework for performing specific surveillance tasks. Under this perspective, it was first described the low-level vision approach for optic flow and binocular disparity, and then it was presented a robotic head which uses this approach to effectively solve the problem of fixation in depth.

In order to evaluate the performance of the system, it is necessary to differentiate the framework instances according to their role in the system. On the one hand, both optic flow and binocular disparity are to be used as prominent features for segmentation; hence, it is important to evaluate the accuracy of the proposed algorithms by using test sequences for which ground truth is available (see *http://vision.middlebury.edu/*). On the other hand, we must evaluate the system performance in relation to the accuracy of the binocular system to correctly change the viewpoint of the cameras.

6.1 Accuracy of the distributed population code

The accuracy of the estimates has been evaluated for a system with $N = 16$ oriented filters, each tuned to $M = 3$ different velocities and to $K = 9$ binocular phase differences. The used Gabor filters have a spatiotemporal support of $(11 \times 11) \times 7$ pixels $\times$ frames and are characterised by a bandwidth of 0.833 octave and spatial frequency $\omega_0 = 0.5\pi$. The Table 3 shows the results for *distributed population code* that has been applied to the most frequently used test sequences. The optic flow was evaluated by using the database described in (Baker et al., 2007) and the disparity was evaluated by using the one described in (Scharstein & Szeliski, 2002); however, in the case of disparity test sequences, the ground truth contains horizontal disparities, only; for this reason, it was also used the data set described in (Chessa, Solari & Sabatini, 2009) to benchmark the 2D-disparity measures (horizontal and vertical).

Distributed population code			
Sequences	Venus	Teddy	Cones
Disparity (%BP)	4.5	11.7	6.4
Sequences	Yosemite	Rubberwhale	Hydrangea
Optic Flow (AAE)	3.19	8.01	5.79

Table 3. Performance of the proposed distributed population code. On the one hand, the reliability of disparity measures has been computed in terms of percentage of bad pixels (%BP) for non-occluded regions. On the other hand, the reliability of optic flow measures has been computed by using the average angular error (AAE) proposed by Barron (Barron et al., 1994).

A quantitative comparison between the proposed *distributed population code* and some of the well-established algorithms in literature has been performed in (Chessa, Sabatini & Solari, 2009). The performances of the stereo and motion modules are shown in Table 3, which substantiates the feasibility of binocular disparity and optic flow estimates for image segmentation; the visual results are shown in Fig. 7.

6.2 Behaviour of the trinocular system

A good perception of the scene's depth is required to properly change the viewpoint of a binocular system. The previous results for disparity estimation have shown to be a valuable cue for 3D perception. The purpose now is to demonstrate the capability of the trinocular head to fixate any object in the field of view of the wide-angle camera. In order to evaluate the fixation in depth algorithm, two different scenarios have been considered: the long-range scenario in which the depth is larger than 50 meters in the line of sight (see Fig. 8), and the short-range scenario in which the depth is in the range between 10 and 50 meters (see Fig. 11).

(a) Cyclopean Image.

(b) Left Image, point A.

(c) Right Image, point A.

(d) Left Image, point B.

(e) Right Image, point B.

(f) Left Image, point C.

(g) Right Image, point C.

Fig. 8. Long-range scenario: Fixation of points A, B and C. A zoom factor of 16x was used in the active cameras. Along the line of sight the measured depths were approximately 80 m, 920 m, and 92 m, respectively.

The angular-position control uses the disparity information to align the binocular images in the Panum area. In order to save computational resources and considering that just a small area around the centre of the image has the disparity information of the target object, the size of the Panum area has been empirically chosen as a square region of 40x40 pixels. Accordingly, the mean value of the disparity in the Panum area is used to iteratively estimate the new Z parameter.

In order to evaluate the performance of the trinocular head, we first tested the fixation strategy in the long-range scenario. In the performed tests, three points were chosen in the cyclopean image (see Fig. 8(a)). For each point, the active cameras performed a version movement according to the coordinate system of the cyclopean image and, inmediately after, the angular-position control started the alignment of the images by changing the pan angles iteratively. Once the images were aligned, a new point in the cyclopean image was provided.

Fig. 9 shows the angular changes of the active cameras during the test in the long-range scenario. In Figs. 9(a) and 9(b) the pan angle of the left and right cameras, respectively, is depicted as a function of time. Fig. 9(c) shows the same variation for the common tilt angle. Each test point of the cyclopean image was manually selected after the fixation in depth of the previous one; consequently, the plots show the angular-position control behaviour during changes in the viewpoint of the binocular system. It is worth noting that the version movements correspond, roughly speaking, with the pronounced slopes in the graphs, while the vergence movements are smoother and therefore with a less pronounced slope.

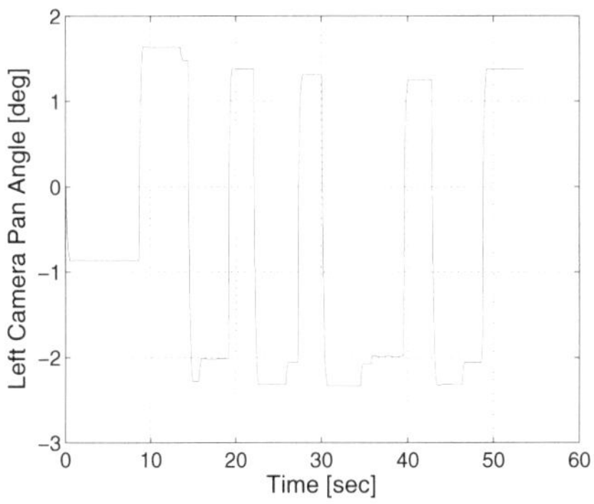

(a) Left camera pan movements.

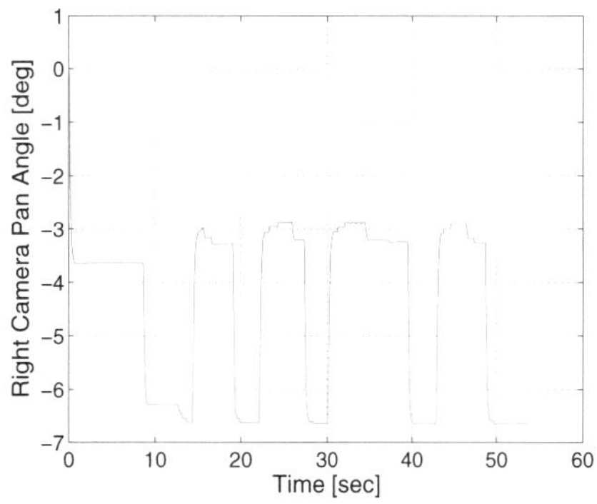

(b) Right camera pan movements.

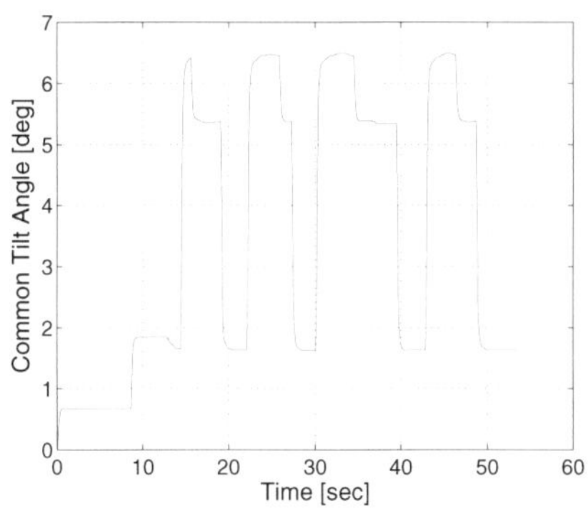

(c) Common tilt movements.

Fig. 9. Temporal changes in the angular position of the active cameras to fixate in depth the points A, B and C in a long-range scenario.

In a similar way, the fixation in depth algorithm was also evaluated in short-range scenarios by using three test points (see Fig. 11). We followed the same procedure used for long-range scenarios and the results are shown in Fig. 10.

From the plots in Figs. 9 and 10 we can observe that small angular shifts were performed just after a version movement; this behaviour is due to two factors: (1) the inverse relationship between the vergence angle and the depth by which for large distances the optical axes of the binocular system can be well approximated as parallel; and (2) the appropriate geometrical description of the system which allows us to properly map the angular position of the active cameras with respect to the cyclopean image. Actually, there are not enough differences between long and short-range scenarios in the angular-position control, because the vergence angles begin to be considerable for depths minor than 10 meters, approximately; it is worth noting that, this value is highly dependent on the baseline of the binocular system.

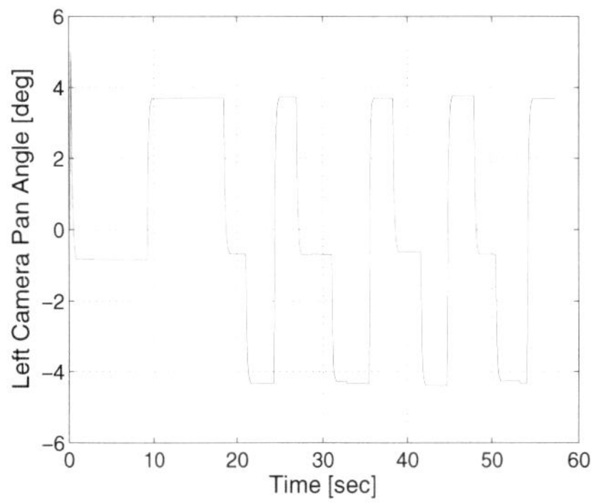

(a) Left camera pan movements.

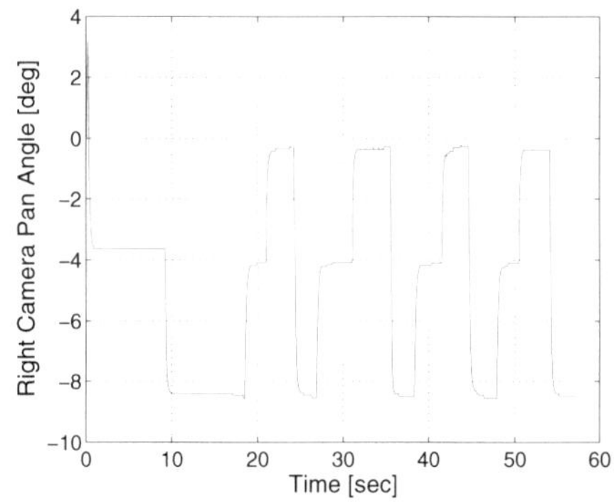

(b) Right Camera pan movements.

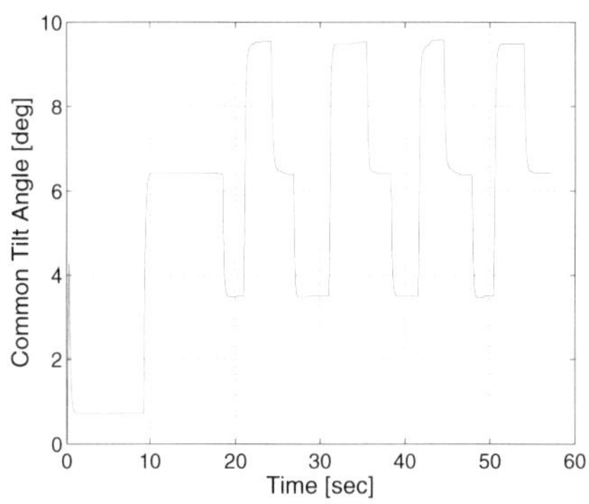

(c) Common tilt movements.

Fig. 10. Temporal changes in the angular position of the active cameras to fixate in depth the points A, B and C in a short-range scenario.

Finally, the justification for using two different scenarios is the field of view of the active cameras. Even though the wide-to-active camera mapping equations do not depend on the field of view of the active cameras, everything else does. It follows that the estimation of optic flow and disparity loses resolution due to narrow fields of view in the active cameras. In order to clarify the system behaviour, it is worth to highlight that the framework always performs the fixation in depth by using the maximum field of view in the active cameras, and immediately after, it changes the field of view of the cameras according to the necessary magnification. In this regard, the adequate definition of the Panum area plays an important role in the framework (see section 4.1). Consequently, Figs. 8 and 11 show the performance of the framework not only in terms of the fixation but also for a proper synchronisation of all processing stages in the system; these images were directly obtained from the system during the experiments in Figs. 9 and 10. Fig. 8 shows the fixation in depth of three test points. The zoom factor of the active cameras in all cases was 16x. The angular-position control estimated the depth along the line of sight for each fixated target and the approximated values were 80 m, 920 m, and 92 m, respectively. Likewise, Fig. 11 shows the fixation in depth of three test points at different zoom factors each one, namely: 4x, 16x, and 4x, respectively. Along the line

(a) Cyclopean Image.

(b) Left Image, point A.

(c) Right Image, point A.

(d) Left Image, point B.

(e) Right Image, point B.

(f) Left Image, point C.

(g) Right Image, point C.

Fig. 11. Short-range scenario: Fixation of points A, B, and C. The different zoom factors used in the active cameras were 4x, 16x, and 4x, respectively. Along the line of sight the measured depths were approximately 25 m, 27 m, and 28 m, respectively.

of sight the measured depths were approximately 25 m, 27 m, and 28 m, for points A, B, and C, respectively.

7. Conclusion

We have described a trinocular active visual framework for video surveillance applications. The framework is able to change the viewpoint of the active cameras toward areas of interest, to fixate a target object at different fields of view, and to follow its motion. This behaviour is possible thanks to a rapid angular-position control of the cameras for object fixation and pursuit based on disparity information. The framework is capable of recording image frames at different scales by zooming individual areas of interest, in this sense, it is possible to exhibit the target's identity or actions in detail. The proposed visual system is a cognitive model of visual processing replicating computational strategies supported by the neurophysiological studies of the mammalian visual cortex which provide the system with a powerful framework to characterise and to recognise the environment, in this sense, the optic flow and binocular disparity information are an effective, low-level, visual representation of the scenes which provide a workable base for segmenting the dynamic scenarios; it is worth noting that, these measures can easily disambiguate occlusions in the different scenarios.

8. References

Adelson, E. & Bergen, J. (1985). Spatiotemporal energy models for the perception of motion, *JOSA* 2: 284–321.

Adelson, E. & Bergen, J. (1991). The plenoptic and the elements of early vision, *in* M. Landy & J. Movshon (eds), *Computational Models of Visual Processing*, MIT Press, pp. 3–20.

Andrade, E. L., Blunsden, S. & Fisher, R. B. (2006). Hidden markov models for optical flow analysis in crowds, *Pattern Recognition, International Conference on* 1: 460–463.

Baker, S., Scharstein, D., Lewis, J., Roth, S., Black, M. & Szeliski, R. (2007). A database and evaluation methodology for optical flow, *Computer Vision, 2007. ICCV 2007. IEEE 11th International Conference on*, pp. 1 –8.

Barron, J., Fleet, D. & Beauchemin, S. (1994). Performance of optical flow techniques, *Int. J. of Computer Vision* 12: 43–77.

Chessa, M., Sabatini, S. & Solari, F. (2009). A fast joint bioinspired algorithm for optic flow and two-dimensional disparity estimation, *in* M. Fritz, B. Schiele & J. Piater (eds), *Computer Vision Systems*, Vol. 5815 of *Lecture Notes in Computer Science*, Springer Berlin / Heidelberg, pp. 184–193.

Chessa, M., Solari, F. & Sabatini, S. (2009). A virtual reality simulator for active stereo vision systems, *VISAPP* .

Daugman, J. (1985). Uncertainty relation for resolution in space, spatial frequency, and orientation optimized by two-dimensional visual cortical filters, *JOSA* A/2: 1160–1169.

Donders, F. C. (1969). Over de snelheid van psychische processen, *Onderzoekingen gedann in het Psychologish Laboratorium der Utrechtsche Hoogeschool: 1868-1869. Tweede Reeks, II, 92-120.*, W. E. Koster (Ed.) and W. G. Koster (Trans.), pp. 412 – 431. (Original work published 1868).

Fleet, D., Wagner, H. & Heeger, D. (1996). Neural encoding of binocular disparity: Energy models, position shifts and phase shifts, *Vision Res.* 36(12): 1839–1857.

Hansard, M. & Horaud, R. (2008). Cyclopean geometry of binocular vision, *Joural of the Optical Society of America A* 25(9): 2357–2369.

Harville, M. (2004). Stereo person tracking with adaptive plan-view templates of height and occupancy statistics, *Image and Vision Computing* 22(2): 127 – 142. Statistical Methods in Video Processing.

Helmholtz, H. v. (1925). *Treatise on Physiological Optics*, Vol. III, transl. from the 3rd german edn, The Optical Society of America, New York, USA.

Hubel, D. H. & Wiesel, T. N. (1968). Receptive fields and functional architecture of monkey striate cortex, *The Journal of Physiology* 195(1): 215–243.

Kapur, J., Sahoo, P. & Wong, A. (1985). A new method for gray-level picture thresholding using the entropy of the histogram, *Computer Vision, Graphics, and Image Processing* 29(3): 273 – 285.

Koenderink, J. & van Doorn, A. (1976). Geometry of binocular vision and a model for stereopsis, *Biological Cybernetics* 21: 29–35.

Kolmogorov, V., Criminisi, A., Blake, A., Cross, G. & Rother, C. (2005). Bi-layer segmentation of binocular stereo video, *Computer Vision and Pattern Recognition, IEEE Computer Society Conference on* 2: 407–414.

Kumar, R. K., Ilie, A., Frahm, J.-M. & Pollefeys, M. (2008). Simple calibration of non-overlapping cameras with a mirror, *Computer Vision and Pattern Recognition, IEEE Computer Society Conference on* 0: 1–7.

Kuon, I. & Rose, J. (2006). Measuring the gap between fpgas and asics, *FPGA '06: Proceedings of the 2006 ACM/SIGDA 14th international symposium on Field programmable gate arrays*, ACM, New York, NY, USA, pp. 21–30.

Lavest, J.-M., Rives, G. & Dhome, M. (1993). Three-dimensional reconstruction by zooming, *Robotics and Automation, IEEE Transactions on* 9(2): 196–207.

Lee, L., Romano, R. & Stein, G. (2000). Monitoring activities from multiple video streams: establishing a common coordinate frame, *Pattern Analysis and Machine Intelligence, IEEE Transactions on* 22(8): 758 –767.

Lenz, R. & Tsai, R. (1988). Techniques for calibration of the scale factor and image center for high accuracy 3-d machine vision metrology, *IEEE Transactions on Pattern Analysis and Machine Intelligence* 10: 713–720.

Lowe, D. G. (1984). *Perceptual Organization and Visual Recognition*, PhD thesis, STANFORD UNIV CA DEPT OF COMPUTER SCIENCE.

Marr, D. (1982). *Vision: A Computational Investigation into the Human Representation and Processing of Visual Information*, Henry Holt and Co., Inc., New York, NY, USA.

Ohzawa, I., DeAngelis, G. & Freeman, R. (1990). Stereoscopic depth discrimination in the visual cortex: neurons ideally suited as disparity detectors, *Science* 249: 1037–1041.

Otsu, N. (1979). A threshold selection method from graylevel histograms, *IEEE Trans. Syst., Madn. & Cybern.* 9: 62–66.

Pauwels, K. & Van Hulle, M. M. (2006). Optic flow from unstable sequences containing unconstrained scenes through local velocity constancy maximization, *British Machine Vision Conference (BMVC 2006)*, Edinburgh, Scotland, pp. 397–406.

Pouget, A., Dayan, P. & Zemel, R. S. (2003). Inference and computation with population codes., *Ann. Rev Neurosci* 26: 381–410.

Rashbass, C. & Westheimer, G. (1961). Disjunctive eye movements, *The Journal of Physiology* 159: 339–360.

Ratha, N. & Jain, A. (1999). Computer vision algorithms on reconfigurable logic arrays, *Parallel and Distributed Systems, IEEE Transactions on* 10(1): 29 –43.

Ridler, T. W. & Calvar, S. (1978). Picture thresholding using an iterative selection method, *Systems, Man and Cybernetics, IEEE Transactions on* 8(8): 630 –632.

Scharstein, D. & Szeliski, R. (2002). A taxonomy and evaluation of dense two-frame stereo correspondence algorithms, *Int. J. of Computer Vision* 47: 7–42.

Shi, J. & Malik, J. (2000). Normalized cuts and image segmentation, *Pattern Analysis and Machine Intelligence, IEEE Transactions on* 22(8): 888 –905.

Stauffer, C. & Grimson, W. (1999). Adaptive background mixture models for real-time tracking, *Computer Vision and Pattern Recognition, 1999. IEEE Computer Society Conference on.*, Vol. 2, pp. 2 vol. (xxiii+637+663).

Stauffer, C. & Grimson, W. (2000). Learning patterns of activity using real-time tracking, *Pattern Analysis and Machine Intelligence, IEEE Transactions on* 22(8): 747 –757.

Theimer, W. & Mallot, H. (1994). Phase-based binocular vergence control and depth reconstruction using active vision, *CVGIP: Image Understanding* 60(3): 343–358.

Tsai, R. (1987). A versatile camera calibration technique for high-accuracy 3d machine vision metrology using off-the-shelf tv cameras and lenses, *Robotics and Automation, IEEE Journal of* 3(4): 323 –344.

Weems, C. (1991). Architectural requirements of image understanding with respect to parallel processing, *Proceedings of the IEEE* 79(4): 537 –547.

Yilmaz, A., Javed, O. & Shah, M. (2006). Object tracking: A survey, *ACM Comput. Surv.* 38.

Zhao, L. & Thorpe, C. (2000). Stereo- and neural network-based pedestrian detection, *Intelligent Transportation Systems, IEEE Transactions on* 01(3): 148 –154.

2

Stereo Matching Method and Height Estimation for Unmanned Helicopter

Kuo-Hsien Hsia[1], Shao-Fan Lien[2] and Juhng-Perng Su[2]
[1]*Far East University*
[2]*National Yunlin University of Science & Technology*
Taiwan

1. Introduction

The research and development of autonomous unmanned helicopters has lasted for more than one decade. Unmanned aerial vehicles (UAVs) are very useful for aerial photography, gas pollution detection, rescue or military applications. UAVs could potentially replace human beings in performing a variety of tedious or arduous tasks. Because of their ubiquitous uses, the theory and applications of UAVs systems have become popular contemporary research topics. There are many types of UAVs with different functions. Generally UAVs can be divided into two major categories, fixed-wing type and rotary-wing type. The fixed-wing UAVs can carry out long-distance and high-altitude reconnaissance missions. However, flight control of fixed-wing UAVs is not easy in low-altitude conditions. Conversely, rotary-wing UAVs can hover in low altitude while conducting surveys, photography or other investigations. Consequently in some applications, the rotary-wing type UAVs is more useful than the fixed-wing UAV. One common type of rotary-wing type UAVs is the AUH (Autonomous Unmanned Helicopter). AUHs have characteristics including of 6-DOF flight dynamics, VTOL (vertical taking-off and landing) and the ability to hover. These attributes make AUHs ideal for aerial photography or investigation in areas that limit maneuverability.

During the past few years, the development of the unmanned helicopter has been an important subject of research. There have been a lot of researches interested in a more intelligent design of autonomous controllers for controlling the basic flight modes of unmanned helicopters (Fang et al., 2008). The controller design of AUHs requires multiple sensor feedback signals for sensing states of motion. The basic flight modes of unmanned helicopters are vertical taking-off, hovering, and landing. Because the unmanned helicopter is a highly nonlinear system, many researchers focus on the dynamic control problems (e.g. Kadmiry & Driankov, 2004; C. Wang et al., 2009). Appropriate sensors play very important roles in dynamic control problems. Moreover, the most important flight mode of autonomous unmanned helicopter is the landing mode. In consideration of the unmanned helicopter landing problem, the height position information is usually provided by global positioning system (GPS) and inertial measurement unit (IMU). The system of the autonomous unmanned helicopter is a 6-DOF system, with 3-axis rotation information provided by IMU and 3-axis moving displacement information provided from GPS.

Oh et al. (2006) brought up the tether-guided method for autonomous helicopter landing. Many researches used vision systems for controlling helicopter and searching landmark (Lin, 2007; Mori, 2007; C.C. Wang et al., 2009). In the work of Saito et al. (2007), camera-image based relative pose and motion estimation for unmanned helicopter were discussed. In the works of Katzourakis et al. (2009) and Xu et al. (2006), navigation and landing with the stereo vision system was discussed. Xu et al. used the stereo vision system for estimating the position of the body. From the work of Xu, it was shown that the stereo vision does work for the position estimation.

For unmanned helicopter autonomous landing, the information of the height is very important. However, the height error of GPS is in general about from 5 to 8 meters, which is not accurate enough for autonomous landing. For example, the accuracy of Garmin GPS 18-5Hz is less than 15 meters (GPS 18 Technical Specifications, 2005). After many times of measurement, the average error of this GPS was obtained to be around 10 meters. Since the height error range of GPS is from 5 to 8 meters, to conquer the height measurment error of GPS, the particular stereo vision system is designed for assisting GPS, and the measurement range of this system is set to be at least 6 m.

Image systems are the common guiding sensors. In the AUHs controll problems, image systems are usually collocated with IMU and GPS in the outdoor environment. The image system has been used on vehicles for navigation, obstacle avoidance or position estimation. Doehler & Korn (2003) proposed an algorithm to extract the edge of the runway for computing the position of airplane. Bagen et al. (2009) and Johnson et al. (2005) discussed the image-guided method with two or more images for guiding the RC unmanned helicopter approaching to the landmark. Undoubtedly multiple-camera system measurement environment is an effective and mature method. However, the carrying capacity of a small unmanned helicopters has to be considered. Therefore the image systems are the smaller the better. A particular stereo vision system is developed for reducing the payload in our application.

In this chapter, we focus on the problem of estimating the height of the helicopter for the landing problem via a simple stereo vision system. The key problem of stereo vision system is to find the corresponding points in the left image and the right image. For the corresponding problem of stereo vision, two methods will be proposed for searching the corresponding points between the left and right image. The first method is searchig corresponding points with epipolar geometry and fundamental matrix. The epipolar geometry is the intrinsic projective geometry between two cameras (Zhang, 1996; Han & Park, 2000). It only depends on the camera internal parameters and relative position. The second method is block matching algorithm (Gyaourova et al., 2003; Liang & Kuo, 2008; Tao et al., 2008). The block matching algorithm (BMA) is provided for searching the corresponding points with a low resolution image. BMA will be compared with epipolar geometry constraint method via experimental results.

In addition, a particular stereo vision system is designed to assist GPS. The stereo vision system composed of two webcams with resolutions in 0.3 mega pixels is shown in Figure 1. To simplify the system, we dismantled the cover of the webcams. The whole system is very light and thin. The resolution of cameras will affect the accuracy of height estimation result. The variable baseline method is introduced for increasing the measuring range. Details will be illustrated in the following sections.

Fig. 1. The stereo vision system composed of two Logitech® webcams

2. Design of stereo vision system

2.1 Depth measuring by triangulation

In general, a 3D scenery projected to 2D image will lose the information of depth. The stereo vision method is very useful for measuring the depth. The most common used method is triangulation.

Consider a point $P=(X, Y, Z)$ in the 3D space captured by a stereo vision system, and the point P projected on both left and right images. The relation is illustrated in Figure 2. In Figure 2, the projected coordinates of point P on the left and the right images are (x_l, y_l) and (x_r, y_r) respectively. The formation of the left image is:

$$\frac{X}{Z} = \frac{x_l}{f} \Rightarrow X = \frac{x_l}{f} Z \tag{1}$$

and the formation of the right image is:

$$\frac{X - b}{Z} = \frac{x_r}{f} \tag{2}$$

From (1) and (2), we have

$$Z = \frac{fb}{(x_l - x_r)} = \frac{fb}{\Delta x} \tag{3}$$

where f is focal length, b is the length of baseline and $\Delta x = (x_l - x_r)$ is the disparity. From (3), the accuracy of f, b and Δx will influence the depth measuring. In the next section, the camera will be calibrated for obtaining accurate camera parameters.

There are three major procedures for stereo vision system design. Fristly, the clear feature points in image need be extracted quickly and accurately. The second procedure is searching for corresponding points between two images. Finally, computing the depth using (3).

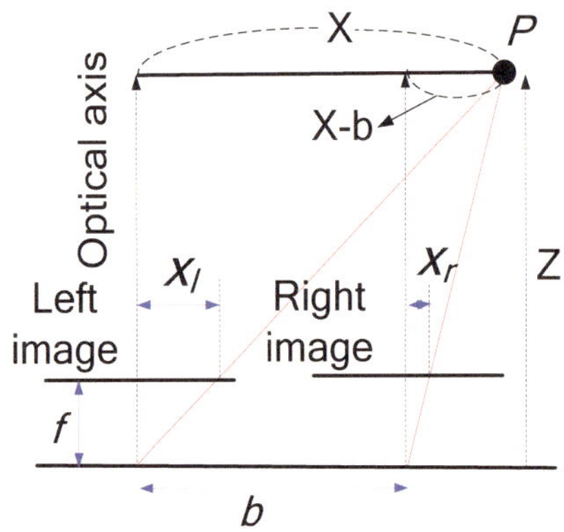

Fig. 2. Geometric relation of a stereo vision system.

2.2 Depth resolution of stereo vision system

The depth resolution is a very important factor for stereo vision system design (Cyganek & Siebert, 2009). The pixel resolution will reduce with the depth. The relations of depth resolution is illustrated in Figure 3.

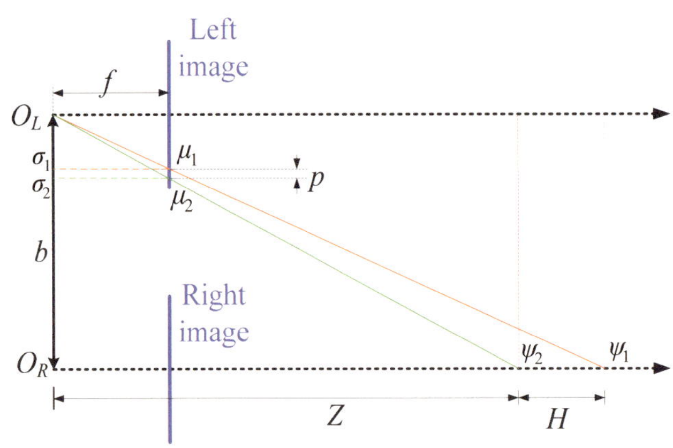

Fig. 3. Geometry relation of depth resolution.

From Figure 3, with the similarity of triangle $\Delta O_L\mu_1\sigma_1$ to $\Delta O_L\Psi_1 O_R$ and $\Delta O_L\mu_2\sigma_2$ to $\Delta O_L\Psi_2 O_R$, we can have the following relations:

$$\frac{\overline{O_L O_R}}{\overline{O_R\psi_1}} = \frac{\overline{O_L\sigma_1}}{\overline{\sigma_1\mu_1}}, \frac{\overline{O_L O_R}}{\overline{O_R\psi_2}} = \frac{\overline{O_L\sigma_2}}{\overline{\sigma_2\mu_2}} \tag{4}$$

where $\overline{O_L O_R} = b$, $\overline{\sigma_1\mu_1} = \overline{\sigma_2\mu_2} = f$, $\overline{O_R\psi_2} = Z$ and $\overline{\psi_2\psi_1} = H$. Then we obtain

$$\begin{aligned} p &= \overline{O_L\sigma_2} - \overline{O_L\sigma_1} \\ &= \frac{fb}{Z} - \frac{fb}{Z+H} \end{aligned} \tag{5}$$

where p is the width of a pixel on image. Next, we can have the following equation by rearranging (5):

$$H = \frac{pZ^2}{fb - pZ} \tag{6}$$

where H is the depth-change when a pixel change in the image, and is called the pixel resolution. Assuming $fb / Z >> H$, the following approximation will be obtained,

$$H \approx \frac{pZ^2}{fb} \tag{7}$$

In addition, the (6) is true with the condition:

$$fb \neq pZ \tag{8}$$

Therefore the limit value of Z is

$$Z = \frac{fb}{p} \tag{9}$$

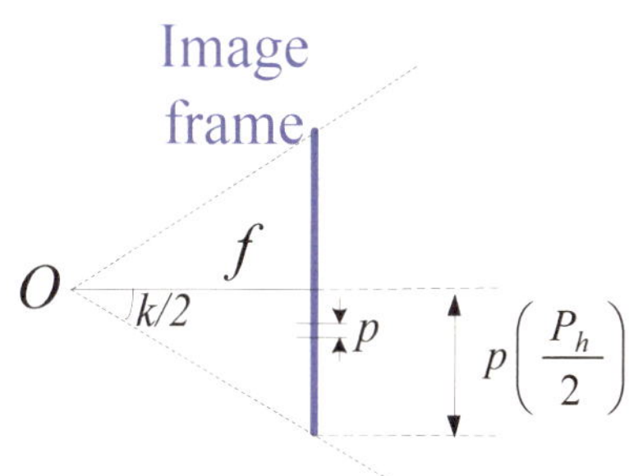

Fig. 4. Geometry relation of f and p.

For single image, the f, b and p are all constants. Thus there is no depth information from a single image. Furthermore, consider Figure 4, and we will have:

$$f = p\left(\frac{P_h}{2\tan(k/2)}\right) \tag{10}$$

where k is the horizontal view angle, P_h is the horizontal resolution of the camera.

Combining (6) with (10), we will have:

$$H = \frac{Z^2}{\left[P_h b / 2\tan(k/2)\right] - Z} \tag{11}$$

From (11), the relation between baseline and pixel resolution are shown in Figure 5. Obviously, the pixel resolution and baseline are in a nonlinear relation. Moreover, they are almost in inverse proportion. The accuracy of system depends on choosing an appropriate baseline. In general, if small pixel resolution is expected, one should choose a larger beseline.

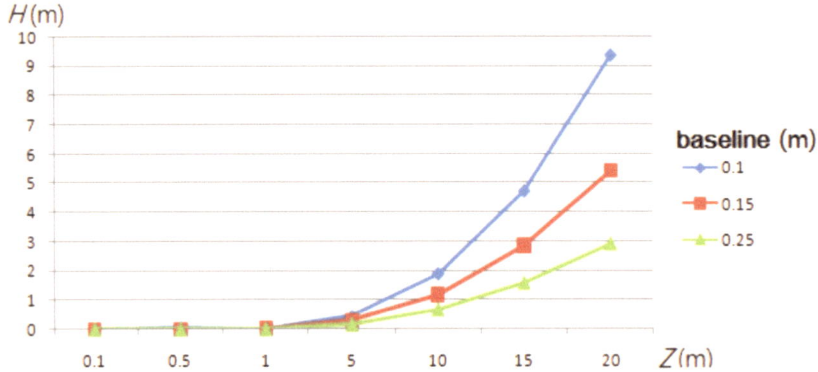

Fig. 5. The pixel resolution H with different baseline for stereo vision system setup.

3. Searching for corresponding points

The stereo vision system includes matching and 3D reconstruction processes. The disparity estimation is the most important part of the stereo vision. The disparity is computed by matching method. Furthermore, the 3D scene could be reconstructed by disparity. The basic idea of disparity estimation is using the pixel intensity of a point and its neighborhood on an image as a matching template to search the most matching area on another image (Alagoz, 2008; Wang & Yang, 2011). The similarity measurment between two images is definded by correlation functions. Based on different matching unit, there are two major categories of matching method which will be discussed. They are area-based matching method and feature-based matching method.

3.1 Area-based matching method

A lot of area-based matching methods have been proposed. Using area-based matching methods, one can obtain the dense disparity field without detecting the image features. Generally, the matching method has good results with flat and complex texture images. Template matching method and block matching method are relatively prevalent methods of the various area-based matching methods. Hu (2008) proposed the adaptive template for increasing the matching accuracy. Another example is proposed by Siebert et al. (2000). This approach uses 1D area-based matching along the horizontal scanline. Figure 6 illustrates the 1D area-based matching. Bedekar and Haralick (1995) proposed the searching method with Bayesian triangulation. Moreover, Tico et al. (1999) found the corresponding points of fingerprints with geometric invariant representations. Another case is area matching and depth map reconstruction with the Tsukuba stereo-pair image (Cyganek, 2005, 2006). In this case, the matching area is 3×3 pixels and the image size is 344×288 pixels (download from http://vision.middlebury.edu/stereo/eval/). The disparity and depth map are reconstructed and the depth information in the 3D scene are obtainded. The results are illustrated in Figure 7.

However, there are still some restrictions for area-based matching method. Firstly, the matching template is established with pixel intensity, therefore the matching performence are depedent on brightness, contrast, and textures. If the brightness is changed a lot or textures are monotonous, the matching performence will not be good. Secondly, the matching results will not be good when the image with depth discontinuity or masking. Last, the computational complexity is very high. Therefore, the feature-based matching methods are developed for improving the defects of area-based matching method.

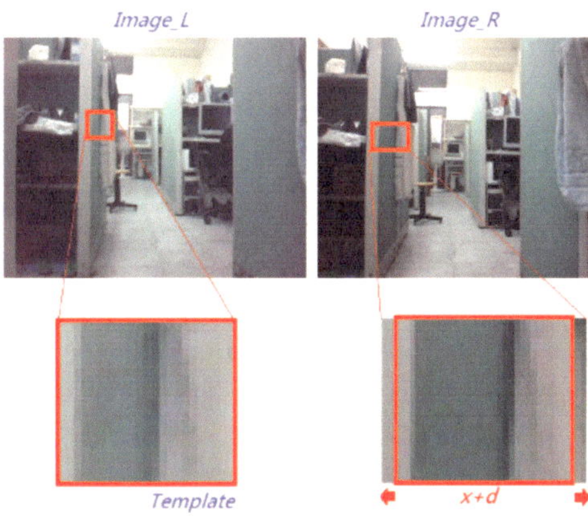

Fig. 6. 1D area-based matching along the horizontal scanline.

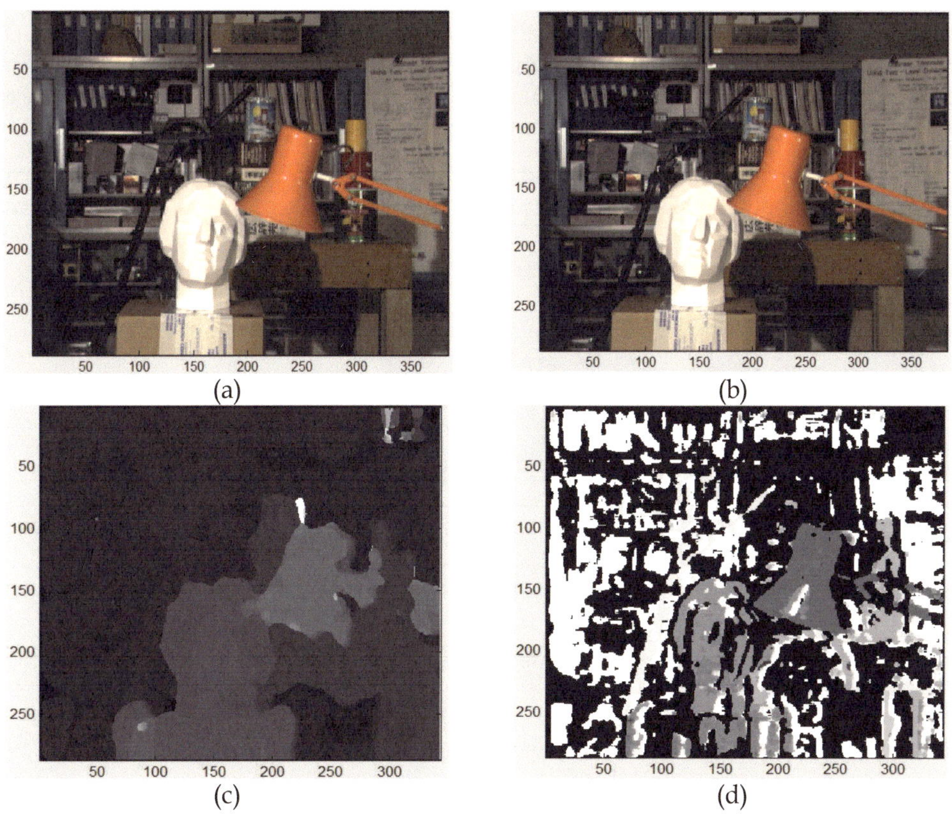

Fig. 7. (a) Left image. (b) Right image. (c) Disparity map. (d) Depth map.

3.2 Feature-based matching method

The feature-based matching method is matching the corresponding points on image with the features of the scene. To highlight the information of space, the features are more easily than the pixel intensity of area. Moreover, feature-based matching method is more robust than area-based method for brightness changing. There are two steps for the feature-based matching method. They are feature extraction and feature matching. The features are usually the lines, corners or planes in the image. The specific operators are utilized for extracting the features.

Many feature-based matching methods have been proposed for searching the feature correlation between the right and left images. Both the intensity and orientation of the features could be the matching templates for searching the correspondence of the features. Therefore, for the depth discontinuity or masking problems, the feature-based matching method can obtain better matching result. In addition, feature-based matching method computes only for the features istead of all pixels, hence the computing load is smaller than area-based matching method. Olson (2002) proposed the matching method based on statistics. This method extracted a few eigenvectors as the matching templates, and it used the maximum-likelihood for template matching.

Moreover, the phase-based image matching is also a very accurate matching method (Muquit et al., 2006). The images are transformed into frequency domain by 2D Discrete Fourier Transforms (2D DFTs). The best matching vector is obtained by computing the phase correlation function. Figure 8 is a simulation of phase-based image matching. The example image is "Cristo Redentor" in Brazil and the image sizes are both 119×127 pixels. Figure 8(c) is the phase correlation of Figures 8(a) and 8(b). From Figure 8(c), we can see that the peak is located at (111, 126), and hence the motion vector is (8, 1).

Similarly, feature-based matching method have two restrictions. Firstly, the dense disparity field could not be obtained. Therefore, it is not easy to reconstruct the complex 3D scene. Secondly, the matching performance is affected by feature extraction results directly. In other words, if the features are too sparse, the matching results will not be good.

(a) (b) (c)

Fig. 8. Case simulation of phase-based image matching.

Since both the area-based and feature-based matching methods have some restrictions, the hybrid matching algorithm has been proposed in recent years. For example, Fuh et al. (1989) combined the optic flow and block matching algorithm for increasing the matching performance. In this chapter, we will combine the feature points and epipolar geometry constraint for reducing the computation.

3.3 Feature points detection

There are usually millions pixels in an image, therefore how to extract the significant feature points is the interesting research topic. Including edge detecting method (Canny, 1986), Tabu search algorithm (Glover, 1989, 1990), neural network (NN) (Takahashi et al., 2006) or Hough transform (Duan et al., 2010) are useful methods for extracting the special features from an image. However, the point sets of lines or edges are still too large. In addition, the searching speeds of most feature extracting algorithms, such as Tabu search, are not fast enough for real-time stereo vision systems. Consequently, the Harris corner detector (Nixon & Aguado, 2008) is proposed for detecting the feature points.

The main principle of Harris corner detector is using the Gaussian filter to detect the cornersresponse of each pixel in the image. Gaussian filter can not only enhance the significant corners, but also remove the undesirable corners. Moreover, it can reduce the probability of misjudgment. Although the Harris corner detector is a very useful tool, it requires a lot of computing time. Therefore the corner detection operations are applied only for the basic rectangle to reduce computation time.

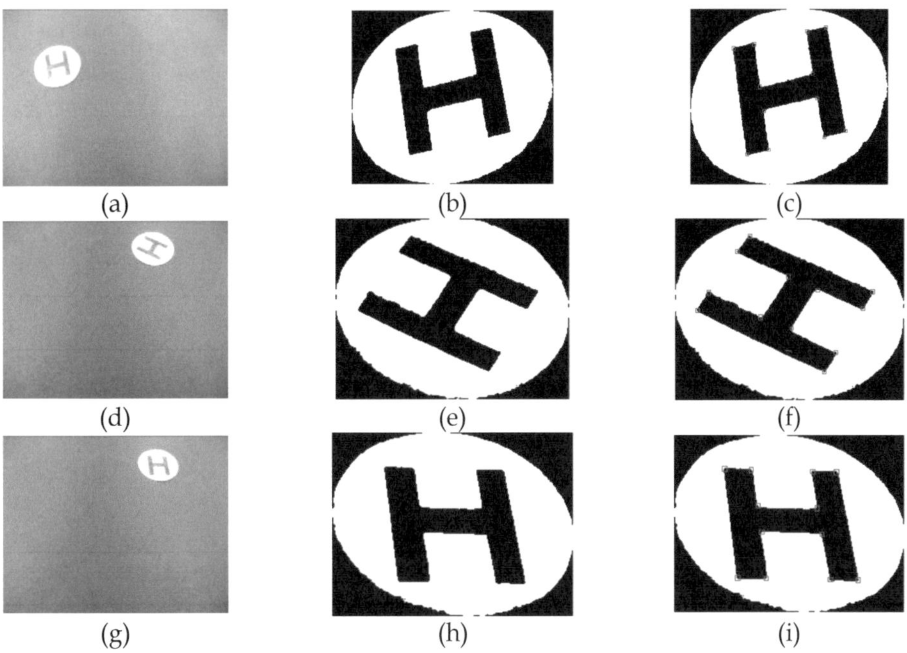

Fig. 9. The simulations of the image processing and feature extraction results. (a) The test image 1. (b) The binary image and the basic rectangle of test image 1. (c) The corner detection result of test image 1. (d) The test image 2. (e) The binary image and the basic rectangle of test image 2. (f) The corner detection result of test image 2. (g) The test image 3. (h) The binary image and the basic rectangle of test image 3. (i) The corner detection result of test image 3.

The following demo example shows the results of landmark image corner detection. The procedure is described as follows.

Step 1. Detect the corners of the image in the basic rectangle.

Step 2. Detect the convex corners of the label 'H'.
Step 3. Label the convex corners and extract the four most outside corners of the label 'H'.
Step 4. Find the intersection of the diagonals, and designate it as the approximate center of the landmark image.

Some examples for demonstrating the Harris corner detector are illustrated in Figure 9. Several advantages can be summarized from Figure 9. The Harris corner detector is very robust for corner detection. Moreover the Harris corner detector could not only detect the edges but also the corners of the object. The detection procedure with the basic rectangle segmentation can greatly enhance the detecting efficiency.

3.4 Epipolar geometry constraints

The epipolar geometry is the intrinsic projective geometry between two cameras. It only depends on the camera internal parameters and relative position. The fundamental matrix F is the algebraic description of the epipolar geometry. Epipolar geometry between two views is illustrated in Figure 10.

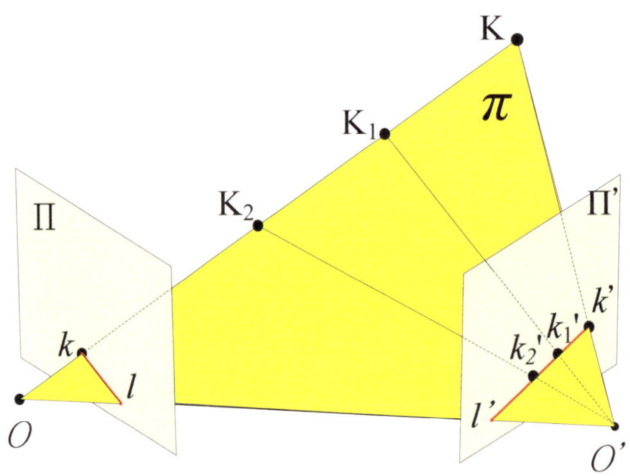

Fig. 10. Epipolar geometry.

From Figure 10, a point K in the 3D space is projected on Π and Π' respectively. The point k in Π corresponds to an epipolar line l' line on Π', and it can be represented as:

$$l' = Fk \tag{12}$$

where F is fundamental matrix, $k = (x, y)$ and $l' = (a\ b\ c)^T$. The point $k' = (x', y')$ lies on l', then:

$$k'^T l' = k'^T Fk = 0 \tag{13}$$

Rewrite equation (13) as:

$$\begin{bmatrix} x' & y' & 1 \end{bmatrix} \begin{bmatrix} f_{11} & f_{12} & f_{13} \\ f_{21} & f_{22} & f_{23} \\ f_{31} & f_{32} & f_{33} \end{bmatrix} \begin{bmatrix} x \\ y \\ 1 \end{bmatrix} = 0 \tag{14}$$

Based on Hartley's 8-point algorithm (Hartley, 1995), (14) can be further represented as:

$$xx'f_{11} + xy'f_{21} + xf_{31} + yx'f_{12} + yy'f_{22} + yf_{32} + x'f_{13} + y'f_{23} + f_{33} = 0 \tag{15}$$

Rewrite (15) as:

$$\begin{bmatrix} x_1x'_1 & x_1y'_1 & x_1 & y_1x'_1 & y_1y'_1 & y_1 & x'_1 & y'_1 & 1 \\ \vdots & \vdots & \vdots & \vdots & \vdots & \vdots & \vdots & \vdots & \vdots \\ x_nx'_n & x_ny'_n & x_n & y_nx'_n & y_ny'_n & y_n & x'_n & y'_n & 1 \end{bmatrix} \begin{bmatrix} f_{11} \\ f_{12} \\ f_{13} \\ f_{21} \\ f_{22} \\ f_{23} \\ f_{31} \\ f_{32} \\ f_{33} \end{bmatrix} = 0 \tag{16}$$

With 8 points, (16) can be solved.

Under epipolar geometry constraint, searching of the corresponding points will be reduced from 2D image to one line.

3.5 Block matching algorithm (BMA)

The key point of the stereo vision is how to search the corresponding points quickly and effectively. Epipolar geometry constraint is the typical skill for finding the corresponding points. However, movement of stereo vision systems will cause the images blurred. Moreover, since the resolution of webcam is pretty low, the task of searching the corresponding points becomes difficult. Here we will apply the block matching algorithm (BMA) for searching the corresponding points.

BMA is a standard technique for encoding motion in video sequences. It aims at detecting the similar block between two images. The matching efficiency depends on the chosen block size and search region. It is not easy to choose an appropriate block size. Usually, bigger blocks are less sensitive to the noise but will spend more computation time. Fast matching methods for searching the corresponding points have been proposed in some representative studies (e.g. Tao, 2008). Tao's method is to match the reference block in the left image and the candidate block on the epipolar line in the right image. Here we need more accurate matching results for finding out the corresponding points. Therefore, the full-search (FS) is used for achieving better searching results.

In Figure 11, the sum of absolute difference (SAD) is used for the block similarity measuring. The first pixel of the chosen block on left image is (x, y) and the block size is N×N. The search region on the right image is defined to be a rectangular with the image width as the width and height of $2k$. Since the left and right cameras of the stereo vision system are placed on the same line, the k can be small in order for reducing the computation time.

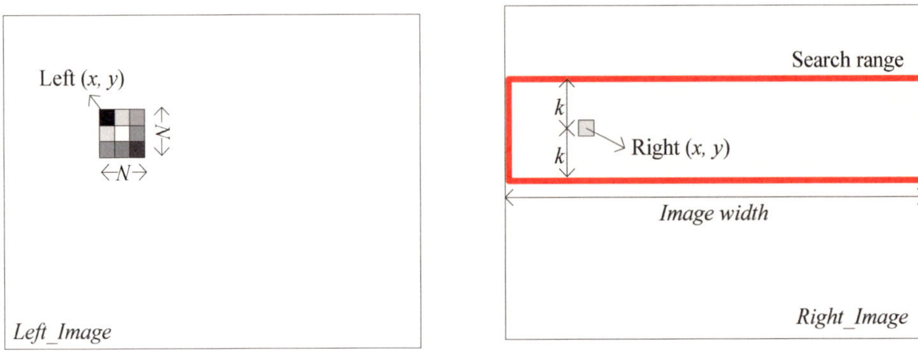

Fig. 11. Block template and the search region.

3.6 Matching cost function

The matching cost function is a norm to represent the degree of correctness of a match. The smaller the matching cost, the higher the correctness. The sum of squared differences (SSD) and normalized cross correlation (NCC) are frequently used matching cost functions other than SAD. Functions of SAD, SSD, and NCC are illustrated as (17-19).

$$SAD(x,y,r,s) = \sum_{i=0}^{i=m}\sum_{j=0}^{j=n}\left|\text{R\_Image}_{(x+i,y+j)} - \text{L\_image}_{((x+r)+i,(y+s)+j)}\right| \tag{17}$$

$$SSD = \sum_{i=0}^{i=m}\sum_{j=0}^{j=n}\left(\text{R\_Image}_{(x+i,y+j)} - \text{L\_image}_{((x+r)+i,(y+s)+j)}\right)^2 \tag{18}$$

$$NCC = \frac{\sum_{i=0}^{i=m}\sum_{j=0}^{j=n}\left(\text{R\_Image}_{(x+i,y+j)}\right)\left(\text{L\_image}_{((x+r)+i,(y+s)+j)}\right)}{\sqrt{\sum_{i=0}^{i=m}\sum_{j=0}^{j=n}\left(\text{R\_Image}_{(x+i,y+j)}\right)^2 \sum_{i=0}^{i=m}\sum_{j=0}^{j=n}\left(\text{L\_image}_{((x+r)+i,(y+s)+j)}\right)^2}} \tag{19}$$

where m and n are the length and width of the block, (x, y) is the position of the block on right image and (r, s) denotes the motion vector.

Comparing these matching cost functions to norms in algebra, the SAD is analogous to an 1-norm in algebra, and the SSD is analogous to a 2-norm in algebra. And the NCC uses an inner-product-like operation. Obviously the computation complexity of SAD is lower than the other matching functions. The SAD is the most frequently used matching cost function in applications since it is one of the more computationally efficient methods. The advantage of SAD has been mentioned in literatures (Humenberger, 2010; Point Grey Research Inc., 2000; Bradski, 2010). In our application, the computation time is an important factor, therefore the SAD will be the matching cost function for searching the correspondence.

4. Camera calibration

A camera calibration is done to build the relationship between the world coordinates and their corresponding image coordinates. Consider a pinhole camera model (David & Ponce, 2002) as shown in Figure 12. The camera parameters can be distributed into intrinsic parameters and extrinsic parameters (Hartley & Zisserman, 2003).

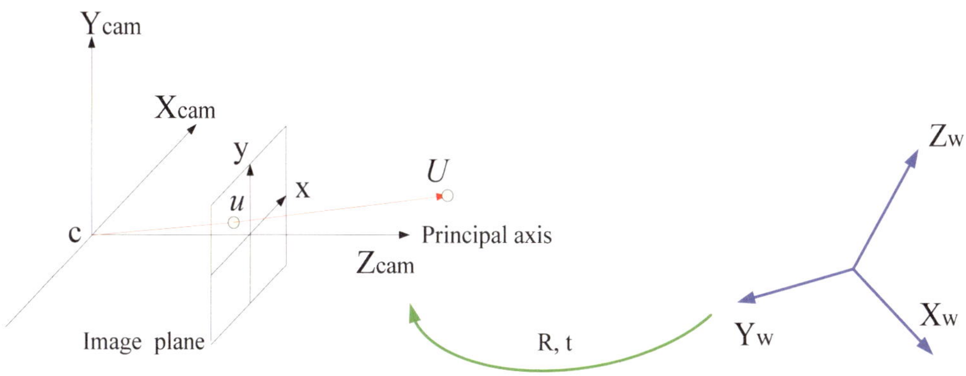

Fig. 12. Pinhole camera model and projective transformation between world and camera coordinate system.

Based on the collinearity equation and the pinhole camera model (Luhmann et al., 2007), the transformation between image point and reference point can be represented as

$$
\begin{aligned}
q &= \frac{L_1X + L_2Y + L_3Z + L_4}{L_9X + L_{10}Y + L_{11}Z + 1} \\
\tilde{q} &= \frac{L_5X + L_6Y + L_7Z + L_8}{L_9X + L_{10}Y + L_{11}Z + 1}
\end{aligned}
\tag{20}
$$

where $(q,\tilde{q})$ is a reference point on the image. Re-arrange equation (20), we will have:

$$
\begin{aligned}
L_1X + L_2Y + L_3Z + L_4 - qL_9X - qL_{10}Y - qL_{11}Z = q \\
L_5X + L_6Y + L_7Z + L_8 - \tilde{q}L_9X - \tilde{q}L_{10}Y - \tilde{q}L_{11}Z = \tilde{q}
\end{aligned}
\tag{21}
$$

Equation (21) is equivalent to

$$
\begin{bmatrix} X & Y & Z & 1 & 0 & 0 & 0 & 0 & -qX & -qY & -qZ \\ 0 & 0 & 0 & 0 & X & Y & Z & 1 & -\tilde{q}X & -\tilde{q}Y & -\tilde{q}Z \end{bmatrix}
\begin{bmatrix} L_1 \\ L_2 \\ L_3 \\ L_4 \\ L_5 \\ L_6 \\ L_7 \\ L_8 \\ L_9 \\ L_{10} \\ L_{11} \end{bmatrix} = 0
\tag{22}
$$

The coefficients $L_1 \sim L_{11}$ are called DLT parameters. In order to solve (22), at least 6 control points are required.

The special orientation of camera can be reconstructed by DLT parameters. The principal point $(q_0, \tilde{q}_0)$ is

$$\begin{aligned} q_0 &= J^2\left(L_1L_9 + L_2L_{10} + L_3L_{11}\right) \\ \tilde{q}_0 &= J^2\left(L_5L_9 + L_6L_{10} + L_7L_{11}\right) \end{aligned} \tag{23}$$

where $J = -\left(\sqrt{K_9^2 + K_{10}^2 + K_{11}^2}\right)^{-1}$, and the principal distance is

$$\begin{aligned} c_q &= \left[J^2\left(L_1^2 + L_2^2 + L_3^2\right) - q_0^2\right]^{1/2} \\ c_{\tilde{q}} &= \left[J^2\left(L_5^2 + L_6^2 + L_7^2\right) - \tilde{q}_0^2\right]^{1/2} \end{aligned} \tag{24}$$

The elements of the rotation matrix R are

$$\begin{aligned} r_{11} &= \frac{J(q_0L_9 - L_1)}{c_q} & r_{12} &= \frac{B(\tilde{q}_0L_9 - L_5)}{c_{\tilde{q}}} & r_{13} &= JL_9 \\ r_{21} &= \frac{J(q_0L_{10} - L_2)}{c_q} & r_{22} &= \frac{B(\tilde{q}_0L_{10} - L_6)}{c_{\tilde{q}}} & r_{23} &= JL_{10} \\ r_{31} &= \frac{J(q_0L_{11} - L_3)}{c_q} & r_{32} &= \frac{B(\tilde{q}_0L_{11} - L_7)}{c_{\tilde{q}}} & r_{33} &= JL_{11} \end{aligned} \tag{25}$$

The position of the camera center will be given by

$$\begin{bmatrix} X_0 \\ Y_0 \\ Z_0 \end{bmatrix} = -\begin{bmatrix} L_1 & L_2 & L_3 \\ L_5 & L_6 & L_7 \\ L_9 & L_{10} & L_{11} \end{bmatrix}^{-1} \begin{bmatrix} L_4 \\ L_8 \\ 1 \end{bmatrix} \tag{26}$$

From (23) -(26), the camera matrix C is obtained.

$$C = KP \tag{27}$$

where K, called four-parameter, is given by:

$$K = \begin{bmatrix} f_x & & p_x \\ & f_y & p_y \\ & & 1 \end{bmatrix} \tag{28}$$

and P, called camera matrix, is given by:

$$P = (R \mid t) \tag{29}$$

where t is the translation matrix.

There are many ways to solve the camera matrix *P*. Least square method, SVD or pseudo-inverse method can be used in the case of an over-determined system. For example, Hartley (1997) used the specific form of Kruppa's Equation and explicitly in terms of singular value decomposition (SVD) of fundamental matrix for calculating the focal length of camera; Zhang (1999) proposed the camera calibration procedure by using specifically model plane; and Heikkilä (2000) proposed the 4-step camera calibration procedure to solve the projective relation between the model plane and the image plane. Here the calibration of the webcam is base on Zhang's procedure for solving camera matrix *P*.

The Matlab® based toolbox developed by Bouguet (2008) is used for camera calibration. The image size is 640×480 pixels and the intrinsic parameters of the webcams and camera matrix are shown in Table 1.

	Left camera		Right camera	
Focal Length	f_x	f_y	f_x	f_y
	688.92	690.84	690.62	690.30
Principal point	293.70	237.91	289.32	242.02

Table 1. Intrinsic parameters of the webcams.

5. Experimental results

In our applications, the stereo system will provide the real-time height information for AUHs. It is necessary that the method should be simple and fast. The local search BMA and epipolar geometry constraints are utilized for searching the stereo corresponding points. There are three parts in this section. In the first part, corresponding points with fundamental matrix are searched and the BMA sreaching are verified. Next, the simulations of height estimation for AUHs are illustrated. The third part of this section demonstrates comparison of our methods with some other methods.

5.1 Measurement results of epipolar geometry constraints and BMA

From Table 1, the corresponding points can be obtained with BMA and epipolar geometry constraint from the obtained parameters of the webcams. As mentioned above, the fundamental matrix can be solved by the left and right camera matrices. An example shows the two images under the epipolar geometry constraint, and the results are shown in Figure 13. Right images are the reference images, and the corresponding points are lying on the epipolar lines of the left images.

Figure 14 demostrates the corresponding points searching results with grid board. Almost all matching points between two images in specific area have been found out. The estimation results of different distances are shown in Figure 15. In the simulations, the measurement range is from 50 cm to 600 cm. Figure 15 shows that the error of estimation is less than 10 cm when the distance is 225 cm and baseline is 10 cm. When the distance is 300 cm with baseline being 15 cm, the estimation error is less than 10 cm. When the distance is 425 cm with baseline being 25 cm, the error of estimation is less than 10 cm. When the distance is 475 cm with baseline being 25 cm, the estimation error is less than 10cm. So we can conclude that the wider the baseline is, the further the measurement distance is.

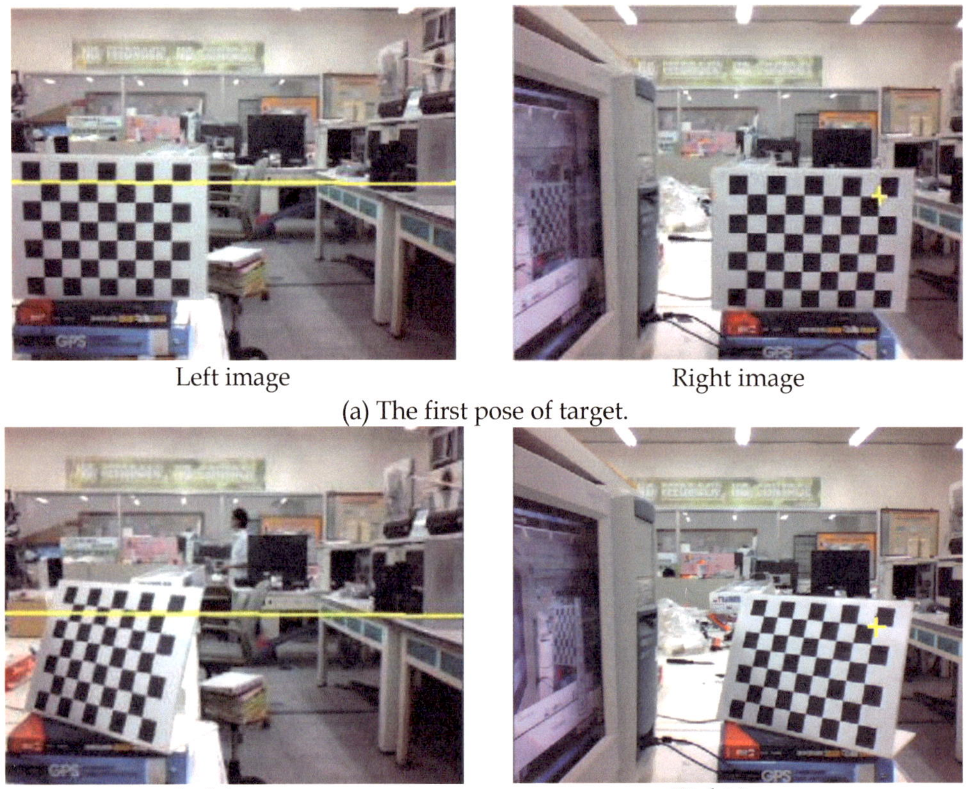

Left image Right image

(a) The first pose of target.

Left image Right image

(b) The second pose of target.

Fig. 13. Epipolar geometry constraint.

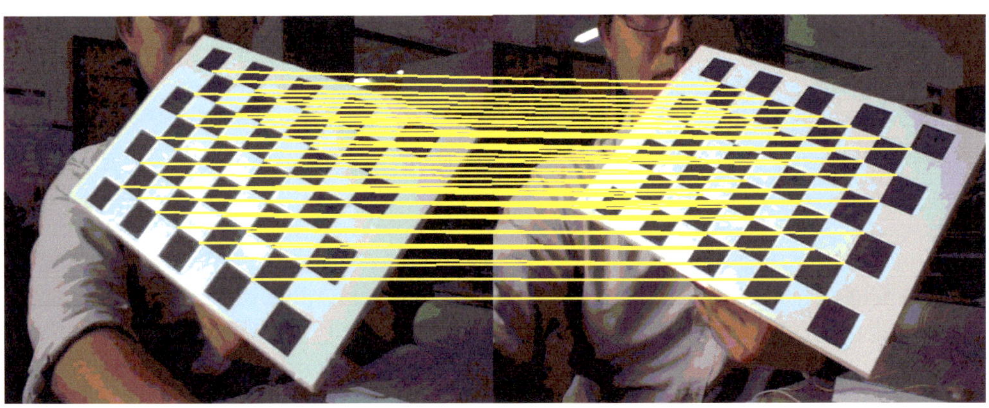

Fig. 14. Searching results of corresponding points with epipolar geometry constraint.

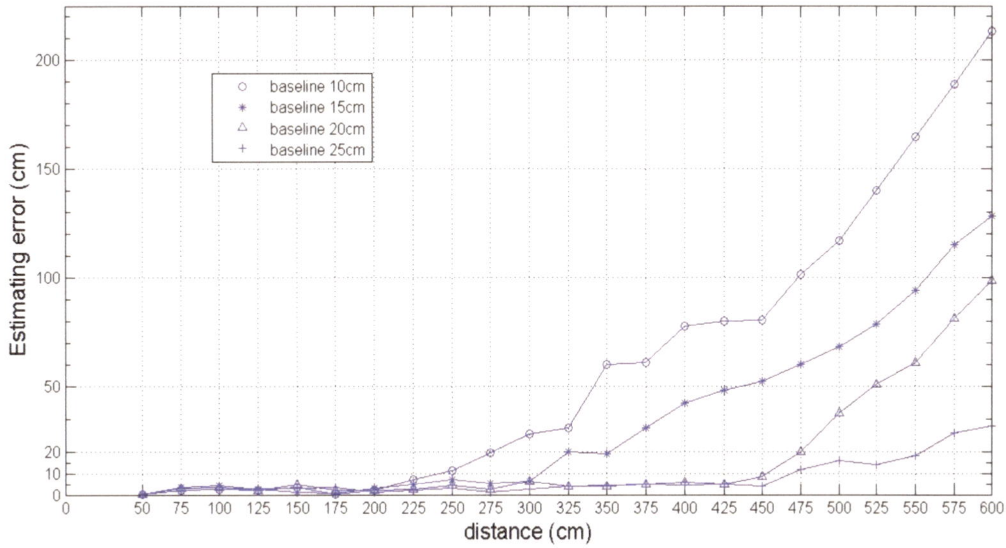

Fig. 15. Estimation errors with epipolar geometry constraint method.

For the BMA, Figure 16 shows that the size of the template block on left image is 5×5 pixels, the searching range is 640×10 pixels and the distance of target is 700 cm. The estimation results of different distances by BMA are shown in Figure 17.

From Figures 15 and 17, we can conclude that the measurement distance increase with baseline increasing. We can also conclude that the range of measurement distance of BMA is more than that of epipolar geometry constraint method, in the sense of the same error tolerance. However, as the measuring range increasing, BMA searching results almost in the same range because of the low resolution of the image, and this causes that the measurement error increases quickly.

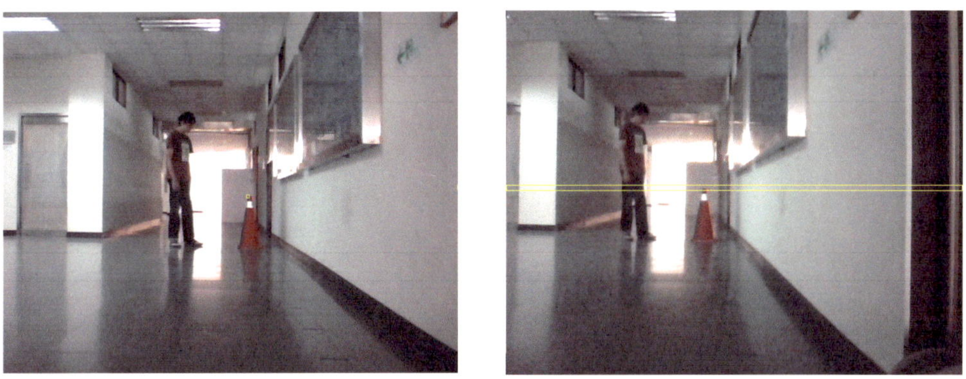

Left image (Baseline 25cm) Right image (Baseline 25cm)

Fig. 16. Searching corresponding points with BMA.

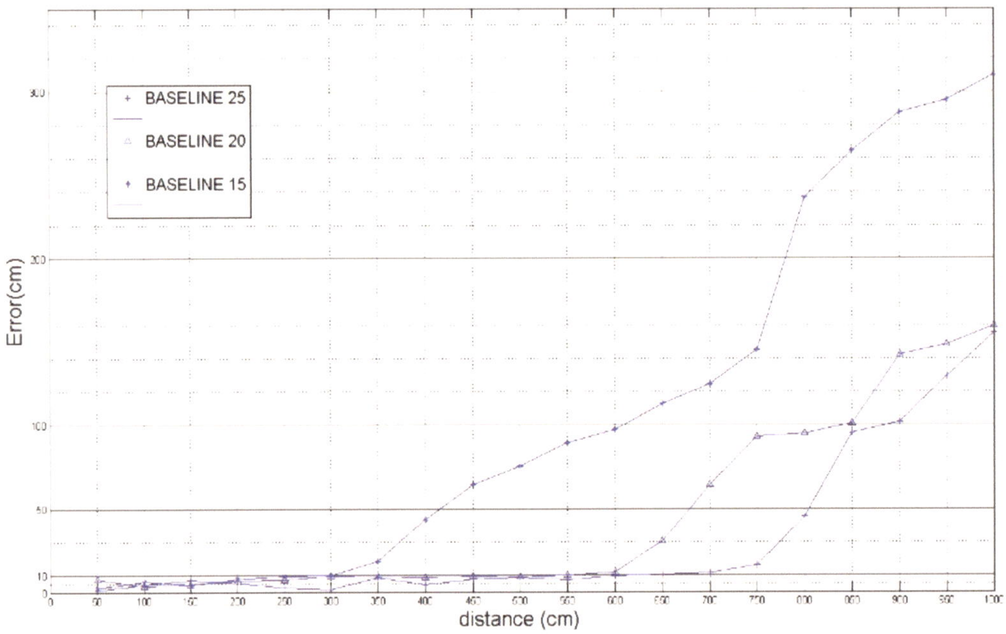

Fig. 17. Estimation errors with BMA.

5.2 The simulations of height estimation for AUHs

In this section, some stereo pairs of aerial photographs are captured for demostrating our methods. Figures 18 through 21 illustrate the stereo pairs captured at different heights including 10.1m, 8.2m, 5.3m and 3.7m with baselines 10cm, 15cm, 20cm and 25cm. The Figure (c) of each figure group is an example for demostrating the height estimation. Figure 22 illustrates the results of image proccessing. Figure 22(b) is the edge image of the landmark and Figure 22(c) is the disparity map.

(a) Baseline 10cm

(b) Baseline 15cm

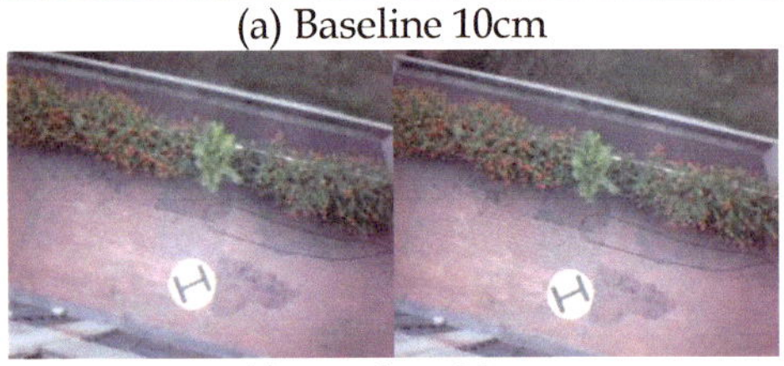

(c) Baseline 20cm

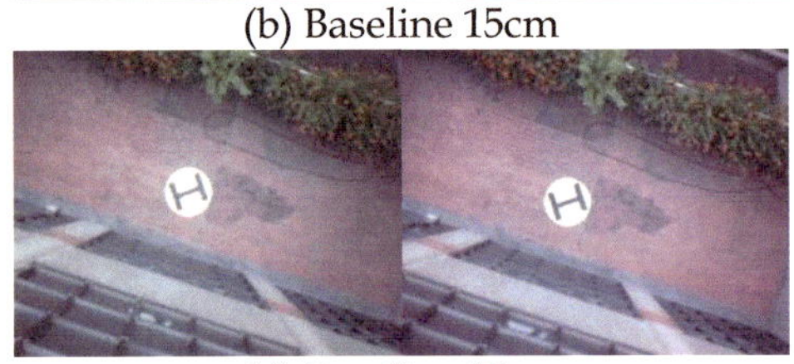

(d) Baseline 25cm

Fig. 18. Stereo pictures captured at height 10.1m.

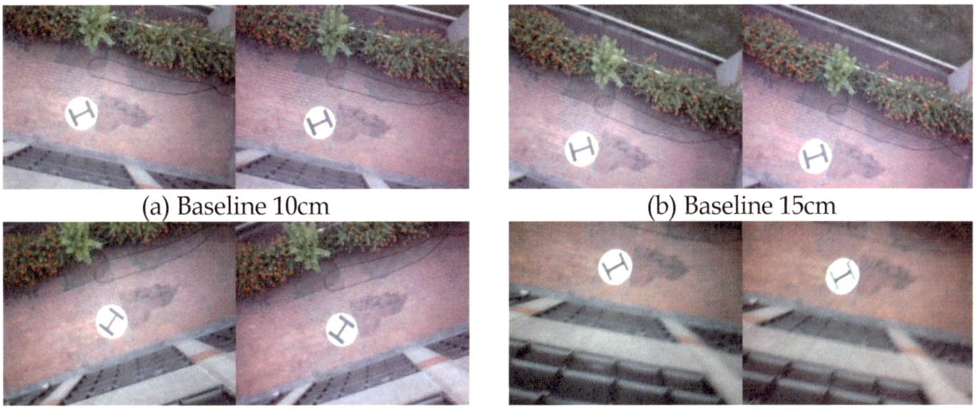

(a) Baseline 10cm (b) Baseline 15cm

(c) Baseline 20cm (d) Baseline 25cm

Fig. 19. Stereo pictures captured at height 8.2m.

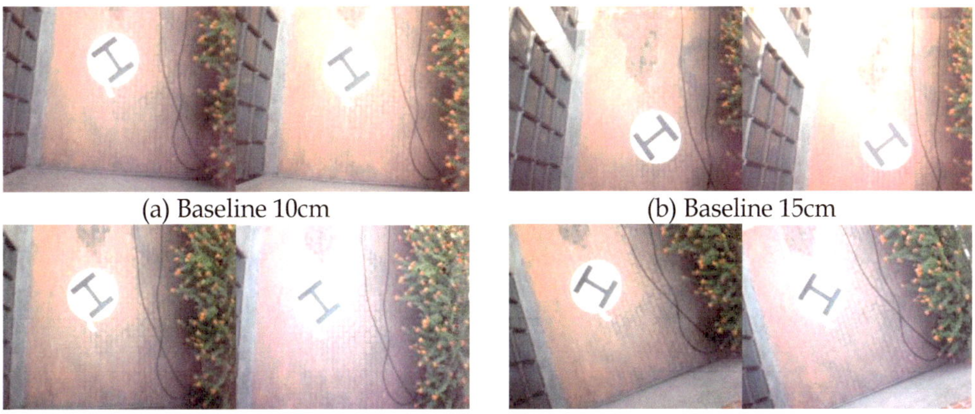

(a) Baseline 10cm (b) Baseline 15cm

(c) Baseline 20cm (d) Baseline 25cm

Fig. 20. Stereo pictures captured at height 5.3m.

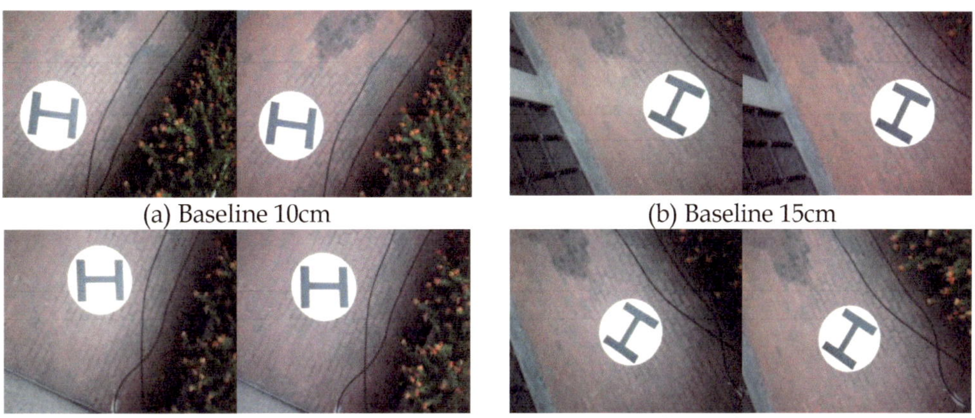

(a) Baseline 10cm (b) Baseline 15cm

(c) Baseline 20cm (d) Baseline 25cm

Fig. 21. Stereo pictures captured at height 3.7m.

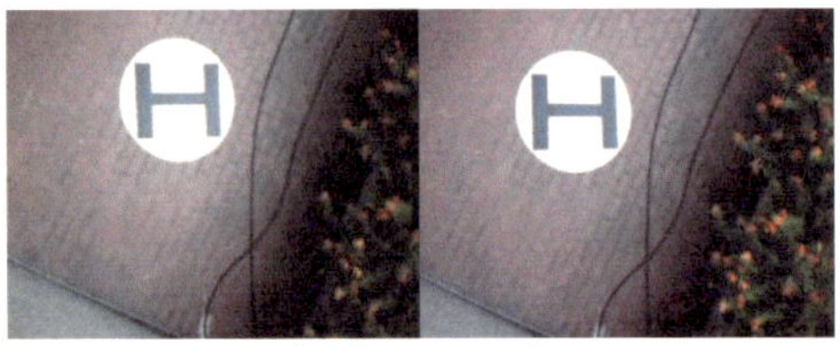

(a) Stereo pair

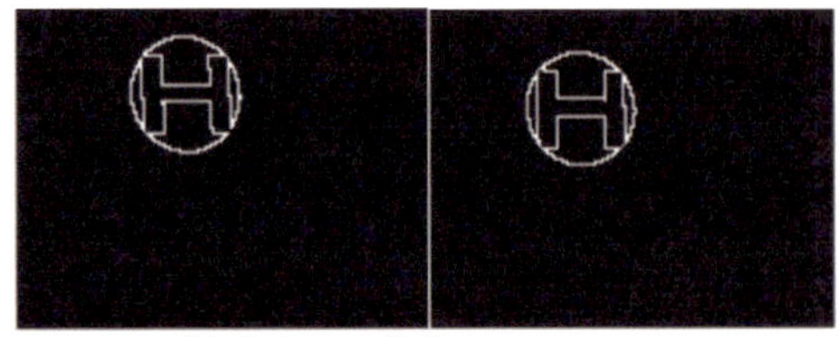

(b) Edges of the stereo pair

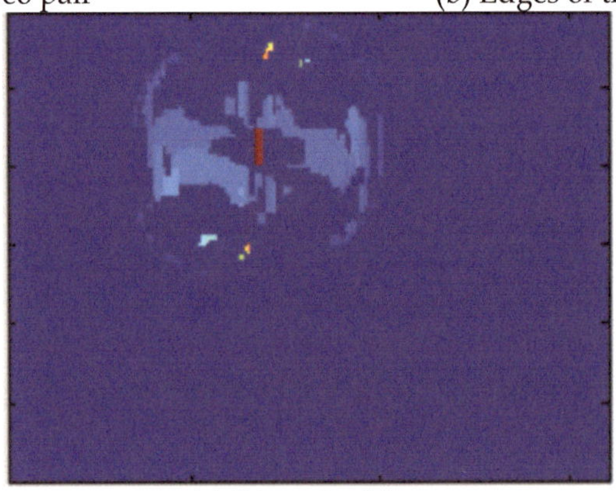

(c) Disparity Map

Fig. 22. Image processing and disparity computing results of Figure 21(b).

The estimating results of the simulations are illustrated in Figure 23. In Figure 23, the x-axis are the length of baselines, the y-axis are the estimation errors, and all the quanties in this figure are in meters. We can find from the figure that the estimation error is decreasing as the baseline increasing. And as the height growing with the baseline unchanged, the estimation error is increasing. When the height is 10.1m and the baseline is 10cm, both the errors of BMA and epipolar geometry constraint are over 2.5m. We can also conclude that the estimation errors by BMA will be less than those by epipolar geometry at the same condition.

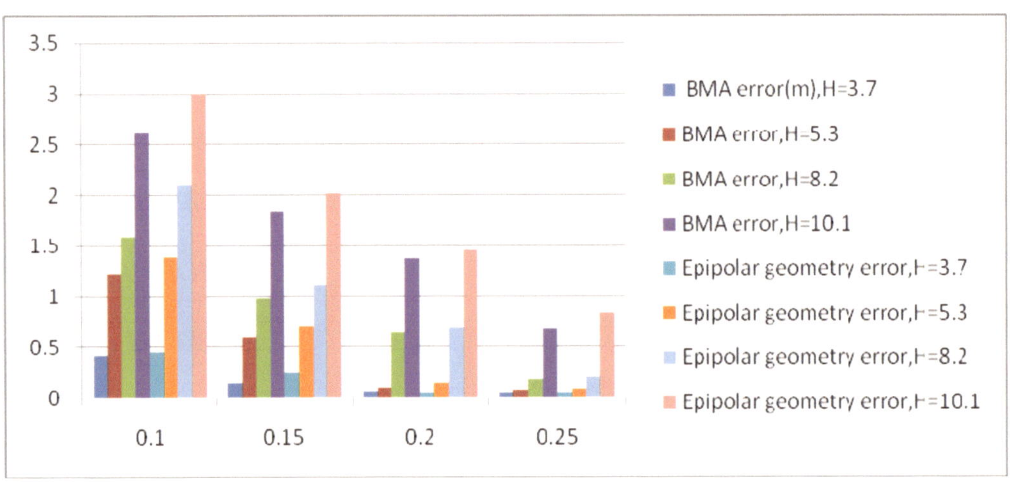

Fig. 23. Estimation results of landing simulations via BMA and epipolar geometry constraint.

5.3 Comparison of our methods with other methods

In Humenberger's work (Humenberger et al., 2010), a comparison of prosessing speed of some real-time stereo vision systems has been made. The proposed methods in this chapter are computed on the platform of CPU. The methods of Point Grey Research Inc. (Point Grey Research Inc., 2000), and Bradski (Bradski, 2010) are also computed on the platform of CPU. Table 2 shows the processing speed of these two systems comparing to our methods with Fig. 22(a) as the test image and SAD as the matching cost function. The size of the test image is 640x480. The processing speed is given in frames per second (fps).

Reference	**Frames per second (fps)**
Point Grey Research Inc.	34.5
Bradski	21.1
Our method 1 (Epipolar Geometry Constraint)	24.6
Our method 2 (BMA)	20.1

Table 2. Computation time of different matching methods.

6. Conclusion

For general purpose of helicopter autonomous flight, GPS is very useful. However, position information provided by GPS is not accurate enough for autonomous landing of an helicopter. The stereo vision system is designed to assist GPS while helicopter is autonomous landing. For small unmanned helicopters, the effective measuring range of 8 m is enough for landing control. The stereo vision system is a very competent sensor for height estimation. On the helicopter autonomous landing problem, stereo vision system could estimate the height of the helicopter. In this chapter, we proposed a low cost stereo vision system which is much cheaper than a GPS. The proposed system can provide acceptably accurate height information for the unmanned helicopter landing control system in certain range.

From the simulation results, it is evident that different baselines will produce different measurement results. The wider the baseline is, the longer that the system can be used for height estimation with acceptable range of error. Comparing the height estimation error of GPS, we can conclude that the system indeed provides more accurate information of height, and it is more useful for the helicopter autonomous landing. To increase the measurement range, one should use cameras of higher resolution and/or increase the baseline.

There are three major works need to be concerned in the future study. Firstly, maybe it is neessary to increase the number of cameras for expanding the camera range. In recent years, multi-view 3D construction technology had made significant progress. The 3D topographical construction with multi-view images that appears to be feasible. The second, the BMA should be further improved. A lot of search methods have been proposed for speeding-up or ameliorating the matching performance. Therefore our approach will be more improved. Finally, the orther matching method (e.g. region matching) will be attempted for better matching performance.

7. References

Alagoz, B.B. (2008). Obtaining depth maps from color images by region based stereo matching algorithms, *Computer Vision and Pattern Recognition*, Vol. 8, No. 4. pp. 1-13.

Bagen, W., Hu, J. & Xu, Y. (2009). A vision-based unmanned helicopter ship board landing system, *International Congress on Image and Signal Processing*, pp. 1–5.

Bedekar, A.S. & Haralick, R.M. (1995). A Bayesian method for triangulation and its application to finding corresponding points, *International Conference on Image Processing*, Vol. 2, pp. 362 - 365.

Bradski, G. & Kaehler, A. (2008). Learning OpenCV, Computer Vision with the OpenCV Library, O'Reilly, 555 pages.

Bouguet, J.Y. (2008). *Camera calibration toolbox for Matlab*, available from: http://www.vision.caltech.edu/bouguetj/calib_doc/

Canny, J.F. (1986). A computational approach to edge detection, *IEEE Transactions on Pattern Analysis and Machine Intelligence*, Vol. 8, No. 6, pp. 679 - 698.

Cyganek, B. (2005) Machine efficient adaptive image matching based on the nonparametric transformations. In: *Computational Science - ICCS 2005, 5th International Conference, Atlanta, GA, USA, May 22-25, 2005, Proceedings, Part I. Volume 3514 of Lecture Notes in Computer Science,* Sunderam, V.S., van Albada, G.D., Sloot, P.M.A., & Dongarra, J. eds., pp. 757 - 765, Springer, ISBN 978-3-540-26032-3, New York.

Cyganek, B. (2006) Matching of the multi-channel images with improved nonparametric transformations and weighted binary distance measures. In: *Combinatorial Image Analysis - Proc. of the 11th International Workshop on Combinatorial Image Analysis (IWCIA 2006), volume 4040 of Lecture Notes in Computer Science*, Reulke, R., Eckardt, U., Flach, B., Knauer, U., & Polthier, K. eds., pages 74–88. Springer, ISBN 978-3-540-35153-5, New York.

Cyganek, B. & Siebert, J.P. (2009). *An Introduction to 3D Computer Vision Techniques and Algorithms*, Wiley, 483 Pages.

David, A. F. & Ponce, J. (2002). *Computer Vision: A Modern Approach*, Prentice Hall, 693 Pages.

Doehler, H.-U. & Korn, B. (2003). Robust position estimation using images from an uncalibrated camera, *Digital Avionics Systems Conference*, Vol. 2, pp. 9.D.2–1–9.D.2–7.

Duan D., Xie M., Mo, Q., Han, Z. & Wan, Y. (2010). An improved Hough transform for line detection, *International Conference on Computer Application and System Modeling*, Vol. 2, pp. V2-354 - V2-357.

Fang, Z., Wu, J. & Li, P. (2008). Cntrol system design and flight testing for a miniature unmanned helicopter. *Proceedings of the 7th World Congress on Intelligent Control and Automation*, pp. 2315-2319.

Fuh, C.S. & Maragos, P. (1989). Region-based optical flow estimation, *Proceedings of IEEE Conference on Computer Vision and Pattern Recognition*, pp.130 - 133.

Glover, F. (1989). Tabu search - Part I, *ORSA Journal on Computing*, Vol. 1, No. 3, pp. 190 - 206.

Glover, F. (1990). Tabu search - Part II, *ORSA Journal on Computing*, Vol. 2, No. 1, pp. 4 - 32.

Gyaourova, A., Kamath C. & Cheung S. (2003). Block matching for object tracking, *Technical Report on Lawrence Livermore National Laboratory*, Report No. UCRL-TR-200271, 13 Pages.

Han, J.H. & Park, J.S. (2000). Contour matching using epipolar geometry, *IEEE Transactions on Pattern Analysis and Machine Intelligence*, Vol.22, pp. 358 - 370.

Hu W.C. (2008). Adaptive template block-based block matching for object tracking, *Eighth International Conference on Intelligent Systems Design and Applications,* Vol. 1, pp. 61 – 64.

Hartley, R.I. & Zisserman, A. (2003). *Multiple View Geometry in computer vision,* Cambridge, 634 Pages.

Hartley, R.I. (1995). In defence of the 8-point algorithm, *5th International Conference on Computer Vision, Cambridge,* pp. 1064-1070.

Hartley, R.I. (1997). Kruppa's equations derived from the fundamental matrix, *IEEE Transactions on Pattern Analysis and Machine Intelligence,* Vol. 19, No. 2, pp. 133 - 135.

Humenberger, M., Zinner, C., Weber, M., Kubinger, W. & Vincze, M. (2010), A fast stereo matching algorithm suitable for embedded real-time systems, *Computer Vision and Image Understanding,* Vol. 114, No.11, pp. 1180 - 1202.

Heikkilä, J. (2000). Geometric Camera Calibration Using Circular Control Points, *IEEE Transactions on Pattern Analysis and Machine Intelligence,* Vol. 22, No. 10, pp. 1066 -1077.

Johnson, A., Montgomery, J. & Matthies, L. (2005). Vision guided landing of an autonomous helicopter in hazardous terrain, *IEEE International Conference on Robotics and Automation,* pp. 3966 - 3971.

Kadmiry, B. & Driankov, D. (2004). A fuzzy gain-scheduler for the attitude control of an unmanned helicopter, *IEEE Transactions on fuzzy systems,* Vol. 12, No. 4, pp. 502 - 515.

Katzourakis, D., Vitzilaios, N.I. & Tsourveloudis, N.C. (2009). Vision aided navigation for unmanned helicopters, *17th Mediterranean Conference on Control & Automation,* pp. 1245-1250.

Liang, T. & Kuo, P. (2008). A novel fast block-matching algorithm for motion estimation using adaptively asymmetric patterns. *International Journal of Innovative Computing, Information and Control (IJICIC),* Vol. 4, No. 8, pp. 2011 - 2024.

Lin, F., Chen, B. M. & Lum, K.Y. (2007). Integration and implementation of a low-cost and vision-based UAV tracking system, *26th Chinese Control Conference,* pp. 731 - 736.

Luhmann, T., Robson, S., Kyle, S. & Harley, I. (2007). *Close Range Photogrammetry: Principles, Techniques and Applications,* Wiley, 528 pages.

Mori, R., Hirata, K. & Kinoshita, T. (2007). Vision-based guidance control of a small-scale unmanned helicopter, *IEEE/RSJ International Conference on Intelligent Robots and Systems,* pp. 2648-2653.

Muquit, M.A., Shibahara, T. & Aoki, T. (2006). A high-accuracy passive 3D measurement system using phase-based image matching, *IEICE Transactions on Fundamentals of Electronics, Communications and Computer Sciences,* Vol. E39-A, No. 3, pp. 686-697.

Nixon, M.S. & Aguado A.S. (2008). *Feature Extraction & Image Processing,* Elsevier, 424 pages.

Oh, S.R., Pathak, K., Agrawal, S.K., Pota, H.R. & Garratt, M. (2006). Approaches for a tether-guided landing of an autonomous helicopter, *IEEE Transactions on Robotics,* Vol. 22, No. 3, pp. 536 - 544 .

Olson, C.F. (1997). Maximum-likelihood image matching, *IEEE Transactions on Pattern Analysis and Machine Intelligence,* Vol. 24, No. 6, pp. 853 – 857.

Saito, S., Bao Y., & Mochizuki T. (2007). Autonomous flight control for RC helicopter using camera image, *SICE Annual Conference,* pp. 1536 - 1539.

Siebert, J.P. & Marshall, S.J. (2000). Human body 3D imaging by speckle texture projection photogrammetry, *Sensor Review,* Vol. 20, No. 3, pp. 218 – 226.

Takahashi, Y., Karungaru, S., Fukumi, M. & Akamatsu, N. (2006). Feature point extraction in face image by neural network, *SICE-ICASE International Joint Conference,* pp. 3783 – 3786.

Tao, T., Koo, J.C. & Choi, H.R. (2008). A fast block matching algorithm for stereo correspondence, *IEEE Conference on Cybernetics and Intelligent Systems*, pp. 38 - 41.

Tico, M., Rusu, C. & Kuosmanen, P. (1999). A geometric invariant representation for the identification of corresponding points, *International Conference on Image Processing*, Vol. 2, pp. 462 - 466.

Wang, C., Lei X., Liang, J., Wu, Y. & Wang,T. (2009). An adaptive system identification method for a micro unmanned helicopter robot. *IEEE International Conference on Robotics and Biomimetics*, pp.1093 - 1098.

Wang, C.C., Lien, S.F., Hsia, K.H. & Su, J.P. (2009). Image-guided searching for a landmark, *Artificial Life and Robotics*, Vol. 14, No. 1, pp. 95 - 100.

Wang, L. & Yang R. (2011). Global stereo matching leveraged by sparse ground control points, *IEEE Computer Society Conference on Computer Vision and Pattern Recognition (CVPR)*, pp. 3033 - 3040.

Xu, C., Qiu, L., Liu, M., Kong, B. & Ge, Y. (2006). Stereo vision based relative pose and motion estimation for unmanned helicopter landing, *IEEE International Conference on Information Acquisition*, pp. 31 - 36.

Zhang, Z. (1996). On the epipolar geometry between two images with lens distortion, *13th International Conference on Pattern Recognition*, Vol.1, pp. 407 - 411.

Zhang, Z. (1999). Flexible camera calibration by viewing a plane from unknown orientation, *7th IEEE International Conference on Computer Vision*, pp. 666 - 673.

GPS 18 Technical Specifications, Revision D (2005), Garmin International, 33 Pages.

Triclops, Technical Manual (2000), Point Grey Research Inc., available from: http://www.ptgrey.com/products/triclopsSDK/triclops.pdf

3

Fast Computation of Dense and Reliable Depth Maps from Stereo Images

M. Tornow, M. Grasshoff, N. Nguyen, A. Al-Hamadi and B. Michaelis
Otto-von-Guericke University of Magdeburg
Germany

1. Introduction

Modern cars and robots act and interact more and more autonomously. Therefore they are equipped with a set of various sensors to monitor their surroundings. Depending on the application of such devices, different aspects of the measurement data are relevant and have to be extracted during post processing. The evenness of the movements depends on the sampling rate of the sensors. Yet for close interaction with people a very reliable information about the environment is necessary.

Autonomous vehicles are very common in work processes as in hospitals or production facilities, but the interaction possibilities are currently very limited. In experimental setups cars can drive fully autonomous and robots can directly interact with a person. The difference between both situations is the availability of computation power needed for an acceptable price.

Nevertheless, the continuous development of electronics provide devices with higher computation power, such as graphic processing units (GPUs) or field programmable gate arrays (FPGAs). The structure of GPUs and FPGAs has to be kept in mind when programming such devices. Therefore an algorithm has to be adapted and optimized or altered respectively, towards this structure, for the individual usage, which results in high design efforts. Combining general purpose CPUs with either GPUs or FPGAs the problems of computation power for embedded systems will be reduced in the near future.

Having an environment which is optimized for the visual perception of the human eye, autonomously acting robots and cars need access to information of the environment, which can be extracted by optical observations of the surroundings. For orientation in a 3-d environment with moving objects a 3-d representation of the surroundings is needed. Using vision based measurement systems the 3-d-information can be gained by mono and multi camera systems (with stereo camera systems as the minimal setup) Favaro & Soatto (2007). Processing stereo images needs complex algorithms, which are running continuously at a high frame rate to provide the necessary information for an accurate perception of the objects in time.

In this chapter a high speed calculation of depth maps from stereo images based on FPGAs is introduced. Therefore several cost functions and post processing methods for increased reliability are evaluated. The implementation should be platform independent for easy adaptation to new FPGA-hardware.

2. Calculation of depth maps of stereo images

The principle of stereophotogrammetry relies on the functionality of the human eye and is very long known and well established. It has been used primarily by architects and for geological surveying in civil engineering. In the beginning analog photographs were analyzed by human operators. At a later stage the analog photographs were digitized to allow a faster analysis by computers, thereby enhancing speed and accuracy.

In stereo photogrammetry a set of two cameras is used to gain 3-d-information about the environment. Therefore the parameters of the camera setup must be estimated with high accuracy and must be held constant during the measurement process. In the standard case of stereophotogrammetry the position of the cameras and the angle between cameras optical axis can be chosen freely, unless parts of the fields of view of both cameras are overlapping. For processing stereo images taken in the standard case of stereophotogrammetry the calibration process (Albertz & Wiggenhagen, 2009, pp. 247) has a high complexity and the correspondence analysis has to cover a wide range. To reduce the calibration effort as well as the range for the correspondence analysis the normal case of stereophotogrammetry as shown in figure 1 is used. In this setup two identical cameras are arranged with parallel optical axis are used, while the image sensors are exactly aligned.

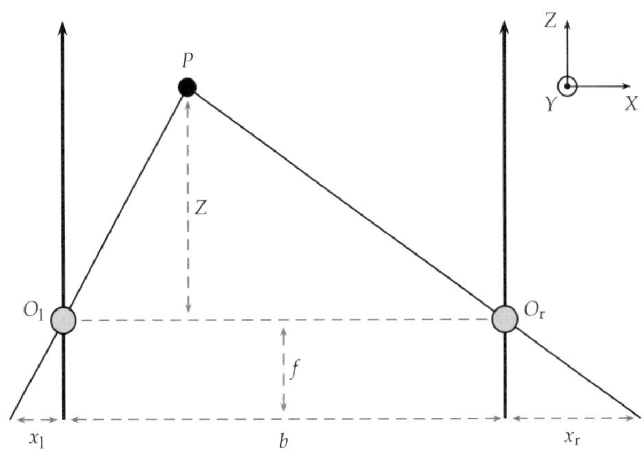

Fig. 1. Normal Case of Stereophotogrammetry

In figure 1 X,Y and Z are the 3-d coordinates of the world coordinations system and x and y are the coordinates of the 2-d image coordination system, with the axes parallel to X and Y. b is the base width, which represents the distance between the perspective centers of both cameras. O_l and O_r are the focal points of both cameras while f is the focal length. Having a scene point P with its representation at x_l, y_l in the left image and $x_r,, y_r$ in the right image its distance Z can be calculated via triangulation. d, known as disparity, is in inverse proportion to Z. It can be determined using equation (1) (Faugeras, 1993, pp. 175).

$$d = x_l - x_r \tag{1}$$

Having c named as the camera constant, the world coordinates can be estimated from digital stereo images using the equations in (2). c is the focal length divided by the pixel size.

$$X = x_l \cdot \frac{b}{d}; Y = y_l \cdot \frac{b}{d}; Z = c_k \cdot \frac{b}{d} \tag{2}$$

Ideally only the base width b and the camera constant c need to be estimated during the calibration process for a stereo camera system arranged in normal case (Trucco & Verri, 1998, p. 140). Yet the lenses, used in cameras, are distorting the images depending on the focal length f. The quality of the lenses has an effect to the representation of the image as well. Thus parameters for the image distortion have to be estimated during the calibration process additionally.

The next step lies in the estimation of the image coordinates. While having x_l and y_l of the left image, the representation x_r of the scene point P in the right image is needed for triangulation, thus the information has to be retrieved by comparing both images. This operation is known as the correspondence problem and is solved by using methods of correspondence analysis.

3. Correspondence analysis

The correspondence analysis is as important as the calibration for generating a dense and reliable depth map. Thus many algorithms have been developed to solve the correspondence problem. Finding the representation of an object in two images taken from a slightly different angle is a very difficult and calculation power consuming task as every pixel of one image has to be compared to every pixel of the other image.

While global methods (Narasimha, 2010, pp. 15) are used to search iteratively for the best depth map of a stereo image pair, with pixel based methods corresponding pixels in both images are searched for. Pixel based methods for correspondence analysis can be divided into feature based and block based algorithms.

Feature based algorithms provide reliable depth maps, yet with a low density. With feature based algorithms characteristic features are determined for both stereo images. Using these features the images are compared and the depth information is extracted. The features are ideally unique to the region (Trucco & Verri, 1998, pp. 145), such as corners, edges and lines. The Speed-Up-Robust-Features (SURF)-algorithm Bay et al. (2008) is an example for a fairly new feature based method and allows a unique and robust identification of blob-like regions using a set of haar-like features, which is independent regarding size and angle. Applying these methods just a few positions have to be compared and computation power can be saved. Thus a lot of high speed stereo algorithms were feature based in the past (Szeliski, 2010, pp. 475). Due to the low resolution of the depth map these algorithms are very useful for high accuracy measurements of 3-d-information in known environments. Yet the representation of unknown objects in changing environments via feature based algorithms is a difficult task, because it can't be ensured that all objects are covered with feature points.

Block based methods (Narasimha, 2010, pp. 15) for correspondence analysis are able to generate relatively dense depth maps, while searching for corresponding blocks for every pixel in the stereo image pair, though with a lower reliability. Dense depth maps have a higher probability to represent all objects in an unknown scene. For block matching algorithms a block taken from the reference image is compared with a set of equally sized blocks of the search image. By varying the number of reference blocks and the number of search blocks the resolution of the depth map can be adjusted. In case of block matching the resolution corresponds directly to the calculation effort.

Applications like driver assistance systems or obstacle detection for autonomic robots need specific processing times that implies real time processing with high requirements. On the other side theses applications need a relative exact measurement and require usually a large measurement range. Close objects have high disparities but are most important for collision avoidance systems. To ensure that all objects in a scene are covered by the depth map a fairly dense depth map is required. This is especially important if the scene is analyzed using statistic methods e.g. grid based approaches. In grid based approaches the environment is represented by cells of a specific size arranged in the so called grid. Each cell contains the occupancy grid. For safety a high reliability is important.

In embedded systems algorithm designers have to deal with massive restrictions according to memory size and calculation power. This is a difficult task for image processing but even more difficult for stereo image processing as two images, taken at the same time, have to be compared. Therefore the usage of simple but effective algorithms is necessary.

3.1 Cost functions for block based algorithms

For comparing reference and search blocks cost functions are used. The traditional criteria such as normalized-cross-correlations-function (NCCF), the sum-of-absolute-differences (SAD) shown in equation (3) and the sum of squared differences (SSD) shown in equation (4) are motivated by signal processing applications. $P_r(i,j)$ is the gray value of a pixel of the reference block at the position i,j. $F(\xi+i,\eta+j)$ is the gray value of a pixel of the search block at the position i,j and displaced by ξ,η.

$$SAD(\xi,\eta) = \sum_{j=0}^{n-1} \sum_{i=0}^{m-1} |P_r(i,j) - F(\xi+i,\eta+j)| \tag{3}$$

$$SSD(\xi,\eta) = \sum_{j=0}^{n-1} \sum_{i=0}^{m-1} (P_r(i,j) - F(\xi+i,\eta+j))^2 \tag{4}$$

By replacing the gray values of the image with the zero mean gray values $\overline{P_r(i,j)}$ and $\overline{F(\xi+i,\eta+j)}$ according to the current block the SSD and the SAD gain robustness regarding brightness variations between both images. The zero mean versions are called ZSAD and ZSSD. The best block combination results in minimal value (ideal zero) for the cost functions SAD, ZSAD, SSD and ZSSD, as they determine the differences between two blocks. In equation (5) the ZNCCF zero-mean-normalized-cross-correlation-function is shown using the same terms as used for the SAD and the SSD. The normalization improves the robustness against image capture variances of both cameras. The values of the ZNCCF range from -1 to 1 due to the normalization. The best fitting block combination can be identified by ZNCCF-values close to 1.

$$ZNCCF(\xi,\eta) = \frac{\sum_{j=0}^{n-1} \sum_{i=0}^{m-1} \left(\overline{P_r(i,j)} \cdot \overline{F(\xi+i,\eta+j)}\right)}{\sqrt{\sum_{j=0}^{n-1} \sum_{i=0}^{m-1} \overline{P_r(i,j)}^2 \cdot \sum_{j=0}^{n-1} \sum_{i=0}^{m-1} \overline{F(\xi+i,\eta+j)}^2}} \tag{5}$$

The Census-transformation (Zabih & Woodfill, 1997, pp. 5) on the other hand is fairly new and motivated by vision systems for robots with strong capabilities for comparing image blocks.

First both images are converted using the Census-transformation shown in fig. 2. A block with an odd number of pixels in horizontal and vertical directions is transformed by comparing the

$$\begin{bmatrix} 41 & 154 & 115 & 211 & 27 \\ 203 & 67 & 21 & 137 & 246 \\ 79 & 167 & (58) & 255 & 1 \\ 135 & 176 & 233 & 20 & 198 \\ 42 & 191 & 39 & 113 & 209 \end{bmatrix} \Longrightarrow \begin{bmatrix} 0 & 1 & 1 & 1 & 0 \\ 1 & 1 & 0 & 1 & 1 \\ 1 & 1 & X & 1 & 0 \\ 1 & 1 & 1 & 0 & 1 \\ 0 & 1 & 0 & 1 & 1 \end{bmatrix} \Longrightarrow 01110\,11011\,1110\,11101\,01011$$

Fig. 2. Census-Transformation of a 5 x 5 px-Block

gray values of all pixels with the pixel at the center (in brackets). If the gray value is lower than the value in the center of the block its value is set to zero otherwise it is set to one. Then these values are assembled to a bit vector of 24 bits and assigned to the position of the center pixel. The block size is usually smaller than the block size used for the correlation. The Census-transformation is only coding the surrounding structure of the pixel, but not the gray values. The Census-transformation is robust against variations of brightness, but shows a sensitivity to high frequency noise.

Next the blocks of the Census-transformed images are compared using the hamming distance. Since the hamming distance of the Census-tranformed images detects differences between reference and search block like the SAD and the SSD, low values indicate good block combinations. Since the correlation of the image blocks is only a binary operation it is suitable for a hardware implementation (Pelissier & Berry, 2010, p. 7) and can be easily realized using combinational logic (Zabih & Woodfill, 1997, p. 7)(Jin et al., 2010, p. 2).

$$\begin{bmatrix} & 115 & \\ & 21 & \\ 79 & (58) & 1 \\ & 233 & \\ & 39 & \end{bmatrix} \Longrightarrow \begin{bmatrix} & 1 & \\ & 0 & \\ 1 & X & 0 \\ & 1 & \\ & 0 & \end{bmatrix} \Longrightarrow 10\,10\,10$$

Fig. 3. Mini-Census-Transformation on a 5 x 5 px-Block

The Mini-Census-transformation (Chang et al., 2010, pp. 3) is optimized for saving computation power by reducing the length of the bit vector for each Census-transformed pixel to 6 bits instead of 24 bits, as shown in fig. 3. This reduces the implementation effort on either, the Census-transformation as well as the calculation of the hamming distance, while its results are nearly as good as the ones using the full Census-transformation.

3.2 Comparison of cost functions

To find the best suited method for calculating depth maps from stereo images several cost functions were evaluated. In Hirschmueller & Scharstein (2009) an overview of the comparison of several methods for stereo matching by usage of the middleburry stereo dataset Scharstein (2011) is given. Therefor different methods, using a set of cost functions, are applied to radio-metrically clean as well as distorted image pairs. These image pairs vary in size from 384 x 288 px to 450 x 375 px with a maximum disparity of 16 px or 64 px including noise and varying brightness. As a result for block based matching the ZNCCF as well as the Census-transformation performed well in most of the tests.

This chapter covers straight block based correlation methods, as iterative methods are not really suitable for real time processing on fast image sequences. First the depth maps of all mentioned cost functions are evaluated by numbers of correct points compared to the ground truth depth map provided within the middleburry datasets. For evaluation of the algorithms the image pairs *Art* and *Dolls* from the middleburry stereo data set from 2005 Scharstein (2011) are used. For meeting our test conditions these images were taken in original resolution 1390 x 1110 px and cut to 1024 x 1024 px (See fig. 4). For both datasets the disparity rises up to 220 px.

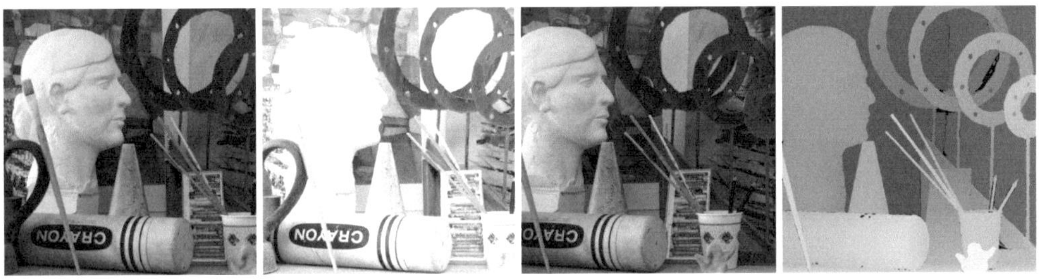

(a) *Art* Left Image (b) *Art* Left Image Brighter (c) *Art* Right Image (d) *Art* Ground Truth

Fig. 4. Middleburry Stereo Dataset *Art*

Following Scharstein & Szeliski (2002) and Scharstein (2011) the calculation of the depth maps uses 5 x 5 px-blocks while the disparity range is extended to 256 px.

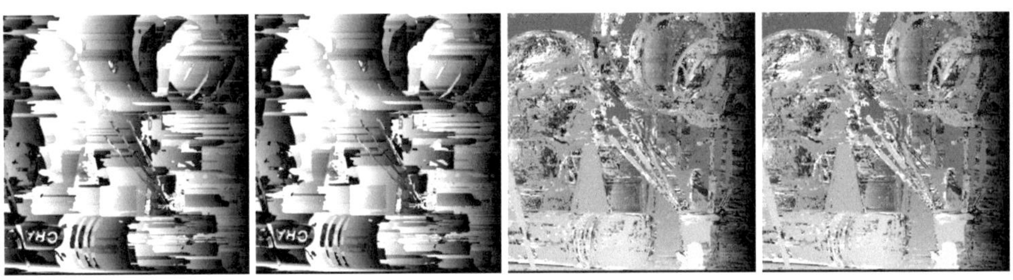

(a) SAD (b) SSD (c) ZSAD (d) ZSSD

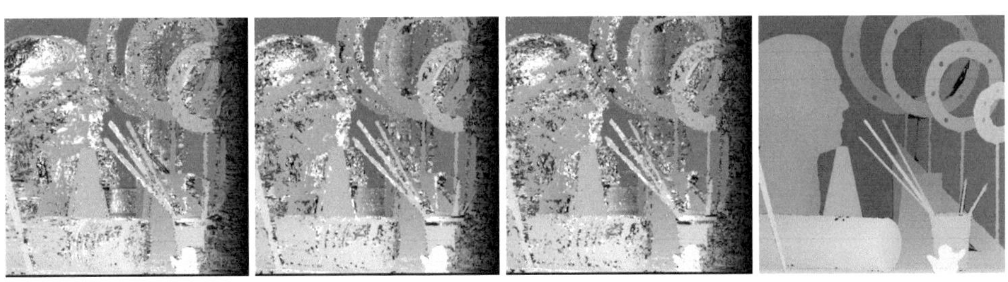

(e) ZNCCF (f) Census (g) Mini-Census (h) Ground Truth

Fig. 5. Disparity Maps for various cost functions of the Stereo Dataset *Art* with unequal exposure time

	Dataset *Art*		Dataset *Dolls*	
Cost Function	Equal Exposure	Unequal Exposure	Equal Exposure	Unequal Exposure
SAD	42 %	0 %	51 %	1 %
SSD	43 %	1 %	52 %	1 %
ZSAD	51 %	34 %	57 %	42 %
ZSSD	51 %	35 %	57 %	43 %
ZNCCF	51 %	41 %	55 %	50 %
Census	55 %	46 %	57 %	52 %
Mini-Census	53 %	44 %	54 %	49 %

Table 1. Accuracy of the Depth Maps for *Art* and *Dolls* Depending on the Cost Function

While the results of all cost functions of the dataset *Art* with equal exposure time are very similar to each other, the results for unequal exposure (see fig. 5) show distinct differences. In table 1 the number of correctly estimated points, points with a maximal difference of the disparity value regarding to the ground truth map, are listed for each of the cost functions for equal as well as unequal exposure. The same information is given for the dataset *Dolls*, yet only in numbers. The best results overall are gained using the Census-transformation, followed by the ZNCCF and the Mini-Census-transformation. The simple SAD and SSD show the worst results. Especially with the unequal exposure these functions are not a good choice. Using the SAD and the SSD cost functions with zero-mean blocks show acceptable results, and when using equal exposure the results are even comparable to the best results of the Census-transformation. The biggest errors of the Census-transformation, the Mini-Census-transformation and the ZNCCF appear at jumps of disparity. These results can be verified by comparing the depth maps with the ground truth data.

3.3 Impact of the block size for correlation

The used block size for the correlation has a major impact on the results. For block based approaches of stereo matching it is assumed that all pixels in a block have the same disparity and the same information is available of both stereo images. The first assumption is violated by all objects which are not aligned parallel to the imaging plane and all blocks with pixel belonging to more than just one object (Faugeras, 1993, p. 191). The second assumption collides with the terms of stereo image acquisition, due to the fact that both images are taken from at least a slightly different angle.

Using small block sizes the impact of perspectively distortion and disparity jumps can be minimized while the probability of disambiguation increases. Big block sizes reduce the probability of disambiguation but leads to blurred edges and flattened small details (Kanade & Okutomi, 1991, p. 1). By multiple applications of block based operations, as for the Census-transformation and the correlation, using windowed hamming distance, the effect is amplified (McDonnell, 1981, p. 2).

Usually the block size is estimated empirically. Empiric evaluation of the block size for the SAD-function is covered by (Kisacanin et al., 2009, pp. 132) and (Scharstein & Szeliski, 2002, p. 18). Kisacanin et al. (2009) compares the number of incorrect assigned pixels by varying the block size from 3 x 3 px to 11 x 11 px evincing the result, that the number of errors is decreasing til a block size of 9 x 9 px is reached and is increasing above a block size of 11 x 11 px due to the low pass filter effect of block based methods. (Scharstein & Szeliski, 2002, p. 18) comes to a similar conclusion while the block size here is varied from 3 x 3 px to 29 x 29 px. For two of the three test images the error is at its minimum between 9 x 9 px and 11 x 11 px.

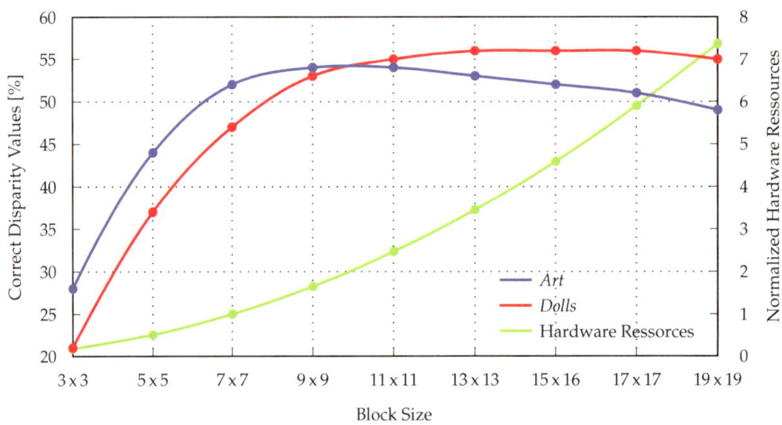

Fig. 6. Results of the Stereo Matching while Varying the Block Size Compared to the Hardware Resources

As the results of both publications are mainly valid for the SAD function, an evaluation for the Mini-Census-transformation was done varying the block size from 3 x 3 px to 19 x 19 px supporting the result of the two publications working with the SAD as the most accurate depth maps here are also calculated using 9 x 9 px and 11 x 11 px for the dataset *Dolls*. But for the dataset *Art* the best results are gained by using a block size of 13 x 13 px. The number of errors increases again past the block size of 17 x 17 px. In figures 10(a) to 10(d) in section 4.8 a choice of depth maps is shown in order to visualize the effect of the block size. In 10(a) with a block size of 3 x 3 px the susceptibility of small block sizes regarding to noise is obvious. Due to a large number of incorrect block assignments the objects are almost indistinguishable. Is the block size increased to the maximum of 19 x 19 px the edges are blurred and details of the objects get lost. Using a block size of 9 x 9 px both effects afore mentioned are noticeable but not prevalent.

To sum up the results are shown in the diagram in fig. 6 for the datasets *Art* and *Dolls*, illustrating that the number of correct block assignments is rising fast for an increased block size with its maximum between the block sizes from 9 x 9 px and 13 x 13 px and falling slowly for bigger block sizes. Additional to the number of correct block assignments for both datasets the diagram shows an estimation for the needed hardware resources. Having squared blocks the hardware resources are rising quadratically. The values are standardized to the needed resources of an hardware implemented stereo analysis for a block size of 7 x 7 px. It is obvious that the stereo image analysis with a block size of 9 x 9 px needs more than one and a half of the resources needed for a block size 7 x 7 px, while the average of the correct block assignment is rising only about 4 % to 5 %. Thus it was decided to use a block size of 7 x 7 px for the processing.

4. Methods for improving the reliability of depth maps from stereo images

Concluding the above section it becomes obvious that straight block based methods always create incorrect block assignments. To improve the reliability of depth maps, using these methods, requires to identify them as incorrect assignments and exclude their values from further processing steps.

The developed methods for improving the depth maps quality are working either iteratively or exclusionary. Iterative methods generate depth maps with higher density and better quality than preclusive methods always resulting in a higher need of computation power, since a lot of intermediate steps have to be calculated as well. Thus the main advantage of excluding incorrect points is that the needed computing power will be reduced, while the main disadvantage is that always correct block assignments are excluded as well. Due to the need of higher computation power iterative methods are not well suited for high speed stereo image processing.

The preclusive methods often work with thresholds to keep the processing as simple as possible. Therefore a good algorithm has to be designed in order to mainly exclude incorrect assignments while ideally no correct assignments are excluded. In this section seven known preclusive methods and their parameters are presented, modified and evaluated for maximizing the number of correct points, while minimizing the number of incorrect points in the depth map. The methods discussed in the next section are: maximal disparity, epipolar lines, thresholds on cost function, first absolute central moment as filter for homogeneous regions, the uniqueness constraint, the continuity constraint, left and right consistency check and multi layer correspondence search.

All methods are introduced shortly and evaluated for their potential of excluding incorrect assigned blocks. Furthermore the best use cases for each method is determined.

4.1 Physical criteria

The physical constrains can exclude many incorrect candidates for block assignments. The maximal disparity is usually applied in restricting regions of searching for correct block assignments. The same can be done for epipolar conditions. Restricting the area to the epipolar line is easily realized for the normal case of stereophotogrammetry but rather difficult in the standard case of the stereophotogrammetry (Faugeras, 1993, pp. 169). Physical constraints will not be further discussed in this work as they can vary according to the camera setup and the application. The maximum disparity is set to 256. For the epipolar condition the normal case of stereophotogrammetry is used and the search area is reduced to an area parallel to the lines of the image sensor.

4.2 Thresholds on the cost function

The threshold on the cost function can be easily applied to the results of the cost function and is a very common method for improving the reliability. In case of the Census-transformation and the Mini-Census-transformation the hamming distance determines the number of differences between the binary vectors. The results of this test are given in table 2. The threshold of 294 bit is the theoretical maximum of the hamming distance for a 7x7 px block and 6 bit Mini-Census vectors. Comparing identical blocks gives a hamming distance of zero. An increasing number of differences between the compared blocks increases the hamming distance until the maximum of 294 bit in this case. By applying a threshold to the cost function the maximal difference between both blocks can be limited (Fua, 1993, p. 2).

	Dataset *Art*			Dataset *Dolls*		
Maximal Hamming Distance	Valid 3-d Points	Correct (Respectively)	Correct (Absolute)	Valid 3-d Points	Correct (Respectively)	Correct (Absolute)
294 Bit	97 %	49 %	48 %	97 %	57 %	55 %
150 Bit	97 %	49 %	48 %	97 %	57 %	55 %
100 Bit	87 %	54 %	47 %	90 %	61 %	55 %
90 Bit	72 %	62 %	45 %	83 %	65 %	54 %
80 Bit	54 %	70 %	38 %	73 %	68 %	50 %
70 Bit	39 %	76 %	30 %	61 %	71 %	43 %
60 Bit	27 %	78 %	21 %	48 %	72 %	35 %
50 Bit	17 %	79 %	13 %	34 %	72 %	24 %

Table 2. Impact of the Threshold on the Maximal Hamming-Distance on the Correlation Result

The calculation of depth maps using various thresholds leads to table 2. All results are given as the ratio of the possible 1048576 point (1024 x 1024 px) in the depth map. The column *"Valid 3-d Points"* represents the rate of the valid 3-d-points. In this case valid 3-d points are identified as valid by applying the threshold on the cost function. The rate of correct block assignments is given by the columns *"Correct (Respectively)"* respectively to the valid 3-d points. Correct points show a difference of maximal one pixel for the disparity according to the ground truth. In column *"Correct (Absolute)"* the rate of all correct block combinations according to the ground truth depth map is given. Thus it is possible to check how many correct values are excluded. The tables 2 to 7 are setup in the same structure.

The results of both images are very similar. Excluding values with a high hamming distance improves the quality of the depth map, by rejecting all points with uncertain matches from the depth map. Yet a very small threshold results in rejecting correct points. In this application a threshold of 90 bits is a good value for restriction via the hamming distance. Here the number of correct points increases about 10% comparing to a threshold of 294 bit (same result when no threshold is given) while only 3% of the correct values are rejected (see table 2). With a threshold of 50 bits for the hamming distance a thin depth map is generated where over 70% of the points are correct according to the ground truth map. The threshold on the hamming distance performs well to rejects incorrect block assignments in case of occlusions due to disparity jumps, but for homogeneous regions its capabilities are limited.

4.3 First absolute central moment for estimating homogeneous regions

Errors in the depth map occur with a high probability in regions of a stereo image pair with low textural information. Especially affected are those areas having the same color or areas covered by large shadows. By identifying blocks belonging to such regions errors can be minimized (van der Mark & Gavrila, 2006, pp. 3).

One possibility to identify regions of low texture information is called *interest operator*,which was introduced by (Moravec, 1977, p. 2) in 1977. The second central moment which complies with the variance is calculated in four directions (vertical, horizontal and both diagonals). The minimum of these four values is used as the variance of the block. This *interest operator* is used in (Konolige, 1997, p.3).

Applying the variance σ^2 to the whole reference block (Falkenhagen, 1994, p. 4) is another method to estimate areas with low textural information. This is demonstrated in equation (6) for a block at the position i, j of an image. The block size is given by W, in either direction, horizontally and vertically. $I_1(x, y)$ gives the intensity of the pixel at position (x, y).

$$\sigma^2 = \frac{1}{(2W+1)^2} \sum_{k=-W}^{W} \sum_{l=-W}^{W} \left(I_1\left(i+k, j+l\right) - \mu\right)^2 \tag{6}$$

μ is the average of the intensity and arises from equation (7).

$$\mu = \frac{1}{(2W+1)^2} \sum_{k=-W}^{W} \sum_{l=-W}^{W} I_1\left(i+k, j+l\right). \tag{7}$$

Calculating the variance, according to equation 6, is not optimized for an FPGA-implementation, due to the number of multiplications and divisions used. While multiplications can be implemented in hardcore embedded multipliers, which are included in most of the current FPGAs, a division is calculated in a resource consuming iterative process (Tornow, 2009, p. 60). To minimize the number of multiplications in a hardware design, the variance known as the second central moment can be replaced by the first absolute central moment (eq. (8)), as used in the opencv library (Willow Garage, 2011, p. 259). In this case the absolute values will prevent that positive and negative differences compensate each other.

$$\overline{\mu} = \frac{1}{(2W+1)^2} \sum_{k=-W}^{W} \sum_{l=-W}^{W} \left|I_1\left(i+k, j+l\right) - \mu\right| \tag{8}$$

If the first absolute central moment is applied to blocks of the same size it can be modified, in order to reduce the needed computation power by substituting the term $1/(2W+1)^2$ with a constant $1/K$. Both divisions can be avoided by multiplying the terms in- and outside the summation with the factor K resulting in the equations (9) and (10). With these steps the number of multiplications is reduced to one.

$$\overline{\mu}_{\text{mod}} = \sum_{k=-W}^{W} \sum_{l=-W}^{W} \left|I_1\left(i+k, j+l\right) \cdot K - \mu_{\text{mod}}\right| \tag{9}$$

$$\mu_{\text{mod}} = \sum_{k=-W}^{W} \sum_{l=-W}^{W} I_1\left(i+k, j+l\right) \tag{10}$$

The first absolute central moment was applied to the reference image to identify and exclude blocks with low textural information. Therefore a block size of 5x5 px was chosen as this block size is used for the Census-transformation. To extend the block size would mean to increase the memory needed, as only 5 lines of the original image are saved in the FPGA-implementation. Different thresholds were used to reduce the number of incorrect points in the depth map. The results are listed in table 3 as ratio of the maximal number of points in the depth map in the same manner as in section 4.2.

Minimal First Absolute Central Moment	Dataset *Art* Valid 3-d Points	Correct (Respectively)	Correct (Absolute)	Dataset *Dolls* Valid 3-d Points	Correct (Respectively)	Correct (Absolute)
0	97 %	49 %	48 %	97 %	57 %	55 %
200	97 %	49 %	48 %	97 %	57 %	55 %
400	89 %	52 %	46 %	95 %	58 %	55 %
500	83 %	54 %	45 %	91 %	59 %	54 %
600	76 %	55 %	42 %	87 %	60 %	52 %
800	65 %	56 %	36 %	79 %	62 %	49 %
1000	57 %	56 %	32 %	71 %	64 %	45 %

Table 3. Impact of the First Absolute Central Moment to the Results of the Correlation

Increasing values lead to a higher reliability especially in areas with low textural information. In regions with occlusions this method is not as effective. For the stereo dataset *Art*, starting with a threshold of 600, a saturation of the reliability can be observed. Increasing the threshold towards higher values will reject correct points. This is not the case for the dataset *Dolls*. A threshold of 500 in this case, is a good compromise for a dense but reliable depth map as only 1–3% of correct points are rejected, while the reliability is increased by 2–5% .

In figures 10(e) and 10(f) the depth maps using thresholds of 500 and 1000 on the first absolute central moment, are shown. The threshold of 500 shows a distinct filter effect in the depth map (see fig. 10(e)). Comparing the depth map with the reference image, it becomes obvious that rejected areas in the depth map correlate with homogeneous colored surfaces in the reference image (see fig. 4(c)). The depth map in fig. 10(f)) shows that the effect of a threshold of 1000 is even stronger. In comparison to fig. 10(e)) the loss of correct block assignments is obvious.

4.4 Uniqueness constraint

The uniqueness constraint was introduced in (Marr & Poggio, 1979, p. 3), it implies that only one disparity value can be assigned to every element of a stereo image. It is substantiated by the physical position of an object which leads to a representation in the reference image and the search image a like alike. Only occlusions by transparent objects violate this constraint (van der Mark & Gavrila, 2006, p. 3).

If the first local minimum C_1 as well as the second local minimum C_2 of the cost function is determined and saved during the correlation process, the uniqueness C_d can be calculated by equation (11).

$$C_d = \frac{C_2 - C_1}{C_1} \tag{11}$$

If the clearance between C_1 and C_2 is small, the uniqueness is small, while the probability of an uncertain result is high. By applying a threshold (see fig 7a)) multi-assignments can be reduced (Hirschmueller & Scharstein, 2009, pp. 6).

Often just the lowest cost function values as shown in fig. 7 are taken for evaluation, due to the complexity of searching for local minima. In this case correct block assignments can be rejected due to a double minima if the eq. 11 (see fig. 7b)) is applied. A double minima

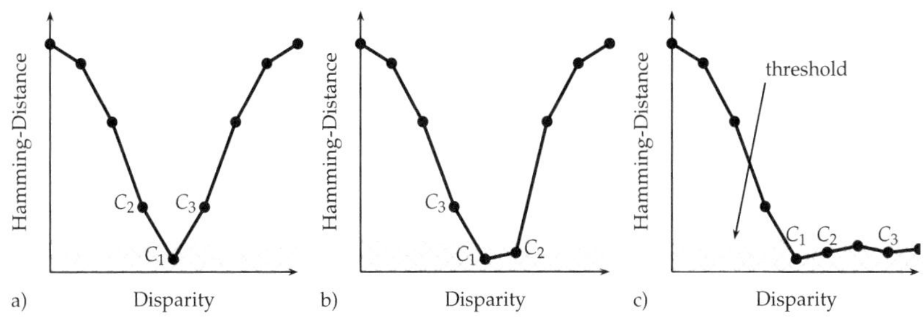

Fig. 7. Problem of Double Minima (inspired by Hirschmueller & Scharstein (2009))

occurs if the real minimum is located between two values. To avoid this problem equation 12 compares the first minimum C_1 and the third minimum C_3 (Hirschmueller & Scharstein, 2009, pp. 6). Uncertain minima are still rejected by this method as shown in fig. 7c). This method performs usually well if the threshold lies between 5–20 % (van der Mark & Gavrila, 2006, p. 4).

$$C_d = \frac{C_3 - C_1}{C_1} \tag{12}$$

The effect of this method was examined by applying thresholds ranging from 0 % to 25 % to

	Dataset *Art*			Dataset *Dolls*		
Minimal Distance	Valid 3-d Points	Correct (Respectively)	Correct (Absolute)	Valid 3-d Points	Correct (Respectively)	Correct (Absolute)
0 %	97 %	49 %	48 %	97 %	57 %	55 %
2 %	93 %	51 %	47 %	93 %	59 %	55 %
5 %	79 %	57 %	45 %	83 %	64 %	53 %
7 %	69 %	63 %	43 %	76 %	68 %	52 %
10 %	58 %	70 %	41 %	66 %	74 %	49 %
15 %	44 %	80 %	35 %	55 %	81 %	45 %
20 %	36 %	87 %	31 %	47 %	86 %	40 %
25 %	29 %	91 %	26 %	41 %	89 %	36 %

Table 4. Impact of the Uniqueness to the Results of the Correlation

the result of the hamming distance, as shown in table 4. By increasing the threshold more correlation results are rejected, while the reliability is increased. In the evaluated range no saturation is reached. A threshold of 7 % gives good results. In *Art* the reliability increases by 14 % while 5 % of the correct values are rejected. For *Dolls* the reliability is increased by 11 % while only 3 % of the correct values are lost.

In the figures 10(g) and 10(h) depth maps using the uniqueness with a threshold of 7 % and 25 % are shown. Especially in areas of occlusions the multi-assignments could be reduced. The depth map regarding the threshold of 25 % shows reduced errors in regions with occlusions as well as homogeneous regions which result in a reliability of 91 % as listed in table 4.

4.5 Left right consistency check

A second method based on the uniqueness assumption is the search for corresponding blocks in both directions (Fusiello et al., 1997, p. 2). First suggestion for such a method are sourced in (Cochran & Medioni, 1992, pp. 5) and (Fua, 1993, p. 2) named as *Two-View-Constraint* and *Validity Test*. In latter publications (Khaleghi et al., 2008, p. 6) and (Zinner et al., 2008, p. 9) it is called *Left-Right-Consistency Checking*.

Carrying out the search for corresponding blocks two times subsequently with two resulting depth maps D_{left} and D_{right} is distinctive for this method. In the first run the left image is the reference image, while the right is the search image and in the second run the roles of both images are reversed. The resulting depth maps are very similar but not identical as visible in the figures 10(m) and 10(n) due to the slightly different angle of view of both images (Fua, 1993, p. 2),(Cochran & Medioni, 1992, pp. 5).

Afterwards the validation of the correspondence search is realized by crosschecking the disparity. At first the disparity $d_{\text{x\_left}}$ at the position x and y in the depth map D_{left} is read out and used to determine the position of the corresponding block in D_{right}.

$$D_{right}(x + d_{\text{x\_left}}, y) = d_{\text{x\_right}} \tag{13}$$

Comparing both disparities gives a clue whether it is a unique block combination. Ideally, regarding the uniqueness assumption the difference between both values has to be zero (Zhao & Taubin (2011)).

$$d_{\text{x\_left}} - d_{\text{x\_right}} = 0 \tag{14}$$

The effect of this method was evaluated, by having an implementation where the difference

	Dataset *Art*			Dataset *Dolls*		
Maximal Difference	Valid 3-d Points	Correct (Respectively)	Correct (Absolute)	Valid 3-d Points	Correct (Respectively)	Correct (Absolute)
256 px	97 %	49 %	48 %	97 %	57 %	55 %
10 px	83 %	55 %	46 %	87 %	61 %	53 %
3 px	81 %	56 %	45 %	85 %	62 %	53 %
2 px	80 %	56 %	45 %	85 %	62 %	53 %
1 px	80 %	56 %	45 %	84 %	63 %	53 %
0 px	77 %	56 %	43 %	80 %	62 %	50 %

Table 5. Impact of the Left Right Consistency Check on the Depth Maps

between both disparity values were be set to any threshold (see. table 5). The number of the points in the depth map as well as the reliability is nearly constant for a difference smaller then 10 pixels and bigger then zero. If a zero-difference between both disparity maps is required a lot of points with a double minima are rejected by this method, hence a threshold of 1 pixel gives the best result as shown in fig. 10(o).

4.6 Applying the continuity constraint

Upon a closer look on a depth map it becomes obvious that incorrect disparity values differ from the neighborhood, especially in homogeneous regions. These nearly homogeneous surfaces follow from the continuity assumption which was introduced by (Marr & Poggio, 1979, p. 3) as a continuous run of disparity values named as *Continuity Constraint*. The *Continuity Constraint* arises from the usually smooth surface of the objects in a scene. This assumtion is violated at object borders.

A similar approach named the *No Isolated Pixel Constraint* Cochran & Medioni (1992) identifies disparity values as isolated if its difference is bigger than 2.5 px according to the average of a 5 x 5 px-block. This method is evaluated subsequently using a block size of 3 x 3 px. The mean of the difference in the Moore-neighborhood is determined, in order to identify a disparity value as isolated. If the difference of a pixel to its neighborhood is bigger than a threshold it is rejected. Fig. 8 shows two example for a threshold of 30.

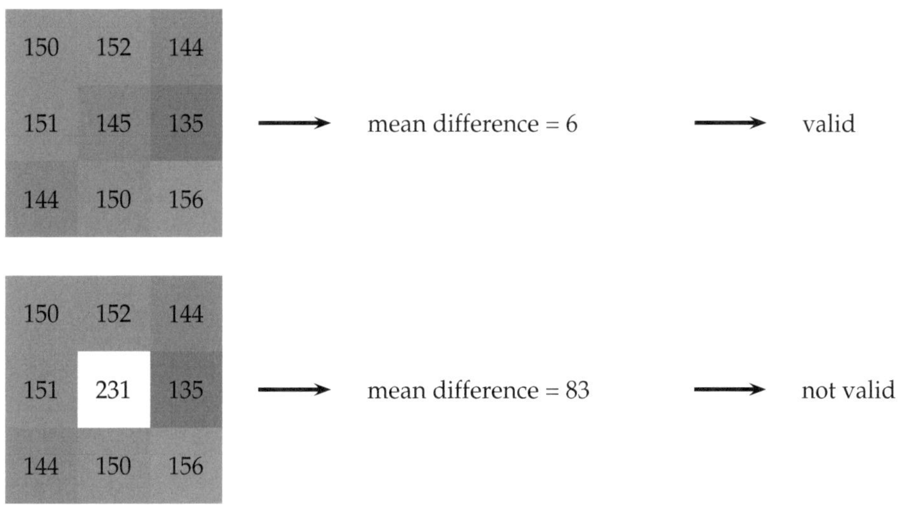

Fig. 8. Validity of Disparity Values according to the Continuity Constraints

The effect of different thresholds using this method is listed in table 6. A decreasing threshold results in a reduced amount of pixels while the rate of correct points in the depth map is rising, since mainly mismatches are rejected. The threshold of 30 is a good choice for keeping a maximum of correct disparity values. Thus the reliability of the depth map rises by 6 % for *Art* and by 4 % for *Dolls* whereas only 2 % for *Art* and 1 % for *Dolls* of the correct values are lost. The resulting depth map is shown in comparison to depth maps without rejected pixels while not allowing a difference in the figures 10(i) and 10(j). The depth map in fig. 10(i) shows the effectivity of this method, which is able to reject mismatches in homogeneous regions as well as regions with occlusions. Setting the threshold to zero a lot of correct disparity values are rejected as well and the depth map is thinner.

The continuity constraint will be used as the final filter in the hardware implementation. It is used for removing isolated pixels, which are left by the other post processing steps as well as outliers in homogeneous regions. While counting the number of valid pixels in their neighborhood it can be easily estimated whether the pixel is isolated or not. Thus in the implementation a pixel identifies as isolated if the number of valid pixels in its neighborhood is less than two, following Cochran & Medioni (1992).

Maximal Mean Difference	Dataset *Art* Valid 3-d Points	Correct (Respectively)	Correct (Absolute)	Dataset *Dolls* Valid 3-d Points	Correct (Respectively)	Correct (Absolute)
255	97 %	49 %	48 %	97 %	57 %	55 %
50	90 %	52 %	47 %	93 %	59 %	55 %
40	87 %	54 %	47 %	91 %	60 %	55 %
30	83 %	55 %	46 %	89 %	61 %	54 %
20	77 %	58 %	45 %	84 %	63 %	53 %
10	66 %	63 %	42 %	76 %	66 %	50 %
0	35 %	83 %	29 %	43 %	84 %	36 %

Table 6. Impact of the Continuity Constraint on the Depth Map

4.7 Multi-layer correspondence search

The effect of different sized correlation blocks is described in section 3.3 whereas small block sizes are good for fine details yet sensitive to noise, while big block sizes create a smooth depth map by blurring edges and details. Due to their size big blocks contain more information and have a greater probability for being unique. Thus they can be used to reduce the ambiguity for smaller blocks by checking if the disparity of a smaller block is within a reasonable range. A similar effect can be gained by changing the resolution of the source images.

Hierarchical methods are widely used to improve the quality of depth maps or to reduce the computation power. All these methods reduce the resolution of the source images and arrange the resulting images in an image pyramid. A common way is to halve the resolution from layer to layer for implementation reasons (Tornow, 2009, pp. 91). In fig. 9 an example for an image pyramid is given. The correspondence search is carried out with the same block size in all three layers. The information covered by a block rises from layer to layer and complies with the effect of different block sizes (Fua, 1993, p. 5). Yet the implementation is more effective for the coarse layers due to smaller block sizes as well as the reduced image size (Falkenhagen, 1994, p. 1).

Nearly all proposed hierarchical methods follow a coarse to fine algorithm, whereas the correlation starts in images with coarse resolution and uses the result as a starting point in the next layer to increase the accuracy. At the second as well as all following layers the range for the disparity search can be reduced to a few pixels. By searching successively throughout all layers the computation power, especially for software implementation can be strongly reduced (Cochran & Medioni, 1992, p. 3), (Sizintsev et al., 2010, pp. 2) and (Zhao & Taubin, 2011, p. 3).

Two different approaches are introduced in (Tornow, 2009, pp. 90) and in Tornow et al. (2006). In contrast to the coarse to fine algorithm the disparity search in Tornow et al. (2006) is realized parallel in all layers. Whereas every layer is used to search only in specific non overlapping disparity ranges. This approach shows good results for very large disparity ranges but leads to a coarse resolution for close objects. The proposed approach in (Tornow, 2009, p. 90) complies with the widely used coarse to fine algorithm. Yet it uses the coarse layers to verify the disparity values found in the highest resolutions. Both approaches are well suited for a

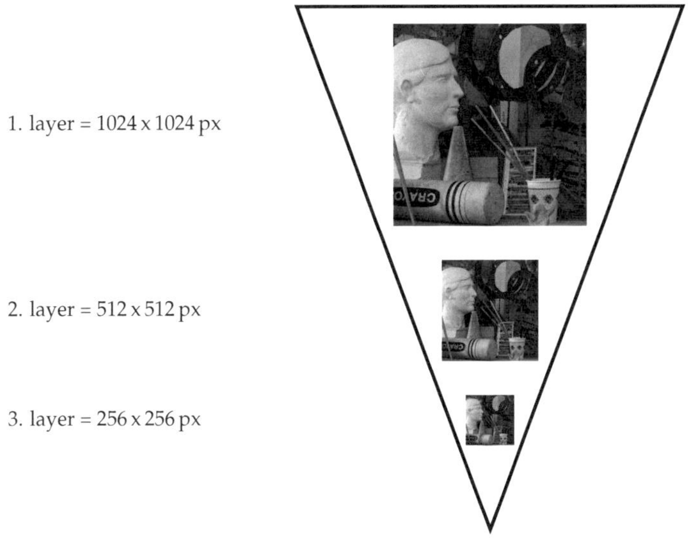

Fig. 9. Example of an Image Pyramid

hardware implementation. As the method presented in (Tornow, 2009, p. 90) is capable of calculating dense depth maps in hardware it is evaluated in the following.

While the block size with 7 x 7 px is constant over all layers, the range for the disparity search is halved layer by layer, starting by 256 px. The resulting disparity maps for layer one is shown in fig. 10(k) and compared to the disparity maps with different block sizes from figures 10(a) to 10(d) the similarity is obvious.

If all disparity values, which can not be verified in the next coarse layers, are rejected, the results in table 7 are gained. The threshold for an optimal result is 1 px for *Art* and 2 px for *Dolls*. Whereas the improvement of the reliability is about 28 % by having only 1 %–2 % of the correct disparities rejected. This proves the efficiency of the algorithm. Higher thresholds are leading to less reliable disparity maps.

	Dataset *Art*			Dataset *Dolls*		
Maximal Difference	Valid 3-d Points	Correct (Respectively)	Correct (Absolute)	Valid 3-d Points	Correct (Respectively)	Correct (Absolute)
256 px	96 %	49 %	47 %	96 %	57 %	55 %
10 px	68 %	68 %	46 %	77 %	71 %	55 %
3 px	60 %	76 %	46 %	69 %	78 %	54 %
2 px	58 %	79 %	46 %	67 %	80 %	54 %
1 px	54 %	83 %	45 %	62 %	83 %	51 %
0 px	25 %	85 %	21 %	30 %	86 %	26 %

Table 7. Impact of the Multi Layer Verification on the Depth Map

Comparing the images in figures 10(k) and 10(l) the effect of the method is visualized. Incorrect block assignments are rejected in case of occlusions as well as homogeneous regions.

The disparity map shown in fig. 10(l) reveals only a few incorrect values and proves the values of table 7.

4.8 Concept of the algorithm

The six presented and evaluated methods for improving the reliability of a disparity map have their strength in different operating ranges. The resulting disparity maps of all methods are shown in fig. 10. The most error-containing regions of an image are regions with homogeneous surfaces and occlusions. Only the multi-layer-verifying performs well for both cases. Mismatches caused by occlusions can be successfully avoided by using the threshold on the hamming distance and the left-right-consistency-check. The first absolute central moment is not suited for treating occlusions. Yet in case of homogeneous regions the minimal hamming distance does not perform well and the first absolute central moment helps avoiding errors in the depth map. The left-right-consistency-check gives average results.

The uniqueness constraint and the continuity constraint give moderate results by rejecting incorrect disparity values in both cases.

To realize a fast hardware implementation those methods giving the best results, while having the least computation power, should be used. As the hierarchical multi-layer correspondence and the left-right-consistency checking require at least two runs of the correlation process, high computation power is needed. The lowest computation power is used by the minimal hamming distance and the uniqueness constraint. The first absolute central moment and the continuity constraint require an average need of computation power.

Parameter	Average	Dense	Reliable
first-abs.-cen.-moment	500	400	800
max. hamming-distance	90 Bit	100 Bit	70 Bit
uniqueness constraint	7 %	2 %	15 %
continuity constraint	30	50	10
Art – valid 3-d-points	40 %	70 %	15 %
correct (respectively)	84 %	64 %	95 %
correct (absolute)	34 %	45 %	14 %
Dolls – valid 3-d-points	54 %	80 %	26 %
correct (respectively)	82 %	67 %	93 %
correct (absolute)	44 %	54 %	24 %

Table 8. Results Using Different Setups

As the threshold on the hamming-distance and the uniqueness can successfully reject a lot of mismatches, the results, especially in homogeneous regions, are not good enough, thus the first absolute central moment and the continuity constraint are added to the set of post-processing. In table 8 three sets of used thresholds and the achieved results of the two stereo data sets are given.

The three different parameter sets are suitable to provide disparity maps with different attributes. The *dense*-setup gives a low reliability of about 60 % with a big amount of

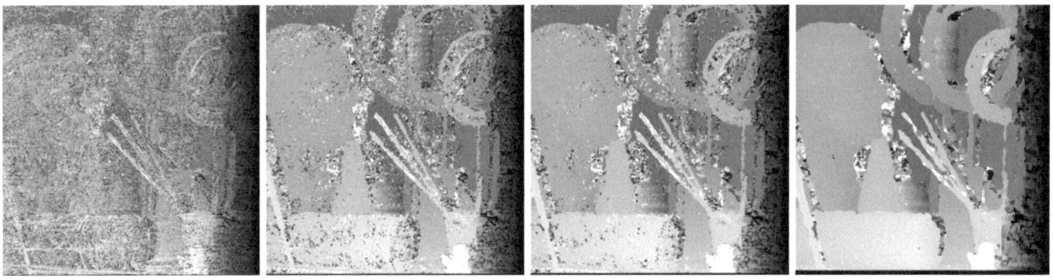

(a) Mini-Census 3 x 3 px (b) Mini-Census 7 x 7 px (c) Mini-Census 9 x 9 px (d) Mini-Census 19 x 19 px

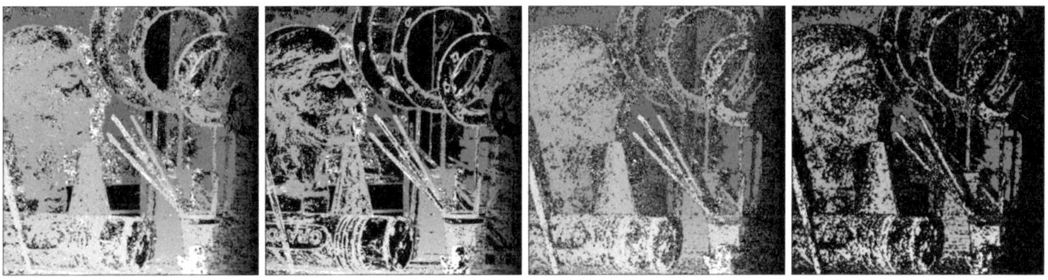

(e) 1st-Abs.-Cen.-Mom. 500 (f) 1st-Abs.-Cen.-Mom. 1000 (g) Uniqueness 7 % (h) Uniqueness 25 %

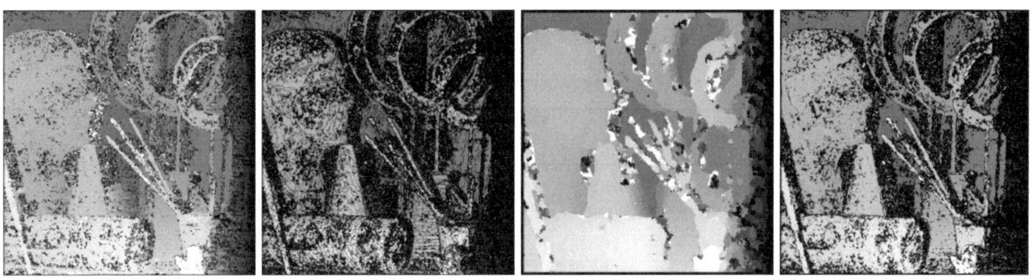

(i) Continuity 30 (j) Continuity 0 (k) Layer 1 (l) Multi-Layer-Approach

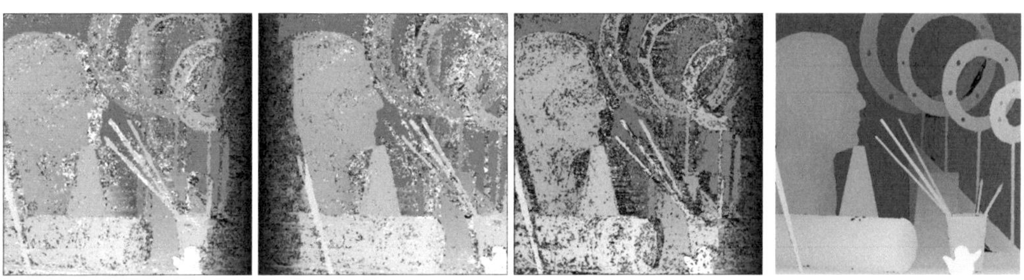

(m) D_{left} (n) D_{right} (o) Left-Right-Cons.-Check (p) Ground Truth

Fig. 10. Depth Maps of Middleburry Stereo Datasets *Art* Using Various Method for Improvement

3-d-points. The *average*- setup with a reliability of more than 80 % is suitable for the most common applications, such as interactive robots, driver assistance systems, facial detection etc. The *reliable*-setup is capable of providing disparity maps with over 90 % reliability and is therefore suitable for safety systems.

Applying these four post-processings very good results of the multi-layer correspondence search are surpassed by using less hardware resources.

In some applications it is more important to have a very dense depth map therefore a certainty value to each 3-d point is added. In such a case the certainty for a 3-d point can be estimated using the methods for quality improvement presented in the sections 4.3 to 4.7 as measurement tools to weight the impact of 3-d-points. In this case fine graduation would give very dense depth maps without any loss of information.

5. FPGA-implementation

Having evaluated all processing steps for a fast, reliable stereo image analysis, suitable for hardware implementation, the next step was to generate a modular and fully platform-independent design using VHDL.

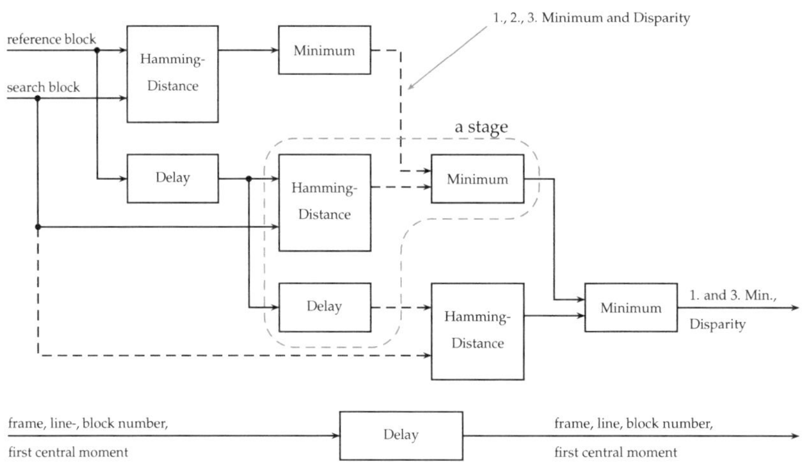

Fig. 11. Structure of the Cascaded Correspondence Analysis

The design is divided into preprocessing, the correlation process and post processing. The preprocessing contains the calculation of the first absolute central moment and the Census-transformation of the stereo image pair, using a block size of 5 x 5 pixels and the determination of the frame-number, line-number and block-number of the synchronization signals. Currently no further filter processes are included. The correlation process is set up by a cascade of processes calculating the hamming distance and determining its first and third minimal value, as shown in fig. 11. The number of blocks for which the hamming distance is calculated in such a stage depends on the required processing speed. For very high speed processing only one block combination of the hamming distance is calculated per stage. The number of stages can be determined by dividing the maximal disparity by the number of hamming distances calculated per stage. For high speed calculations with a disparity range of 256 pixels, when only one block combination is processed per stage, 256 stages are needed. If only a lower speed is required, for example 16 block combinations can be processed in one stage, thus only 16 stages are needed.

(a) Aloe D:71 % R:89 % (b) Baby3 D:42 % R:80 % (c) Books D:40 % R:77 % (d) Dolls D:55 % R:84 %

(e) Laundry D:35 % R:73 % (f) Moebius D:55 % R:91 % (g) Reindeer D:56 % R:84 % (h) Rocks1 D:81 % R:93 %

Fig. 12. Depth map of Different Datasets of the Middlebury Stereo Dataset (D: Density – Rate of Valid Pixels; R: Reliability – Rate of Correct Disparity Values)

In the post processing the four chosen methods of improvement the reliability of the depth map are applied: Starting with the first absolute central moment, followed by the uniqueness constraint and the maximal hamming distance, ending with the continuity constraint. The last step of the post processing is to calculate the depth map from the disparity map using equation (2) according to the calibration information.

Using Alteras NIOS-II processor the processing can be compiled as a hardware coded function available from within the processor. Thus small sets of block combinations could be processed on demand by the processor.

The configuration was tested with a Terasic DE3-board containing an Altera Stratix III EP3SE260 using a set of PhotonFocus cameras MV-D1024 Photonfocus (2008) connected via CameraLink-interface. Using the hardware implementation the datasets Aloe, Art(D: 44 % R: 80 %), Baby3, Books, Dolls, Laundry, Moebius, Reindeer and Rocks1 of the Middlebury stereo dataset were processed (see fig. 12) using the parameter set *average*. The overall density is 53 % and the overall reliability is 84 %. Correct points show a difference of maximal one pixel for the disparity regarding the ground truth.

6. Comparison with the state of the art of stereo image analysis systems

An implementation on different FPGA-platforms could be realized due to a fully platform independent design. In table 9 five different versions are listed. The implementations for the Virtex 6, the Stratix IV and the Cyclone III, which are marked with an *, were only tested with a timing simulation. Just the implementations using Stratix III was tested on real hardware. The hardware approach introduced in section 5 is running with a disparity range of 256 px at a maximal pixel clock of 189 MHz on an Altera Stratix III FPGA. With an image size of 1024 x 1024 px a frame rate of 180 Hz could be reached.

Reference	Hardware Platform	Image Size	Max. Disparity	Frame Rate
Hirschm. & Scharst.(2009)	CPU – 2,6 GHz Xenon	450 x 375 px	64 px	0.5 fps
Zinner et al. (2008)	CPU – 2,0 GHz Core 2 Duo	450 x 375 px	50 px	13 fps
Sizintsev et al. (2010)	GPU – GeForce GTX 280	640 x 480 px	256 px	32 fps
Zhao & Taubin (2011)	GPU – GeForce GTX 280	1024 x 768 px	256 px	36 fps
Khaleghi et al. (2008)	DSP – ADSP-BF561	160 x 120 px	30 px	20 fps
Chang et al. (2007)	DSP – TMS320C6414T-1000	384 x 288 px	16 px	50 fps
Jin et al. (2010)	FPGA – Virtex 4	640 x 480 px	64 px	230 fps
Pelissier & Berry (2010)	FPGA – Cyclone III	1024 x 1024 px	64 px	160 fps
Masrani & MacLean (2006)	FPGA – Stratix	640 x 480 px	128 px	30 fps
Zhang et al. (2011)	FPGA – Stratix III	1024 x 785 px	64 px	60 fps
This work	FPGA – Stratix III	1024 x 1024 px	256 px	180 fps
This work	FPGA – Stratix III	1024 x 1024 px	64 px	205 fps
This work	FPGA – Stratix IV*	1024 x 1024 px	256 px	198 fps
This work	FPGA – Virtex 6*	1024 x 1024 px	256 px	241 fps
This work	FPGA – Cyclone III*	1024 x 1024 px	180 px	124 fps

Table 9. State of the Art of Stereo Image Analysis Systems (* Only Timing Simulation)

A comparison with state of the art implementations on major platforms for on-line processing is shown in table 9. By comparing different approaches it has to be taken into account, that every platform has its own advantages and disadvantages. PC-based solutions are generally used when the focus lies on the quality of the depth map or when the frame rate is not important. The approach of Hirschmueller & Scharstein (2009) provides a slow processing speed, but with a high accuracy. It covers a comparison of several different methods. Zinner et al. (2008) is optimizing a PC-based software implementation, it reaches 13 Hz while the maximum disparity is 50 px. This article shows clearly that image size and disparity range are a trade off to the frame rate. Both approaches use rather small image sizes. DSP based solutions (Khaleghi et al. (2008) and Chang et al. (2007)) for the stereo image analysis are even more limited on image size and disparity range as well as their frame rate. Due to DSPs which are optimized for processing one dimensional signals. Image processing on DSP is still important for smart phones. The proposed algorithm is suitable for GPU-implementation as well, but the Census-transformation is optimized for hardware implementation. Thus the different cost functions should be used.

For high speed applications massive parallelizing is necessary. This can be realized either by hardware implementation or using SIMD-processors, as used in current GPUs. The advantage of GPUs is the possibility of high speed processing with at least double precision floating point units, using a high speed memory connection. Furthermore GPUs can be programmed using standard programming languages like C and are fairly cheap. On the other hand some of the cheap GPUs are limited in accuracy, due to their main field of application as graphic cards.

While rendering images is an easy task on GPUs, optimizing other algorithms to GPUs is still a difficult task.

Sizintsev et al. (2010) uses an adaptive coarse to fine algorithm and a left right consistency check. The system is capable of a frame rate of 32 Hz on images with a resolution of 640 x 480 px, while the disparity range is 256 px wide. By skipping some parts of the algorithms for the sake of improving the quality, 113 Hz are possible. The overall rate of points with a higher difference to the ground truth than one pixel is 15.8 %. The approach of Zhao & Taubin (2011) reaches 36 Hz with an algorithm optimized to measure moving parts in stereo images. After using a foreground detection a multi-resolution stereo matching is applied. The overall error rate lies by 14.5 %. Generally GPUs are capable of calculating dense and reliable depth maps, yet the power consumption and the waste heat of GPU based systems are a major draw back for usage in embedded systems. Nevertheless the combination of small GPUs with embedded microprocessors are on their way.

FPGAs are well suited for embedded systems as they have a very low power consumption combined with a high processing speed. Programming FPGAs is a time consuming process which requires special knowledge, regarding digital electronics and hardware programming. The implementation of complex iterative algorithms in FPGAs is possible but it is often less effective, as FPGA implementations are data flow driven with concurrent processing of algorithm parts. Thus iterative algorithms which provide depth maps with the highest quality must be highly adapted as they are heavily control flow oriented. Straight algorithms perform very well on FPGAs due to their data flow orientation.

The quality of the disparity maps in the presented work is, with 84 %, in the same dimension as other state of the art FPGA-solutions. The rate of incorrect points in the depth map ranges from 14 % in Zhang et al. (2011) to 17 % in Jin et al. (2010). The hardware resources required are fairly low compared to other approaches. Zhang et al. (2011) requires about 95000 logic blocks and 3.77 MBit of memory at the Stratix III, compared to ca 40000 logic blocks and 321 kBit of memory needed for the presented solution with a disparity range of 64 px (disparity range of 256 px used for this research ($\Rightarrow$ ca 147000 logic blocks and 328 kBit). The approach presented by Jin et al. (2010), implemented on a Virtex 4, needs 51000 logic elements and 322 memory blocks compared to the Virtex 6 implementation with a disparity range of 64 px with 39000 logic elements and 23 memory blocks (disparity range of 256 px $\Rightarrow$ ca 149000 logic blocks 23 memory blocks). Both approaches provide more dense disparity maps due to a more complex post processing but with a lower disparity range. Only the Cyclone III implementation of Pelissier & Berry (2010) requires 61000 logic blocks and 131 kBit which are less than the 72000 logic blocks and the 322 kBit of memory needed by our approach, yet without post processing for improving the depth map quality and a disparity range of 64 px.

7. Conclusion

In this chapter an algorithm for generating dense and reliable disparity maps of stereo images, suitable for high speed processing, using an FPGA, is presented. Therefore several cost functions, as well as post processing steps to increase the reliability, are evaluated. The algorithm uses a correlation with the hamming distance, having a block size of 7 x 7 px, on Census transformed images. In four post processing steps incorrect points of the disparity map are rejected and the reliability as well as the quality is increased up to 84 % of correct pixels for the *average*-setup. By choosing a set of parameters either a very dense or a very reliable (*reliable*-setup: 95 % reliability) depth map can be calculated. Due to its modular setup the implementation can be easily adapted for optimizing either speed or needed FPGA-resources. In Stratix III the implementation runs with a maximal frame rate of 180 Hz, having a resolution of 1024 x 1024 px and a disparity range of 256 px.

The main applications for high speed stereo analysis are autonomous robots and driver assistance systems as well as in line quality controls and sensor systems for automation. Especially for grid based analysis of the surroundings, used for vicinity observations for autonomous vehicles and driver assistance systems, a dense and reliable depth map can be provided. Hence the problem of splitting one object into several objects due to the low resolution of the depth map can be overcome. This method can be adapted to be well suited to various applications due to a modular setup where processing speed, which is not needed, can be easily used to reduce the size of the implementation. Additional post processing steps, iterating on the results given by the presented algorithm, could increase both, the reliability and the density of the depth map, while requiring more hardware resources as well.

8. References

Albertz, J. & Wiggenhagen, M. (2009). *Guide for Photogrammetry and Remote Sensing*, 5. edition edn, Wichmann.

Bay, H., Andreas, Ess, A., Tuytlaars, T. & Gool, L. V. (2008). Surf: Speeded up robust features, *Computer Vision and Image Understanding(CVIU)*, Vol. Vol. 10, pp. pp. 346–359.

Chang, N., Lin, T.-M., Tsai, T.-H., Tseng, Y.-C. & Chang, T.-S. (2007). Real-time DSP implementation on local stereo matching, *Proceedings of the IEEE International Conference on Multimedia and Expo*, pp. p. 2090–2093.

Chang, N. Y.-C., Tsai, T.-H., Hsu, B.-H., Chen, Y.-C. & Chang, T.-S. (2010). Algorithm and architecture of disparity estimation with mini-census adaptive support weight, *IEEE Transactions on Circuits and Systems for Video Technology* Bd. 20(6): pp. 792–805.

Cochran, S. D. & Medioni, G. (1992). 3-d surface description from binocular stereo, *IEEE Transactions on Pattern Analysis and Machine Intelligence* Bd. 4(10): pp. 981–994.

Falkenhagen, L. (1994). Depth estimation from stereoscopic image pairs assuming piecewise continuos surfaces, *Proceedings of the European Workshop on Combined Real and Synthetic Image Processing for Broadcast and Video Production*, pp. 115–127.

Faugeras, O. (1993). *Three-Dimensional Computer Vision : A Geometric Viewpoint*, MIT Press, Cambridge, MA.

Favaro, P. & Soatto, S. (2007). *3-D Shape Estimation and Image Restoration*, Springer-Verlag.

Fua, P. (1993). A parallel stereo algorithm that produces dense depth maps and preserves image features, *Machine Vision and Applications* Bd. 6(1): pp. 35–49.

Fusiello, A., Roberto, V. & Trucco, E. (1997). Efficient stereo with multiple windowing, *Proceedings of the 1997 Conference on Computer Vision and Pattern Recognition (CVPR '97)*, pp. 858–863.

Hirschmueller, H. & Scharstein, D. (2009). Evaluation of stereo matching costs on images with radiometric differences, *IEEE Transactions on Pattern Analysis and Machine Intelligence* Bd. 31(9): pp. 1582–1599.

Jin, S., Cho, J., Pham, X. D., Lee, K. M., Park, S.-K., Kim, M. & Jeon, J. W. (2010). FPGA design and implementation of a real-time stereo vision system, *IEEE Transactions on Circuits and Systems for Video Technology* Bd. 20(1): pp. 15–26.

Kanade, T. & Okutomi, M. (1991). A stereo matching algorithm with an adaptive window: Theory and experiment, *Proceedings of the 1991 IEEE International Conference on Robotics and Automation (ICRA '91)*, Vol. 2, pp. 1088–1095.

Khaleghi, B., Ahuja, S. & Wu, Q. M. J. (2008). An improved real-time miniaturized embedded stereo vision system (mesvs-ii), *IEEE Computer Society Conference on Computer Vision and Pattern Recognition Workshops (CVPRW '08)*, pp. 1–8.

Kisacanin, B., Bhattacharyya, S. S. & Chai, S. (2009). *Embedded Computer Vision*, Springer-Verlag, London.

Konolige, K. (1997). Small vision systems: Hardware and implementation, *Proceedings of the International Symposium on Robotics Research*, pp. 111–116.

Marr, D. & Poggio, T. (1979). A computational theory of human stereo vision, *Proceedings of the Royal Society of London. Series B* Bd. 204(1156): pp. 301–328.

Masrani, D. K. & MacLean, W. J. (2006). A real-time large disparity range stereo-system using FPGAs, *in* P. Narayanan, S. Nayar & H.-Y. Shum (eds), *Computer Vision - ACCV 2006*, Vol. 3852, Springer-Verlag, Berlin Heidelberg, pp. 42–51.

McDonnell, M. (1981). Box-filtering techniques, *Computer Graphics and Image Processing* Bd. 17(1): pp. 65–70.

Moravec, H. (1977). Towards automatic visual obstacle avoidance, *Proceedings of the 5th International Joint Conference on Artificial Intelligence*, p. 584.

Narasimha, R. (2010). *Depth Recovery from Stereo Matching Using Coupled Random Fields*, PhD thesis, UNIVERSITÃL' DE GRENOBLE.

Pelissier, F. & Berry, F. (2010). Design of a real-time embedded stereo smart camera, *in* J. Blanc-Talon, D. Bone, W. Philips, D. Popescu & P. Scheunders (eds), *Advanced Concepts for Intelligent Vision Systems*, Vol. 6474, Springer-Verlag, Berlin Heidelberg, pp. 344–356.

Photonfocus (2008). User manual – MV-d1024 series CMOS area scan cameras, Online-Source.

Scharstein, D. (2011). Middlebury stereo datasets, Online-Source.
URL: `http://vision.middlebury.edu/stereo/data`

Scharstein, D. & Szeliski, R. (2002). A taxonomy and evaluation of dense two-frame stereo correspondence algorithms, *International Journal of Computer Vision* Bd. 47(1): pp. 7–42.

Sizintsev, M., Kuthirummal, S., Samarasekera, S. & Kumar, R. (2010). GPU accelerated realtime stereo for augmented reality, *Proceedings of the 5th International Symposium 3D Data Processing, Visualization and Transmission (3DPVT '10)*.

Szeliski, R. (2010). *Computer Vision : Algorithms and Applications*, Springer-Verlag, London.

Tornow, M. (2009). *Untersuchung und Entwicklung von Algorithmen zur Stereobildauswertung fuer die Erfassung von Objekten im Umfeld von Fahrzeugen und Realisierung einer Hindernisdetektion in Echtzeit mittels einer Hardwareimplementierung auf einem FPGA*, Dissertation, Otto-von-Guericke-University, Magdeburg.

Tornow, M., Kazubiak, J., Kuhn, R. W., Michaelis, B. & Schindler, T. (2006). Hardware approach for real time machine stereo vision, *Journal of systemics, cybernetics and informatics* Bd. 4(1): pp. 24–34.

Trucco, E. & Verri, A. (1998). *Introductory Techniques for 3-D Computer Vision*, Prentice Hall, Upper Saddle River, NJ.

van der Mark, W. & Gavrila, D. M. (2006). Real-time dense stereo for intelligent verhicles, *IEEE Transactions on Intelligent Transportation Systems* Bd. 7(1): pp. 38–50.

Willow Garage (2011). The opencv 1.x c reference manual, Online-Quelle.

Zabih, R. & Woodfill, J. (1997). A non-parametric approach to visual correspondence, *IEEE Transactions on Pattern Analysis and Machine Intelligence*.

Zhang, L., Zhang, K., Chang, T. S., Lafruit, G., Kuzmanov, G. K. & Verkest, D. (2011). Real-time high-definition stereo matching on FPGA, *Proceedings of the 19th ACM/SIGDA international symposium on Field programmable gate arrays (FPGA '11)*, pp. 55–64.

Zhao, Y. & Taubin, G. (2011). Real-time stereo on GPGPU using progressive multi-resolution adaptive windows, *Image and Vision Computing* Bd. 29(6): pp. 420–432.

Zinner, C., Humenberger, M., Ambrosch, K. & Kubinger, W. (2008). An optimized software-based implementation of a census-based stereo matching algorithm, *Advances in Visual Computing*, Vol. 5358, Springer-Verlag, Berlin Heidelberg, pp. 216–227.

4

Real-Time Processing of 3D-TOF Data in Machine Vision Applications

Stephan Hussmann, Torsten Edeler and Alexander Hermanski
Institute for Machine Vision Technology (Ma.Vi.Tec),
West Coast University of Applied Sciences
Germany

1. Introduction

In machine vision applications, Time-of-Flight (TOF) sensors like the Photonic Mixer Devices (PMD Technologies [PMD], 2011) became considerable alternatives to common 3D sensing devices. Because of the enormous progress in TOF-vision systems, nowadays 3D matrix cameras can be used for many applications such as robotic, automotive, industrial, medical and multimedia applications. Due to the increasing demand of safety requirements in the automotive industry it can be assumed that the TOF-camera market will grow and the unit price of these systems in the mass production will drop down to ca. 100 € (Hussmann & Hess, 2006).

Many 3-D sensing techniques have been developed in the past decades. A good review can be found in (Jarvis, 1983). 3-D sensing methods can be divided into several categories:

1. *Shape-from Techniques*: These monocular approaches recover relative depth from texture, from shading, from contours, from motion, etc.; resulting in surface orientations with respect to a viewer-centered coordinate system (Hu & Stockman, 1989). These techniques must deal with correspondence problems. Furthermore problems arise due to the uncertainty of the position at which each image is taken and due to dynamic changes that may occur in the time between two images.
2. *Stereo*: This method simulates the two eyes of a human. It uses multiple visual sensors (two cameras, for example) to estimate stereo disparity and then recover depth (Hu & Stockman, 1989). Stereo cameras introduce physical restrictions due to the need for camera separation. Further, stereo cameras depend on texture matching from both camera images for range estimation. This produces a rather sparse and unevenly distributed data set. Due to the allocation problem dynamic tracking of objects is not an easy task (Hussmann & Liepert, 2009).
3. *Structured Light*: In illuminating the scene, natural ambient light is replaced by an artificial light source, which can be of any structure (pattern) that is convenient for the task. Using structured light by itself is not an independent approach to 3-D sensing; the underlying means is still a monocular (shape-from) or binocular (stereo) method, but under different illumination conditions and hence facing rephrased problems. If a single light beam or a single light plane is used as the light source, the underlying method is direct sensing through triangulation. If a uniform grid of light is the light source, the underlying method is stereo and analysis of textures and contours. The

major advantage of using structured light over ambient light is that features in the images are better defined. Image features are easier to detect; their relationships are more regular (parallel, equi-spaced, etc.) following the property of the generating pattern of the light source; and they prominently reveal the surface geometry that humans can readily use to interpret the scene (Hu & Stockman, 1989).

4. *Direct Sensing* (time-of-flight): One representative of this approach is a laser range finder, which can sense the depth to any surface point in its field of view. Laser range finders are often used for navigation tasks (Nuechter et al., 2003). The major disadvantage of these systems is the use of mechanical components and that they do not deliver 2D intensity images and range data at the same time. Another representative are TOF cameras (Blanc et al., 2004 ; Schwarte et al., 1997), which combine the advantage of active sensors and camera based approaches as they provide a 2D image of intensity and exact distance values in real-time. They do require a synchronized light source. Compared to *Shape-from Techniques* and *Stereo* TOF cameras can deal with prominent parts of rooms like walls, floors, and ceilings even if they are not structured. In addition to the 3D point cloud, contour and flow detection in the image plane yields motion information that can be used for e.g. car or person tracking (Hussmann et al., 2008). *Structured light* applications can have a higher measurement accuracy as TOF-cameras, however the measurement range is then limited.

Inspection task in machine vision applications e.g. quality control of bulk materials are very complex due to their real-time requirements. New accurate and fast algorithms for 3D object recognition and classification are needed as the inspection time is always decreasing. As now commercial 3D-TOF cameras are available at a reasonable price the number of machine vision applications using this technology is expected to increase significantly. Hence in this book chapter the use of 3D-TOF cameras for machine vision applications is investigated.

The state of the art is to use a four-phase shift algorithm for TOF cameras to correctly determine the range value (Blanc et al., 2004; Lange & Seitz, 2001; Ringbeck & Hagebeuker, 2007). The optical signal is sampled four times per period at equidistant intervals. The corresponding sampling points permit unique determination of all relevant parameters of the incoming optical echo's waveform. The sample points are not acquired during only a single period but summed over several hundreds or thousands of periods, which considerably increases the signal-to-noise ratio and hence, finally, the accuracy of the measurement. Due to acquisition of the four subsequent phase images, fast object motion leads to distance uncertainties in situations where corresponding phase images do not properly align with respect to object points.

In (Hussmann & Edeler, 2010) we presented a pseudo-four-phase-shift algorithm for 3D-TOF photonic mixer device (PMD) cameras, which only has to capture two phase images and thereby doubles the frame rate. Hence distance uncertainties by fast moving objects will be reduced. In (Hussmann et al., 2011a) we presented a simple motion compensation algorithm for constant lateral motion such as measure objects on a conveyor belt, which can be processed with the maximum frame rate of currently available commercial TOF cameras. However this algorithm was based on the state of the art 4-phase shift algorithm.

In this book chapter we will combine the two proposed algorithms and evaluate their performance in comparison to the state-of-the-art algorithm. The book chapter is structured as follows. In section 2 we derive the basics of PMD TOF vision systems and subsequently present algorithms used for real-time processing of PMD TOF data. In section 3 experiments are conducted to investigate the real-time performance of the proposed algorithm for machine vision applications. Concluding remarks will summarize the chapter.

2. Real-time processing of PMD TOF data

2.1 Operating principle of PMD TOF sensors

Fig. 1 shows the cross section of a typical PMD pixel comprising two symmetrical, transparent photo gates. The photons of the received optical echo P_{opt} enter the p-doped substrate through these gates and are generating charge carrier (electron/hole-pairs). The gates are isolated from the p-doped substrate by a SiO_2 - or Si_3N_4 – isolation layer (channel stop) and bounded on the left and right side by n^+ - diffusion readout gates. The photo gates are controlled by the modulation voltage u_m and the offset voltage U_0. The schematic potential distribution in the p-doped substrate between the photo gates is shown in Fig. 1 for a negative modulation voltage u_m.

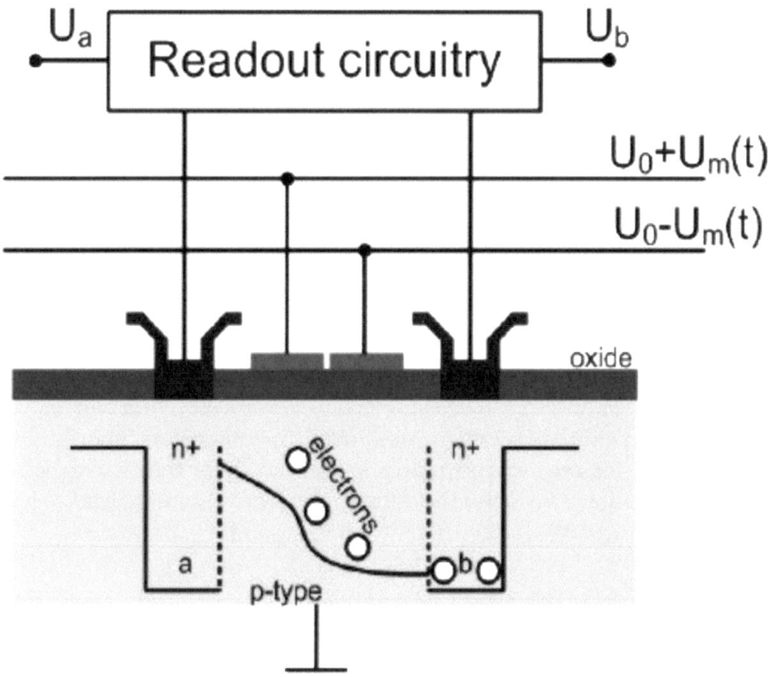

Fig. 1. Cross section of a typical PMD pixel

A PMD pixel may be understood as a modulation controlled photo charge distributer (photonic mixer). In principle the PMD pixel works like a seesaw for electrons while controlling its motion by means of polarity and slope of the seesaw. If no modulated light is received the photo generated charges symmetrically drift to both readout gates *a* and *b*. If modulated light is received the photo generated charges drift only to readout gate *b*, when the modulated light and the modulation voltage have a phase difference of 180° (see Fig. 1). If the phase difference is 0° the photo generated charges drift only to readout gate *a*.

State of the art is to use continuous wave (CW) modulation with square waves for TOF cameras with a typical modulation frequency of 20 MHz (Hussmann et al., 2008). Hence the modulation voltages can be easily generated digitally with a high accuracy and stability using programmable logic devices (PLDs) such as complex programmable logic devices (CPLD) or field programmable gate arrays (FPGA). For the illumination source infrared (IR) - light emitting diodes (LEDs) are used. The low-pass characteristic of the IR-LEDs leads to an attenuation of the square waves' harmonics for larger frequencies. This results in an optical output that gradually looks sinusoidal for frequencies larger than 5-10 MHz (see Fig. 2). This has to be taken into account if CW modulation with square waves is used.

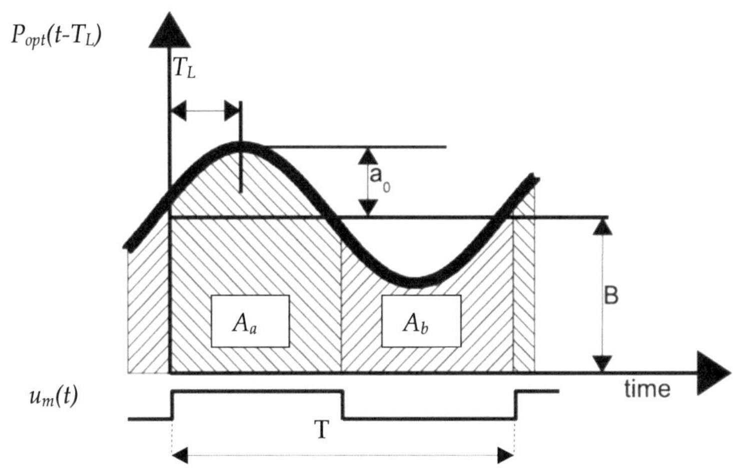

Fig. 2. Correlation process between the received optical echo P_{opt} and the modulation voltage u_m for a single modulation period

The readout gates a and b are each connected to an integration capacitor. Hence the corresponding voltages U_a and U_b can be expressed as a correlation function between the optical echo $P_{opt}(t\text{-}T_L)$ and the modulation voltage $u_m(t)$ over the integration time T_{int}. Fig. 2 illustrates the correlation process for one single modulation period T for the two signals $P_{opt}(t\text{-}T_L)$ and $u_m(t)$. The modulation voltage $u_m(t)$ and the optical echo $P_{opt}(t\text{-}T_L)$ are defined as follows:

$$u_m(t) = \begin{cases} 1, & \text{for } 0 \le \text{t-N} \cdot \text{T} \le \text{T}/2 \\ 0, & \text{for T}/2 < \text{t-N} \cdot \text{T} \le \text{T} \end{cases}, \qquad \text{N} = 0,1,2\ldots \tag{1}$$

$$P_{opt}(t - T_L) = a_0 \cdot \cos(\omega t - T_L) + B \tag{2}$$

T_L is the 'time-of-flight' time for the light (camera-object-camera), B is the received average incident light (background light and DC component of the modulated light source) and a_0 is the amplitude of the received modulated light. U_a and U_b are then proportional to the areas A_a and A_b as shown in Fig. 2. If the complete integration time T_{int} (corresponds to several hundreds or thousands of periods) has taken into account, U_a and U_b can be written as:

$$U_a(T_L) = K \cdot \int_0^{T_{int}} P_{opt}(t - T_L) \cdot u_m(t)dt = K \cdot \frac{T_{int}}{T}\left[\frac{T \cdot a_0}{\pi}\sin(T_L) + \frac{T \cdot B}{2}\right] = K \cdot \frac{T_{int}}{T} \cdot A_a(T_L) \tag{3}$$

and

$$U_b(T_L) = K \cdot \int_0^{T_{int}} P_{opt}(t - T_L) \cdot u_m(t - T/2)dt = K \cdot \frac{T_{int}}{T}\left[-\frac{T \cdot a_0}{\pi}\sin(T_L) + \frac{T \cdot B}{2}\right] = K \cdot \frac{T_{int}}{T} \cdot A_b(T_L) \quad (4)$$

The conversion gain K converts the received optical energy into a voltage. The integration time T_{int} does not have to be necessarily a multiple of the single period time T as the number of periods integrated over the integration time is in the range of hundreds to thousands. Looking at Fig. 2 it can be noticed that U_a and U_b are always a positive voltage. To remove the influence of the background light B the difference of ΔU_{ab} has to be determined:

$$\Delta U_{ab}(T_L) = U_a - U_b = K \cdot \frac{T_{int}}{T} \cdot \left(A_a(T_L) - A_b(T_L)\right) \quad (5)$$

The autocorrelation function ΔU_{ab} corresponds to the distance value of a PMD pixel. The sum of U_a and U_b corresponds to all received and converted photons. Hence this sum is equivalent to the grey level value of standard CCD/CMOS video cameras (amplitude image):

$$\sum U_{ab} = U_a + U_b = K \cdot \frac{T_{int}}{T} \cdot (A_a + A_b) = K \cdot \int_0^{T_{int}} P_{opt}(t - T_L)dt = B \quad (6)$$

It has to be mentioned that in this book chapter only infrared light is used as an IR-filter is mounted on top of the sensor chip. Using an IR-filter reduces the effects of the background illumination B on the distance resolution. Hence the amplitude image in this book chapter could be also called "infrared amplitude image". However without the IR-filter the TOF camera would behave like a standard 2D-camera and therefore we still use the word "amplitude image".

Equation (5) and (6) demonstrate the advantage of the PMD technology compared to other 3-D sensing techniques. The PMD pixel is a TOF vision system with inherent suppression of uncorrelated light signals such as sun light or other modulated light disturbances (neon tubes, high frequency illumination modules etc.). More advantages of a PMD TOF vision system are the acquisition of the amplitude value and range data in each pixel without high computational cost and any moving components as well as the monocular setup.

2.2 State-of-the-art range image calculation using 4-phase shift algorithm

As mention in the last section the range value corresponds to ΔU_{ab}. The amplitude of the received optical echo P_{opt} varies with the measure object reflectivity coefficient and the distance. Hence the amplitude of the output voltage ΔU_{ab} is also affected by these changes. To overcome the amplitude dependency of the output voltage of ΔU_{ab} state of the art is to use a 4-phase shift algorithm (Blanc et al., 2004; Lange & Seitz, 2001; Ringbeck & Hagebeuker, 2007). In (Hussmann & Liepert, 2009) the following equation to calculate the phase difference φ_0 without any dependency on the received optical echo's amplitude is derived:

$$\varphi_0 = \arctan\left(\frac{\Delta U_{ab}(270°) - \Delta U_{ab}(90°)}{\Delta U_{ab}(0°) - \Delta U_{ab}(180°)}\right) \quad (7)$$

The range value R can now be calculated by taken into account the modulation frequency f_{mod} and the physical constant for the speed of light c ($3{\cdot}10^8$ m/s). N represents the ambiguity in range estimation when $\varphi_0 > N \cdot 360°$. For example if a modulation frequency of 20 MHz is used the ambiguity range is 7,5 m. If the distance to an object is now 9 m, $N = 1$ and the distance measured by the camera is 1.5 m.

$$R = \frac{c}{2 \cdot f_{\mathrm{mod}}} \cdot \left(\frac{\varphi_0}{360°} - N \right) \quad with\ N = 0,1,2,3... \tag{8}$$

2.3 Real-time range image calculation using pseudo 4-phase shift algorithm

Looking at equation (3) and (4) it can be noticed that U_a and U_b have a phase difference of 180° (T/2) to each other:

$$U_a(T_L + \frac{T}{2}) = K \cdot \frac{T_{\mathrm{int}}}{T} \left[\frac{T \cdot a_0}{\pi} \sin(T_L + \frac{T}{2}) + \frac{T \cdot B}{2} \right] = K \cdot \frac{T_{\mathrm{int}}}{T} \left[-\frac{T \cdot a_0}{\pi} \sin(T_L) + \frac{T \cdot B}{2} \right] = U_b(T_L) \tag{9}$$

Hence the output voltage ΔU_{ab} can be expressed as:

$$\Delta U_{ab}(T_L) = U_a(T_L) - U_b(T_L) = U_a(T_L) - U_a(T_L + T/2) \tag{10}$$

Equation (10) shows that a PMD pixel delivers two phase values ($U_a(T_L)$ and $U_a(T_L+T/2)$) at one image capture. Therefore equation (7) can be simplified to:

$$\varphi_0 = -\arctan\left(\frac{\Delta U_{ab}(90°)}{\Delta U_{ab}(0°)} \right) \tag{11}$$

The range value R can now be calculated by using equation (8). Equation (11) demonstrates the advantage of the pseudo 4-phase shift algorithm. Only two image captures instead of four are required to calculate the phase difference φ_0. Hence the frame rate of PMD TOF sensors is doubled without changing the integration time T_{int}. A typical frame rate of TOF PMD cameras is 50 Hz. The pseudo 4-phase shift algorithm increases this frame rate to 100 Hz and hence is well suited for real-time machine vision applications. A more detailed description of the pseudo 4-phase shift algorithm can be found in (Hussmann & Edeler, 2010).

2.4 Real-time arctangent calculation using a reconfigurable processor system

The most time critical operation of the phase difference φ_0 calculation is the arctangent function (see equation (7) and equation (11)). As the arctangent function is called for each individual pixel to determine the range value, the processing time increases with the number of present pixels. The hardware algorithm proposed in (Hussmann et al., 2011b) to calculate the arctangent value for 3D-TOF PMD cameras in real-time is realized as a custom functional unit on the reconfigurable functional unit (RFU) of a reconfigurable processor in a FPGA. This algorithm replaces the state-of-the-art CORDIC arctangent function commonly used in microcontrollers. This significantly decreases the processing time to determine the range image of the 3D vision system.

As the arctangent function is symmetrical only the angles from zero to ninety degree have to be calculated. A further reduction of the angle range down to 45° can be achieved by taken into account the following:

$$\varphi_0 = \begin{cases} \arctan\left(\dfrac{y_0}{x_0}\right), & \text{for } y_0 \le x_0 \\ 90° - \arctan\left(\dfrac{x_0}{y_0}\right), & \text{for } y_0 > x_0 \end{cases} \tag{12}$$

With a desired range resolution of 1 mm (equivalent to an angle resolution of 0.048°) and an angle range of 45°, a LUT with 2048 elements is needed. As shown in table 1 this LUT is filled with the distance values of the according phase angles. As can be seen in table 1 the distance resolution between each LUT entry is smaller than 1 millimeter. Furthermore it can be seen that at the end of the LUT the distance resolution is better as at the start of the LUT.

i	tan(angle) = i / 2048	angle in degree	distance in mm
0	0	0	0
1	0.000488	0.027976	0.6
2	0.000977	0.055953	1.2
3	0.001465	0.083929	1.7
4	0.001953	0.111906	2.3
5	0.002441	0.139882	2.9
6	0.002930	0.167858	3.5
7	0.003418	0.195834	4.1
8	0.003906	0.223811	4.7
9	0.004395	0.251786	5.2
...			
2042	0.997070	44.915948	935.7
2043	0.997559	44.929973	936.0
2044	0.998047	44.943992	936.3
2045	0.998535	44.958005	936.6
2046	0.999023	44.972010	936.9
2047	0.999512	44.986008	937.2

Table 1. Lookup table of the hardware algorithm

Looking at equation (12) a comparator, a hardware divider, a subtraction device and a LUT is needed to determine the distance value. The comparator checks if $y_0 > x_0$, the hardware divider calculates y_0 / x_0 or x_0 / y_0 depending on the comparator result, the division result is used as index for the LUT and finally the LUT delivers, depending on the comparator result, the distance value directly or this value has to be subtracted from the distance value 1.875 m (equivalent to 90°).

The hardware algorithm and the state-of-the-art CORDIC algorithm is implemented into a FPGA (Altera Stratix EP1S10F780C6) using a clock frequency of 50 MHz. Therefore the CORDIC algorithm takes 340 ns and the proposed hardware algorithm 160 ns respectively for a standard arctangent calculation. Compared to the execution time of 800 μs for the arctangent C-function *atan2()* from the "math.h" library on the NIOS II processor, a speed-up factor of 2,353 for the CORDIC algorithm and 5,000 for the proposed hardware algorithm is achieved. Hence the total processing time of one range image of a TOF camera with 204 x 204 pixels (PMD[vision]® CamCube 2.0) takes for the CORDIC algorithm 14.15 ms and for the proposed hardware algorithm 6.66 ms respectively.

The maximum frame rate of the used commercial camera (PMD[vision]® CamCube 2.0) is 25 fps, which corresponds to a capture time of 40 ms per image. Using the proposed algorithm with the total processing time for the arctangent function of 6.66 ms leaves enough time to process the range calculation in real-time. To our knowledge there is no other hardware algorithm with the same performance. The proposed approach will significantly reduce the system costs of TOF cameras as state-of-the-art is to use high performance microcontroller in combination with a FPGA. This is an important achievement as the current 3D TOF cameras are too expensive for common machine vision applications. A more detailed description of the hardware algorithm can be found in (Hussmann et al., 2011b).

2.5 Real-time motion artifact compensation

Distance uncertainties typically occur where objects or the camera itself move while the consecutive phase images are taken. They arise from unmatched phase values during the demodulation process. The faster the objects move or the higher the integration time the higher are the distance uncertainties. In (Hussmann et al., 2011a) a compensation algorithm for constant lateral motion is proposed as this is a typical motion in machine vision applications. One industrial example is 3D dimension measurement of objects on a conveyor belt (luggage handling systems, quality control of food or beverages etc.).

The lateral motion of objects on a conveyor belt has to be corrected in only one direction (moving direction of the conveyor belt). This can be done by subtracting the captured amplitude images $\Sigma U_{ab}(0°)$, $\Sigma U_{ab}(90°)$, $\Sigma U_{ab}(180°)$ and $\Sigma U_{ab}(270°)$, and subsequent thresholding using a fixed threshold s to get the binary image B_1-B_3:

$$B_n = \left|\Sigma U_{ab}(0°) - \Sigma U_{ab}(n \cdot 90°)\right| > s \quad \text{, with n} = 1,2,3 \tag{13}$$

Fig. 3 shows typical binary images of an object moving on a conveyor belt. It can be seen that the width of the white area increases linear with the capture time. To compensate the motion the width of the area has to be determined and the amplitude images have to be moved accordingly before the distance image is calculated. The proposed method is computational not expensive and can be easily integrated into an FPGA as shown in (Hussmann et al., 2011a). Hence the motion compensation can be realized in real-time. It has to be noticed that the sensor must be calibrated to make sure that every pixel has a uniform behaviour when exposed with the active light source of the TOF camera (multiplicative shading calibration). A more detailed description of the motion artefact compensation algorithm and the calibration method can be found in (Hussmann et al., 2011a).

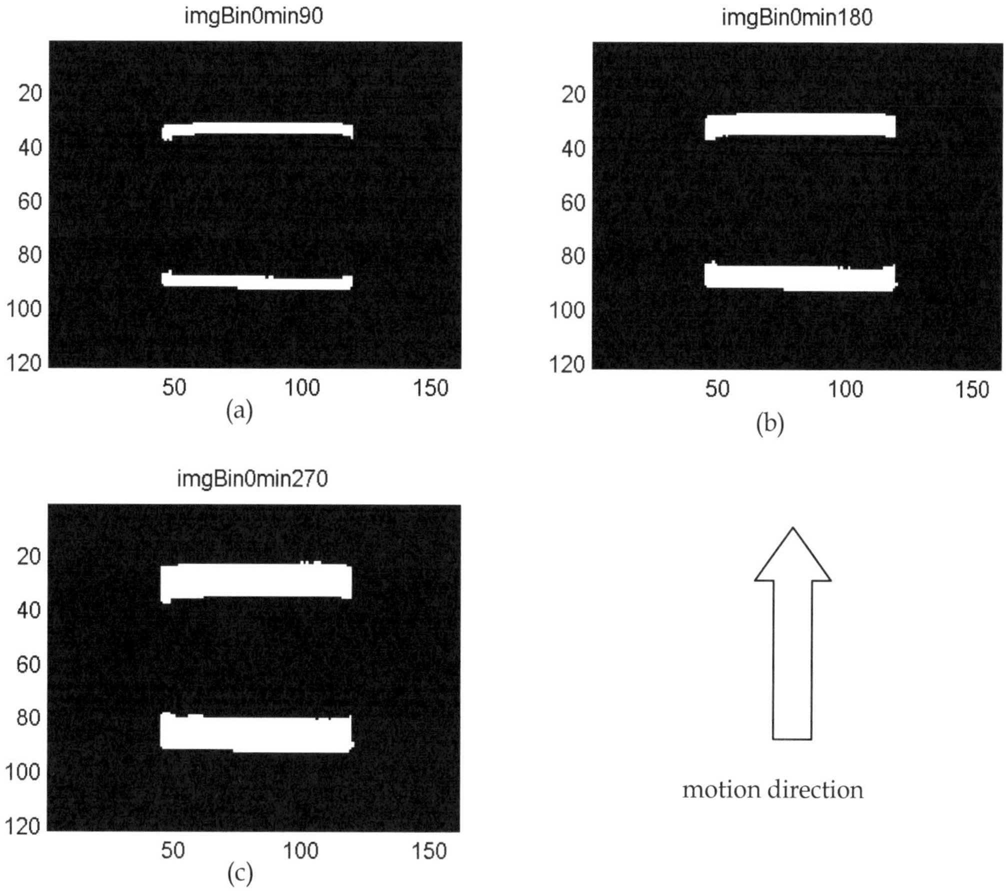

Fig. 3. Binarized difference images of the application in section 3
(a) Binary image of difference image $\Sigma U_{ab}(0°)$-$\Sigma U_{ab}(90°)$
(b) Binary image of difference image $\Sigma U_{ab}(0°)$-$\Sigma U_{ab}(180°)$
(c) Binary image of difference image $\Sigma U_{ab}(0°)$-$\Sigma U_{ab}(270°)$

3. Experiments

3.1 Experimental setup

In Fig. 4 the laboratory setup is shown. A measure object (10 cm x 10 cm x 12 cm) is placed on a conveyor belt, which runs at a speed of 1 m / s. A PMD TOF camera (PMD[vision]® CamCube 2.0) with 204 x 204 pixels is placed 103 cm above the conveyor belt. The raw data (U_a and U_b) of the PMD camera for the four different phases τ ($\tau = 0°$, $\tau = 90°$, $\tau = 180°$ and $\tau = 270°$) are captured and the proposed motion compensation algorithm in (Hussmann et al., 2011a) combined with the pseudo-four-phase-shift algorithm proposed in (Hussmann & Edeler, 2010) is investigated offline using Matlab.

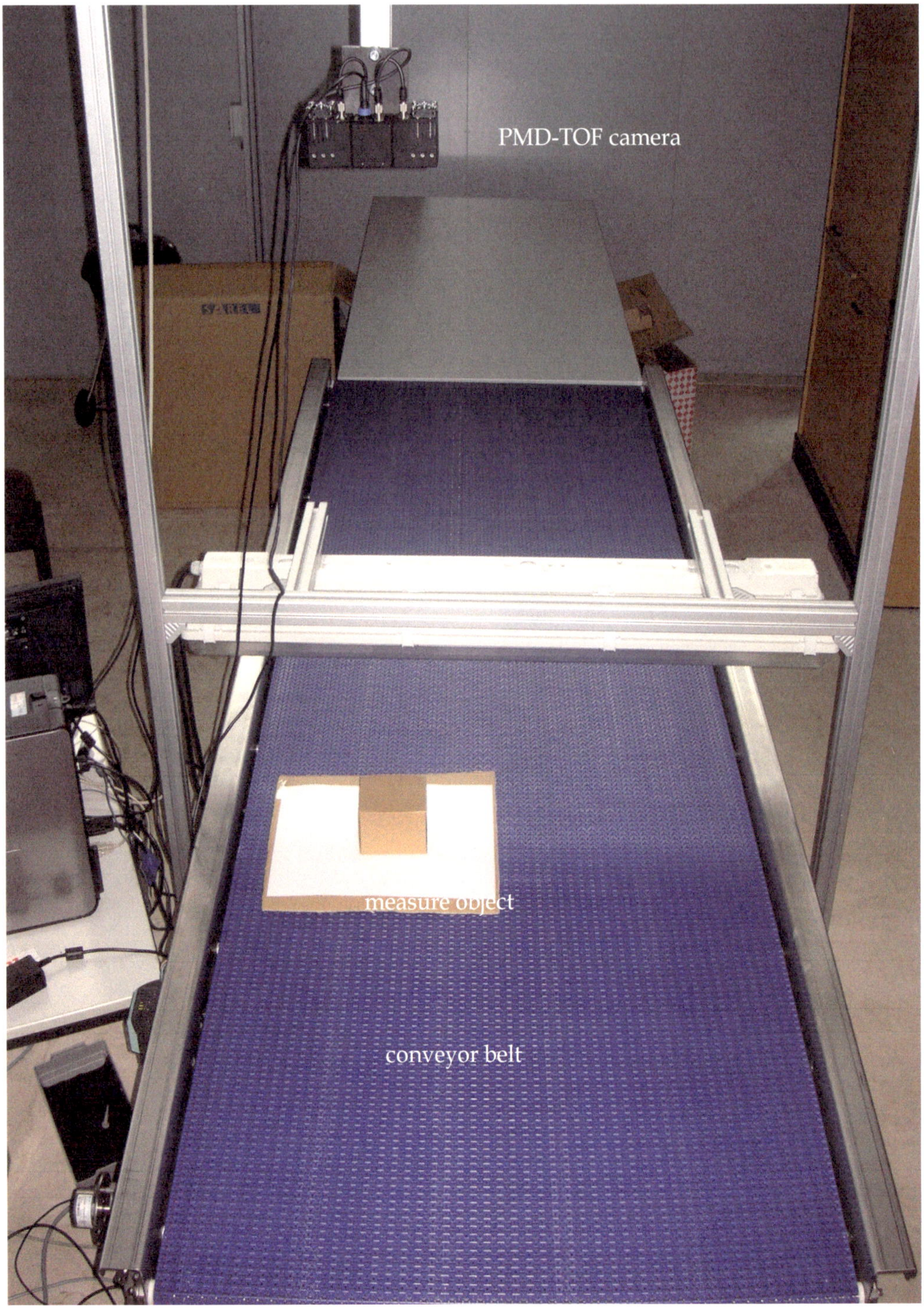

Fig. 4. Laboratory setup

3.2 Experimental results

Fig. 5 shows the amplitude images of the four different phases. The amplitude image is calculated using equation (6). The displacement of the measure object is difficult to notice hence the subtraction results between the different amplitude images are shown in Fig. 6.

The displacement of the object during the acquisition of the four phase images can be clearly seen in Fig. 6. After thresholding and moving of the phase images as proposed in section 2.5, the corrected distance image can be calculated.

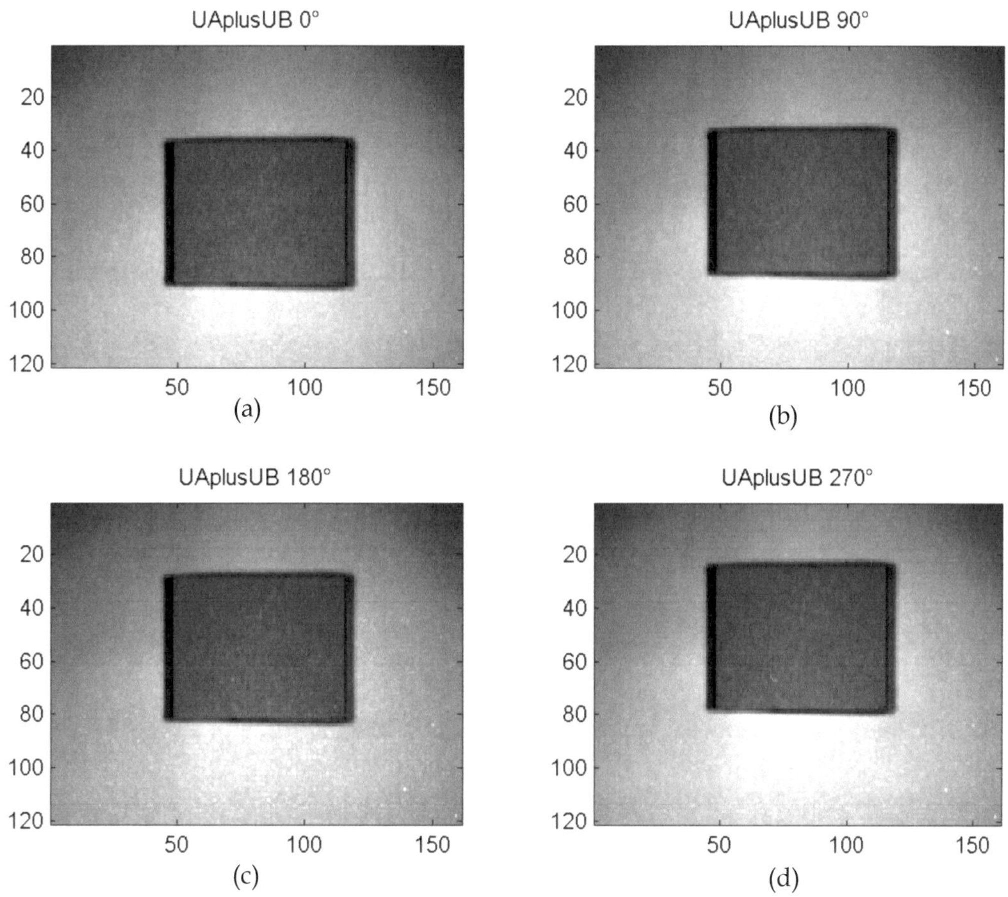

Fig. 5. Amplitude images of the four different phases:
(a) Amplitude image of $\Sigma U_{ab}(0°)$ ("ground truth")
(b) Displaced amplitude image of $\Sigma U_{ab}(90°)$
(c) Displaced amplitude image of $\Sigma U_{ab}(180°)$
(d) Displaced amplitude image of $\Sigma U_{ab}(270°)$

Fig. 7 - Fig. 10 illustrate the influence of the calibration method and the motion artefact compensation algorithm proposed in (Hussmann et al., 2011a) on the distance image using the state-of-the-art 4-phase shift algorithm and the pseudo 4-phase shift algorithm proposed in (Hussmann & Edeler, 2010) respectively. Fig. 7 shows the distance image without motion compensation using the state-of-the-art 4-phase shift algorithm. It can be seen that the distance at the object edges are not calculated correctly (area 1 and area 3) and that the standard deviation is larger than in the other areas (see table 2). Furthermore it can be noticed that the distance image in Fig. 7 (a) is noisier than in Fig. 7 (b) due to the calibration method proposed in (Hussmann et al., 2011a).

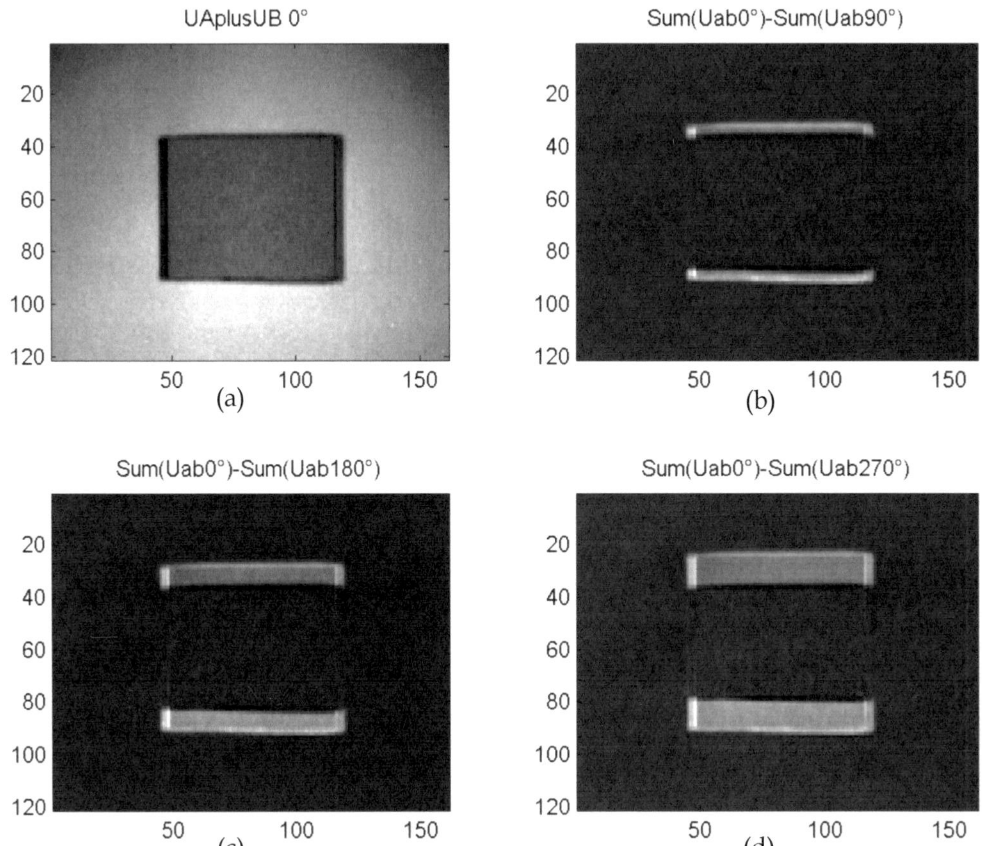

Fig. 6. Subtraction results of the amplitude images:
(a) Amplitude image of $\Sigma U_{ab}(0°)$ ("ground truth")
(b) Difference image $\Sigma U_{ab}(0°)$-$\Sigma U_{ab}(90°)$
(c) Difference image $\Sigma U_{ab}(0°)$-$\Sigma U_{ab}(180°)$
(d) Difference image $\Sigma U_{ab}(0°)$-$\Sigma U_{ab}(270°)$

Fig. 8 shows the distance image without motion compensation using the pseudo 4-phase shift algorithm. Again it can be seen that the distance at the object edges are not calculated correctly (area 1 and area 3) and that the standard deviation is larger than in the other areas (see table 2). However the distorted edges are smaller as only two image captures are needed to calculate the distance. It can be also noticed that the distance image in Fig. 8 (a) is noisier than in Fig. 8 (b) due to the calibration method proposed in (Hussmann et al., 2011a).

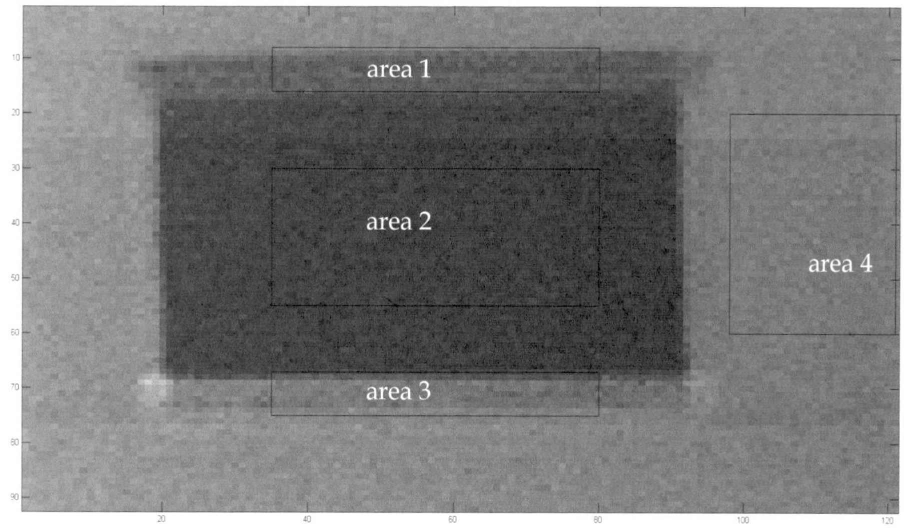

(a)

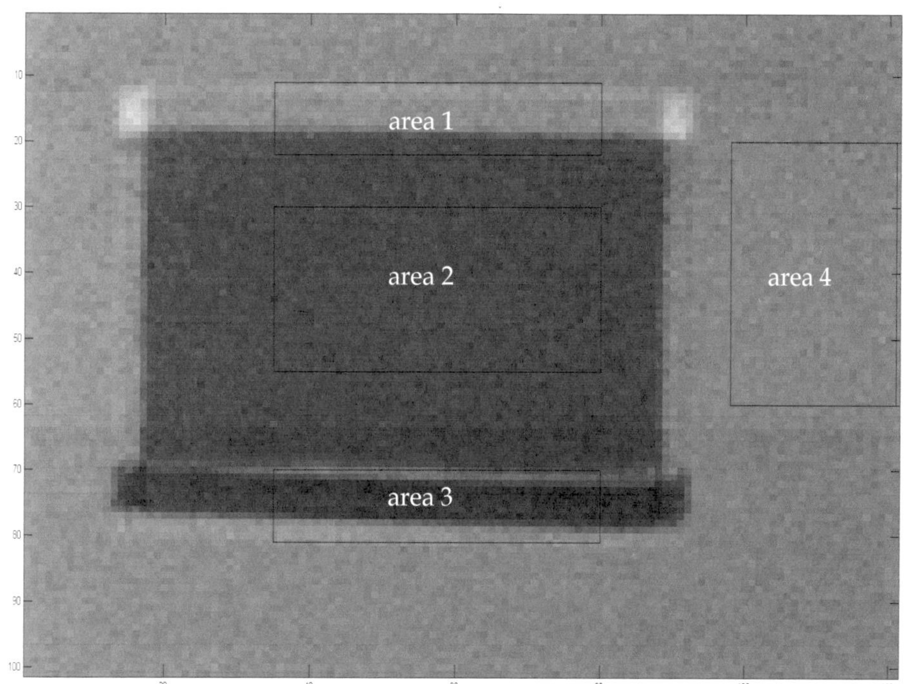

(b)

Fig. 7. Distance image using state-of-the-art 4-phase shift algorithm without calibration (a) and with calibration (b).

(a)

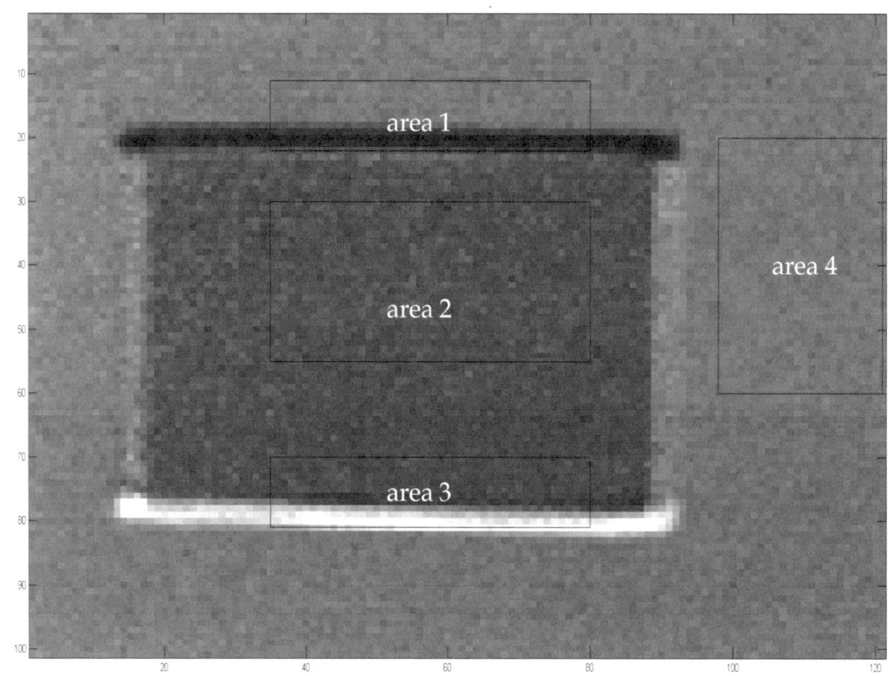

(b)

Fig. 8. Distance image using pseudo 4-phase shift algorithm without calibration (a) and with calibration (b).

Fig. 9 shows the distance image with motion compensation using the 4-phase shift algorithm. The distance image has clear edges without any distance uncertainties and the object dimensions can be calculated correctly. However it can be noticed again that the distance image in Fig. 9 (a) is noisier than in Fig. 9 (b) due to the calibration method proposed in (Hussmann et al., 2011a).

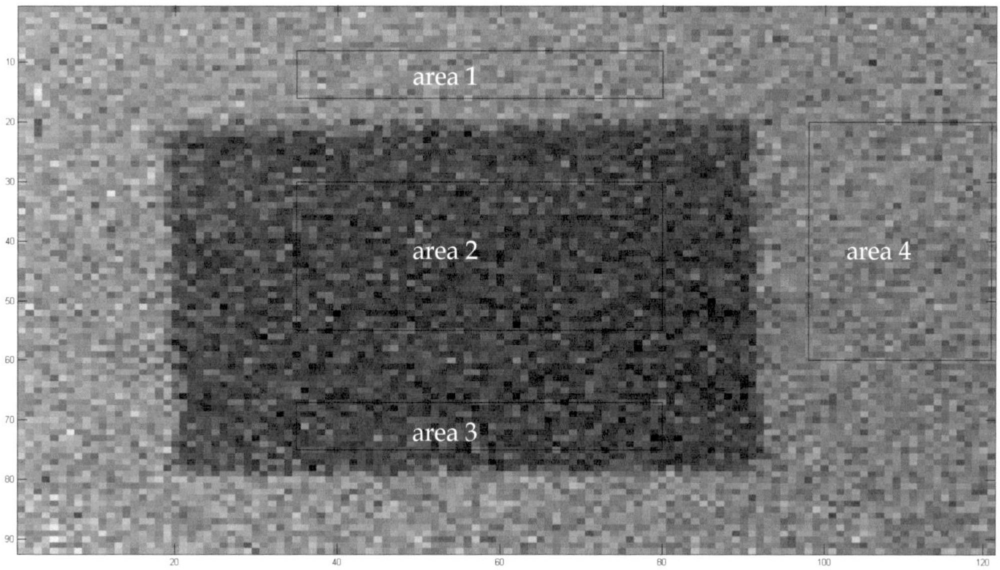

(a)

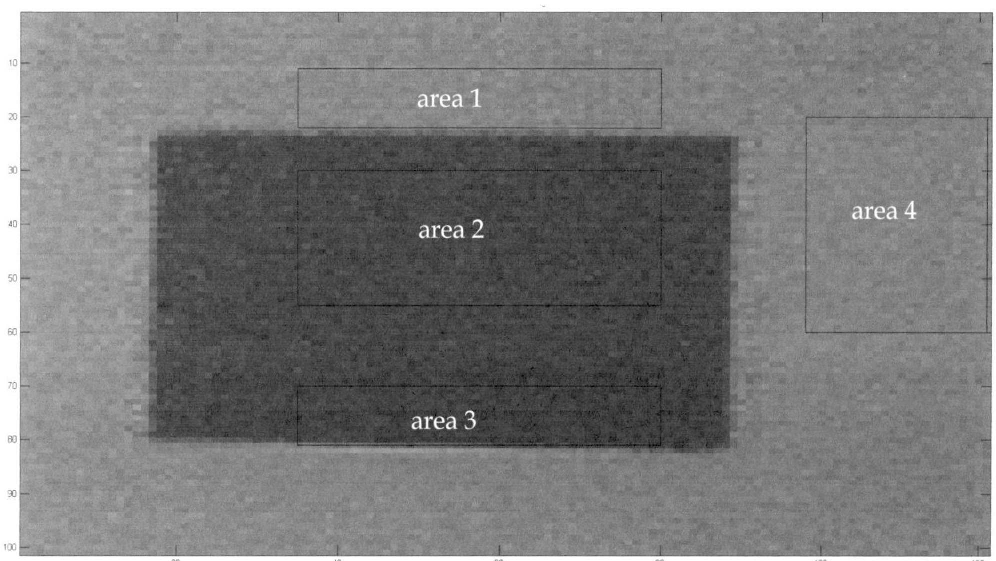

(b)

Fig. 9. Distance image using motion compensation algorithm based on the 4-phase shift algorithm without calibration (a) and with calibration (b).

Fig. 10 shows the distance image with motion compensation using the pseudo 4-phase shift algorithm. The distance image has also clear edges without any distance uncertainties and the object dimensions can be calculated correctly. However it can be noticed again that the distance image in Fig. 10 (a) is noisier than in Fig. 10 (b) due to the calibration method proposed in (Hussmann et al., 2011a).

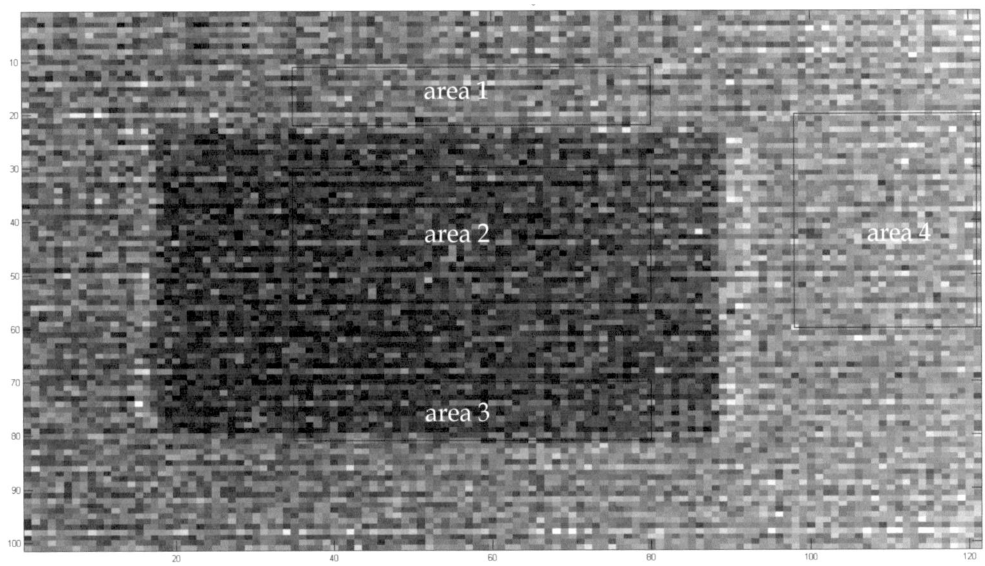

(a)

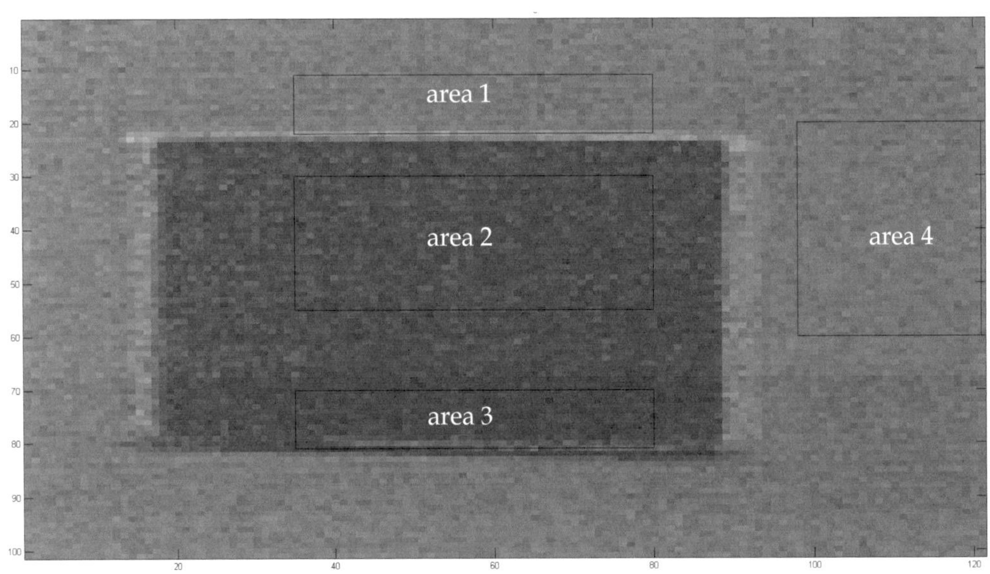

(b)

Fig. 10. Distance image using motion compensation algorithm based on the pseudo 4-phase shift algorithm without calibration (a) and with calibration (b).

Distance image (marked area)	Mean value and Standard deviation in cm			
	State-of-the-art 4-phase shift algorithm	*Pseudo 4-phase shift algorithm*	*Motion compensation algorithm (4-phase shift algorithm)*	*Motion compensation algorithm (pseudo 4-phase shift algorithm)*
area 1 uncalibrated	100.2/1.8	99.0/4.5	104.5/2.3	100.7/4.2
area 1 calibrated	102.0/3.9	127.6/4.4	104.2/0.8	131.0/1.4
area 2 uncalibrated	95.1/0.8	93.1/4.5	95.4/2.7	93.1/4.3
area 2 calibrated	95.5/0.8	125.9/1.0	95.8/0.9	125.8/0.9
area 3 uncalibrated	101.6/2.4	96.6/5.7	95.3/3.0	93.7/4.4
area 3 calibrated	97.5/5.1	130.3/9.0	95.1/1.6	125.2/1.4
area 4 uncalibrated	102.7/0.6	104.9/3.7	103.1/2.2	105.1/3.8
area 4 calibrated	103.5/0.6	130.7/0.7	103.7/0.6	130.5/0.7

Table 2. Spatial noise of the distance images in Fig. 7 – Fig. 10

The mean value and standard deviation across all pixels within the marked areas in Fig. 7 – Fig. 10 has been calculated and the results are shown in table 2. It can be clearly seen that the proposed motion compensation algorithm (with calibration) in (Hussmann et al., 2011a) combined with the pseudo-four-phase-shift algorithm proposed in (Hussmann & Edeler, 2010) has almost the same standard deviation for the background (area 4) and the object (area 2) as the state-of-the-art 4-phase shift algorithm and the motion compensation algorithm (with calibration) in (Hussmann et al., 2011a). However the mean value of the proposed motion compensation algorithm (with calibration) has an offset due to the calibration. Hence this offset has to be calibrated as well. Anyhow the proposed combination algorithm is working at the double frame rate, which is a desired feature for machine vision applications.

4. Conclusion

In this chapter we highlighted the advantages of the PMD technology for machine vision applications compared to other range measurement systems. The equations needed for the design of such a system are derived and demonstrate the simplicity of the extraction of the range information. A PMD camera delivers absolute geometrical dimensions of objects without depending on the object surface, - distance, -rotation and –illumination. Hence PMD TOF vision systems are rotation-, translation- and illumination invariant.

The major advantage of the PMD technology is the delivery of an evenly distributed range and intensity images because each pixel calculates a range and intensity value. The PMD technology has an inherent suppression of uncorrelated light signals such as sun light or other modulated light disturbances. However if those light sources saturate the sensor, the range information is lost. More advantages of the PMD technology are the acquisition of the intensity and range data in each pixel without high computational cost and any moving components as well as the monocular setup. All these advantages lead to a compact and economical design of 3D TOF vision system with a high frame rate. This vision system can not only be used for machine vision applications but also for many other applications such as robotic, automotive, medical and multimedia applications.

In this chapter experimental results of a modified motion artefact compensation algorithm for PMD TOF vision system for a typical machine vision application are presented. Distance uncertainties at the object edges are greatly reduced. However a calibration (multiplicative shading correction) has to be done before to achieve this performance. The experimental results show that the proposed modified algorithm is working at the double frame rate compared to the original motion artefact compensation algorithm with almost the same performance. For real-time machine vision applications it is very important to have a high frame rate. The proposed algorithm will be more suited as only two image captures are needed instead of four to calculate the distance image.

The aim of the proposed motion compensation algorithm is to remove the displacement between the same target object points in the amplitude images due to the target movement. Subsequently the compensated amplitude images are used to calculate the distance image. Hence the target object must not have a uniformly reflecting surface. For the same reason the object orientation or form also does not affect the performance of the proposed algorithm.

It has to be noticed that the proposed modified motion artefact compensation algorithm (and also the original algorithm) only works for constant lateral movements. Speed changes are not taken into account, which would result in distance uncertainties at the object edges. But still the proposed method is able to determine the 3D dimension measurement of fast moving objects with a higher precision than the state-of-the art 4-phase-shift algorithm.

The performance of the proposed modified motion artefact compensation algorithm is investigated offline using Matlab. It has been shown that the algorithm is working at the double frame rate compared to the original motion artefact compensation algorithm. The maximum frame rate of the used commercial camera (PMD[vision]® CamCube 2.0) is 25 fps, which corresponds to a capture time of 40 ms per image. If the proposed algorithm would be implemented into the PMD camera, the frame rate would increase to 50 fps and a capture time of 20 ms respectively. If the hardware arctangent function in section 2.4, which has a total processing time for one range image of 6.66 ms, would be implemented as well, there would be 13.34 ms left for the remaining processing. This time is long enough to calculate the range data in real-time. Hence it can be concluded that the proposed algorithms are well suited for machine vision applications.

5. Acknowledgment

This work was supported in part by the European Union (European Regional Development Fund - EFRE) and in part by the federal state of Schleswig Holstein in Germany (Zukunftsprogramm Wirtschaft). The authors are grateful for the financial support.

6. References

Blanc, N., Oggier, T., Gruener, G., Weingarten, J., Codourey, A. & Seitz, P. (2004). Miniaturized smart cameras for 3D-imaging in real-time, *Proc. of the IEEE Sensors*, vol.1, pp. 471-4

Hu, G. & Stockman, G. (1989). 3-D Surface Solution Using Structured Light and Constraint Propagation, *IEEE Trans. Pattern Anal. Machine Intell.*, 11(4), pp. 390-402

Hussmann, S. & Hess, H. (2006). Dreidimensionale Umwelterfassung, *Trade Journal: "Elektronik automotive"*, WEKA Publisher House, Issue 8, ISSN 1614-0125, pp. 55-59

Hussmann, S., Ringbeck, T. & Hagebeuker, B. (2008). A performance review of 3D TOF vision systems in comparison to stereo vision systems, In: *Stereo Vision* (Online book publication), I-Tech Education and Publishing, Vienna, Austria, ch. 7, ISBN 978-953-7619-22-0, pp. 103-120

Hussmann, S. & Liepert, T. (2009). 3D-TOF Robot Vision System, *IEEE Trans. on Instrumentation and Measurement,* 58(1), pp. 141-146

Hussmann, S. & Edeler, T. (2010). Pseudo 4-phase shift algorithm for performance enhancement of 3D-TOF vision systems, *IEEE Trans. on Instrumentation and Measurement,* 59(5), pp. 1175-1181

Hussmann, S., Hermanski, A. & Edeler, T. (2011). Real-Time Motion Artifact Suppression in TOF Camera Systems, *IEEE Trans. on Instrumentation and Measurement,* 60(5), pp. 1682-1690

Hussmann, S., Knoll, F. & Edeler, T. (2011). Real-time image Processing of TOF range images using a reconfigurableprocessor system, *Proc. SPIE Vol.8085, Videometrics, Range Imaging and Applications XI*, pp. 808507 (8)

Jarvis, R. A. (1983). A perspective on range finding techniques for computer vision, *IEEE Trans. Pattern Anal. Machine Intell.*, 5(2), pp. 122-139

Lange, R. & Seitz, P. (2001). Solid-state time-of-flight range camera, *IEEE Journal of Quantum Electronics*, vol. 37, no. 3, pp. 390–397

Nuechter, A., Surmann, H. & Hertzberg, J. (2003). Automatic model refinement for 3D reconstruction with mobile robots, *Proc. of the 4th IEEE Intl. Conference on Recent Advances in 3D Digital Imaging and Modeling*, pp. 394–401

PMD Technologies, http://pmdtec.com (last accessed August 2011)

Ringbeck, T. & Hagebeuker, B. (2007). A 3D Time of flight camera for object detection, *Proc. of the 8th Conf. On Optical 3-D Measurement Techniques*, Zürich, Online-publication:

(http://www.pmdtec.com/fileadmin/pmdtec/downloads/publications/200705_PMD_ETHZuerich.pdf)

Schwarte, R., Xu, Z., Heinol, H., Olk, J., Klein, R., Buxbaum, B., Fischer H. & Schulte, J. (1997). New electro-optical mixing and correlating sensor: facilities and applications of the photonic mixer device (PMD), *Proc. SPIE, vol. 3100,* pp. 245-53

5

Rotation Angle Estimation Algorithms for Textures and Their Implementations on Real Time Systems

Cihan Ulas, Onur Toker and Kemal Fidanboylu
Fatih University,
Turkey

1. Introduction

In this chapter, rotation angle estimation algorithms for textures and their real time implementations on a custom smart camera called FU-SmartCam is introduced (Ulas et al., 2007) and improved. In the textile industry, weft-straightening is a fundamental problem which is closely related to the rotation angle estimation. Earlier weft-straightening machines used simple sensors and hardware; however, with the increased complexity of fabric types and demand for faster and more accurate machines, the whole industry started to switch to smart camera systems. Three basic methods, which are based on FGT constellation, polar transformation, and statistical features, are proposed and their performances are evaluated. As an improvement to statistical based method, we introduce a neural network based approach to choose optimum weights for the statistical features. Moreover, a comparison between FU-SmartCam and a commercial one called Tattile Smart Camera is given. Experimental results show that the introduced algorithms provide satisfactory performance, and can be used in real time systems.

Weft-straightening operation is a well-known problem in the textile industry. After the fabric is washed, before it goes to the drying process, weft-straightening must be done. Namely, rotation and deformations in the fabric must be measured and corrective action must be taken. In principle, this can be done by a human operator at relatively low speeds. An experienced operator can both detect the rotation and/or deformation in the fabric with naked eye and take corrective action by sending the proper commands to the motor drivers. Primitive weft-straightening machines used relatively simpler optical sensors and hardware. That is, instead of using CCD cameras and embedded systems to analyze the captured images in real-time, earlier systems and their sensors were based on interference and other optical/physical properties of light. Seiren Electronics' DENSIMATIC is one of such example (Seiren Electronics). However, speed and accuracy can be improved considerably by using machine vision systems. With the increased complexity of fabric types and demand for faster and more accurate processing, use of advanced machine vision algorithms with CCD cameras and embedded systems started to appear in commercial products. ElStraight manufactured by Erhardt + Leimer Company (Erdhard + Leimer) is a well-known example that uses four cameras (Tattile Smart Cameras) for weft-straightening as shown in Fig. 1.

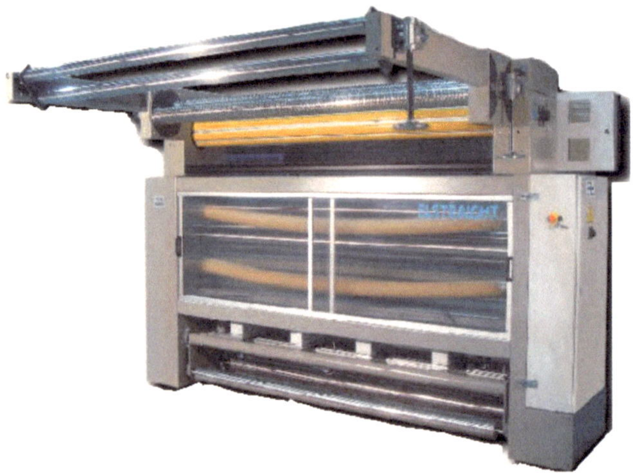

Fig. 1. Weft-straightening machine from by Erhardt+Leimer company (Erhardt+Leimer).

There are several known pattern recognition algorithms to identify different patterns in an image under the presence of translation and rotation. Some of the relevant research papers are (Tuceryan and Jain, 1998), (Loh and Zisserman, 2005), (Josso et al., 2005), and (Araiza et al., 2006), and the references therein. However, in the weft-straightening problem, we have a known texture which is subject to translation and rotation, and the problem is to estimate the rotation angle only. In a typical industrial setup, the width of a fabric equals to a couple of meters, and there are four to six equally spaced sensors, each measuring the local rotation angle. By interpolating these rotation angle measurements, it is possible to estimate mild deformations and curvatures in the fabric. Basically, we have a known 2-D periodic or almost periodic signals if the textile irregularities are taken into account. The problem is to estimate the rotation angle from discretized and windowed version of the rotated texture under the presence of camera noise and quantization errors.

Rotation angle estimation is not a new subject in computer vision. LI et al. proposed a robust rotation angles estimation algorithm from image sequences using annealing m-estimator (Li et al, 1998). They call the method robust since the proposed method can deal with the outliers. Their aim of proposing a rotation angle estimation algorithm was to solve the motion estimation problem. In (Kim Yul and Kim Sung, 1999), another method based on Zernike moments is proposed to estimate rotation angles of the circular symmetric patterns. Since circular symmetric objects have similar eigenvalues in both directions, the principal axes cannot be used for rotation angle estimation. Therefore, they introduce a robust method which uses the phase information of Zernike moments. Recently, a rotation angle estimation algorithm based on wavelet analysis is proposed for textures (Lefebvre et al., 2011). The key point is to find the rotation angle that best concentrates the energy in a given direction of a wavelet decomposition of the original image.

A typical texture image and its 10 degree rotated version are given in Fig. 2. In Fig. 3, a hypothetical deformation of the fabric is shown, which is exaggerated for a better illustration. Measurement of four local rotation angles can be interpolated to estimate the actual deformation.

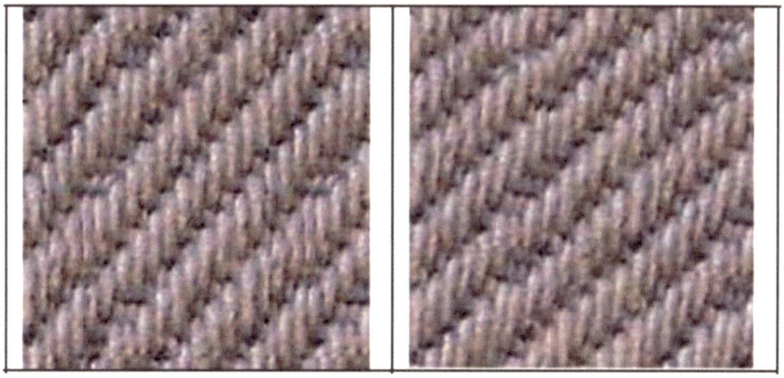

Fig. 2. A typical fabric image (Left) and its rotated and translated version (Right) are shown.

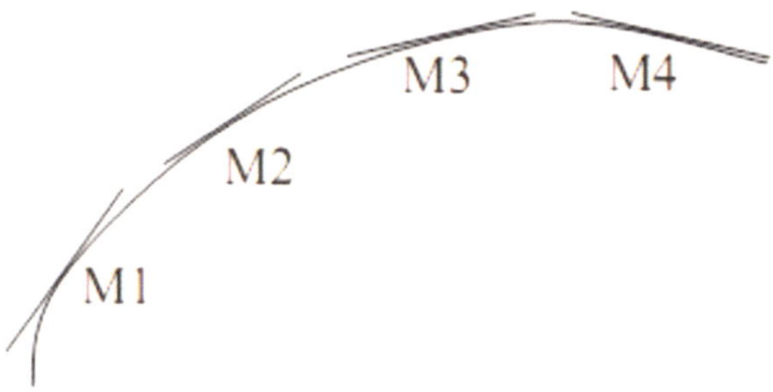

Fig. 3. Measurement of four local rotation angles is used to estimate the fabric deformation.

In real applications, deformations are much smaller than the exaggerated deformation curve shown above.

In this study, in order to solve the rotation angle estimation problem, three algorithms, which are based on "FGT Constellation", "Polar Transformation", and "Statistical Parameters" are proposed. In addition, neural networks based approach is used to choose optimum weights for statistical parameters. All of these methods are dedicated to solve weft-strengthening problem in the textile industry.

In Section 2, the proposed methods, FGT constellation, Polar Transformation, and Statistical Parameters and its extension to neural networks are discussed. Their performance analysis is given in Section 3. Finally, some concluding remarks are made in Section 4.

2. Rotation angle estimation algorithms

2.1 Polar transform approach

Polar transformation approach is related to the computation of the autocorrelation, $R_t(x,y)$, of the texture, $t(x,y)$. For an MxN image, the autocorrelation function is also an image and can be written as;

$$R_t(x,y)=\sum_{i=1}^{M}\sum_{j=1}^{N} t(i,j)\; t(i+x,i+y) \tag{1}$$

where i and j are the dummy variables for summation. This eliminates translation dependence. Rotation and translation of the texture, $t(x,y)$ only rotates $R_t(x,y)$. To estimate pure rotation in $R_t(x,y)$, one can look at its polar representation, $R_{t,polar}(r,\theta)$. It is easy to see that pure rotation around the origin in the Cartesian space corresponds to translation in the θ direction in the polar space. Therefore, now the problem is reduced to the estimation of the shift in the y direction of $R_{t,polar}(r,\theta)$. A simple correlation analysis can be used to find the value of d for which the correlation between $R_{t,polar}(x,y+d)$ and $R_{t-rotated,polar}(x,y)$ is maximum:

$$\arg\max_{d} < R_{t,polar}(x,y+d), R_{t-rotated,\, polar}(x,y) > \tag{2}$$

This requires a search over d, and for each d value, the computation of the inner product requires $O(n^2)$ floating point operations. As an alternative, it can be considered taking 2D Fourier Transform of $R_{t,polar}(x,y)$, which converts translation in the second coordinate to a linear phase shift in the second coordinate. A simple graphical approach can be used to estimate the proportional constant in the linear phase shift, and hence estimate the rotation angle.

Preliminary tests indicate that both variations of this approach are computationally demanding, but give accurate angle estimates. For more information about the polar transform approach based rotation angle estimation one can look at (Sumeyra, 2007).

2.2 FGT Constellation approach

The FGT-Constellation approach also involves computation of the autocorrelation, $R_t(x,y)$, of the texture, $t(x,y)$. However, following this a thresholding is done, and a "constellation" like image is obtained. Basically, peaks of $R_t(x,y)$, will appear as bright spots in the thresholded image. If the texture is rotated, as shown in Fig. 4, one can see that the bright points also rotate in the same way and the same amount as shown in Fig. 5. Then the problem turns into finding the brightest point position on the thresholded image by searching in the first quadrant of the coordinate axis (see Fig. 6). An illustrative video which shows the operation of this algorithm can be found in the following link; www.fatih.edu.tr/~culas/rotationestimation/video1.avi.

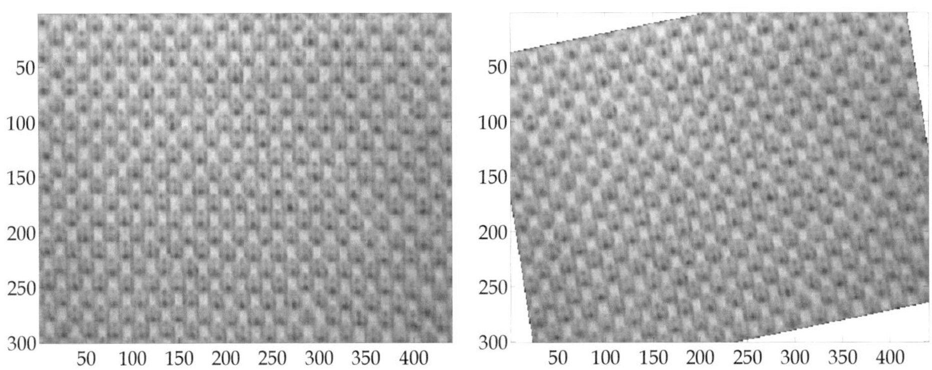

Fig. 4. Left: A picture of texture is having a size of 300 x 450 pixels. Right: The texture is rotated about 10 degrees in the counter clockwise direction.

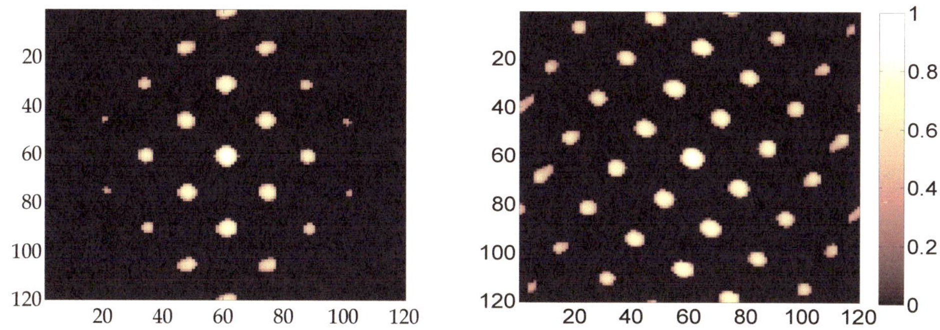

Fig. 5. Left: FGT constellation of the original image texture. Right: The FGT constellation of the rotated image. When the picture is rotated, the bright spots also rotate by the same amount and in the same direction; however, new small spots may appear in the constellation.

Also, preliminary tests on the .NET platform using the DirectX framework showed the feasibility of this approach both computational-wise, and performance-wise. This algorithm is later implemented on the FU-SmartCam, and for 64x64 image size, we were able to get a couple of estimates per second with about 1 degree or better accuracy. However, significant improvement can be achieved if larger images are used. To overcome the computationally demanding autocorrelation computation, which is done by floating point FFT, we tested 2-D autocorrelation computation via *GF(p)* transformation (GT), where p is a large prime satisfying

$$p > 2N^2 2^{2m} \tag{3}$$

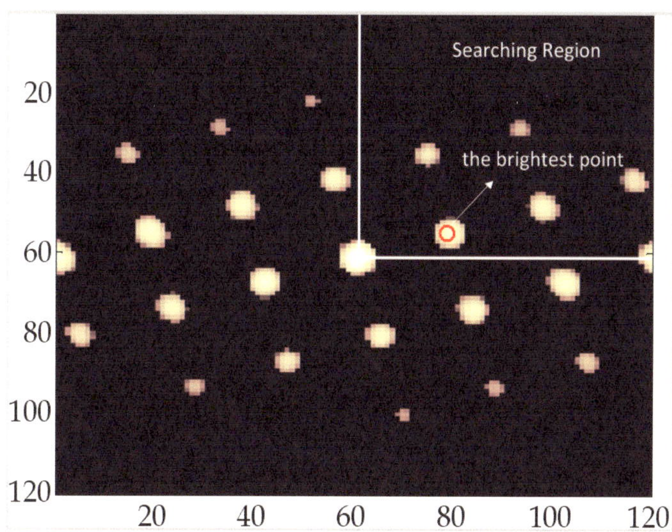

Fig. 6. Brightest point search region.

where is the image size, and m is the number of bits used to store the intensity information. Note that, to be able to perform fast GT by using the well-known "divide and conquer" type approach of FFT, we need the prime to be of the form,

$$p = 2^{u+1}k + 1 \quad (4)$$

where $N = 2^u$ is the image size. We have successfully tested this approach in .NET platform using the DirectX framework: We have selected, $p = 1073741953$ and used $g = 10$ instead of $e^{i2\pi/128}$. Preliminary tests indicate significant speedup, because instead of floating point operations, only 64 bit integer operations are needed, and the structure of the code is very similar to FFT code structure. The theoretical FGT approach, in principle can also be applied for the computation of the autocorrelation in the Polar Transform method. However, there seems to be no simple and fast way of doing Cartesian to polar transformation without using computationally expensive floating point operations.

2.3 Extraction of statistical features

In order to avoid the computational difficulty of autocorrelation computation, and polar transformation, we use a completely different method based on the computation of several statistical features from the fabric. Parameters varying significantly with rotation are considered as suitable for the rotation estimation. In addition, the parameter changes should be preferably linear or almost linear with the rotation, and for real-time applications, they should be easy to compute. For this purpose, five statistical features are proposed for a texture image. These are given as;

- 1-D linear model
- 2-D linear model
- Means of the standard deviations parallel to the x - axis
- Means of the standard deviations parallel to the y - axis
- Means of the standard deviations along the diagonal axes.

These features affect the overall system performance. In performance evaluation tests, we show the benefit of using large number of statistical parameters instead of using small set of features. Statistical parameters based approach is computationally more attractive. The reason is that a look up table is generated based on the reference fabric and this is stored in the memory. Then for a rotated image, these statistical features are computed, and then using the nearest neighborhood method or weighted 1-norm distance metric, best matching rotation angle is estimated.

Computation of the statistical features is explained starting from following subsection. Then two scenarios for rotation estimation based on statistical features are analyzed,

- By using only 2-D model parameters,
- By using all features.

2.3.1 2-D modeling

We model each pixel value with the following equation.

$$\hat{t}(x,y) = \alpha t(x, y - d_y) + \beta t(x - d_x, y) \quad (5)$$

where α and β are the model parameters. Apart from these parameters, there are also two variables which are d_x and d_x. These variables, d_x and d_x, correspond to shifts/periods in x and y directions respectively. These parameters can be determined by using a trial and error depending on the fabric type. Hence, there are a total of 4 parameters. One possible method to determine α and β given the values of d_x and d_x is the minimization of the following cost function.

$$f = \sum_{x=1}^{M}\sum_{y=1}^{N}[-\hat{t}(x,y) + \alpha t(x,y-d_y) + \beta t(x-d_x,y)]^2 \tag{6}$$

To minimize this cost function, derivatives with respect to α and β are computed, and set to zero;

$$\frac{\partial f}{\partial \alpha} = 0 \qquad \frac{\partial f}{\partial \beta} = 0 \tag{7}$$

After the solving these two equations, one can get α and β as,

$$\alpha = \frac{A_{22}B_1 - A_{12}B_2}{A_{11}A_{22} - A_{21}A_{12}}, \quad \beta = \frac{A_{12}B_1 - A_{11}B_2}{A_{12}A_{21} - A_{11}A_{22}}, \tag{8}$$

where A_{11}, A_{12}, A_{21}, A_{22}, B_1, and B_2 and are:

$$A_{11} = \sum_{x=1}^{M}\sum_{y=1}^{N} t^2(x-d_x,y), \qquad A_{12} = \sum_{x=1}^{M}\sum_{y=1}^{N} t(x-d_x,y)\cdot t(x,y-d_y), \tag{9}$$

$$A_{21} = \sum_{x=1}^{M}\sum_{y=1}^{N} t^2(x-d_x,y), \qquad A_{22} = \sum_{x=1}^{M}\sum_{y=1}^{N} t(x-d_x,y)\times t(x,y-d_y), \tag{10}$$

$$B_1 = \sum_{x=1}^{M}\sum_{y=1}^{N} t(x,y-d_y)\cdot t(x,y) \qquad B_2 = \sum_{x=1}^{M}\sum_{y=1}^{N} t(x,y)\times t(x-d_x,y). \tag{11}$$

Variation of the model parameters with respect to rotation angle is shown in Fig. 7. The texture image is rotated from -30 to +30 with 0.5 degree steps and its corresponding 2-D model parameters are computed. We observe that these parameter variations are almost linear when the distance values are $d_x = 2$ and $d_y = 1$.

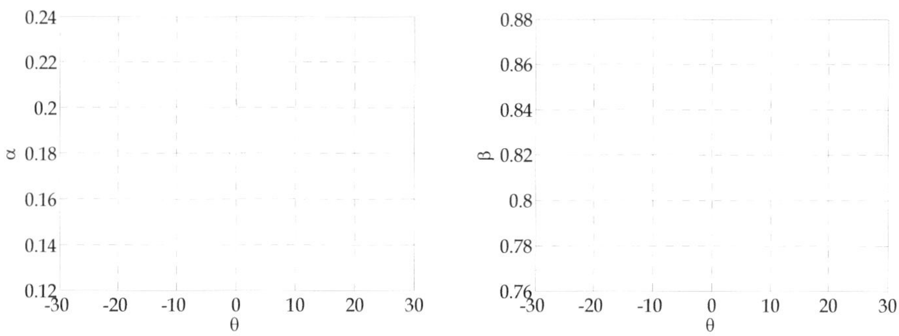

Fig. 7. 2-D model parameters versus rotation angle. The rotation angle is incremented from -30 to +30 with 0.5 degree steps.

2.3.2 1-D modeling

The 1-D model parameter approximation is similar to 2-D model parameter approximation. The following 1-D equation can be used for this type of modeling,

$$\hat{t}(x,y) = \gamma t(x - d_x, y - d_y) \tag{12}$$

In this case, γ has the following equality,

$$\gamma = \frac{\sum_{x=1}^{M}\sum_{y=1}^{N} t(x - d_x, y - d_y)\, t(x,y)}{\sum_{x=1}^{M}\sum_{y=1}^{N} t(x - d_x, y - d_y)\, t(x - d_x, y)}. \tag{13}$$

Variation of the 1-D model parameter, γ, with respect to rotation angle is shown in Fig. 8. The texture image is rotated from -30 degrees to +30 degrees with 0.5 degree steps and its corresponding 2-D model parameter is plotted versus the rotation angle. As it is seen, variation is not linear for the distance values $d_x = 2$ and $d_y = 1$ and this is a high undesirable feature.

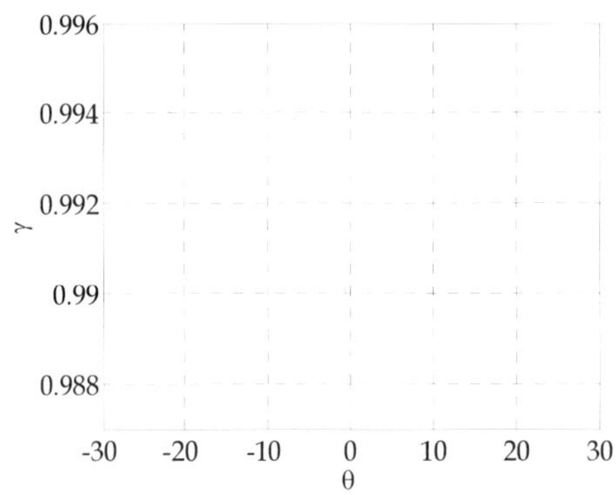

Fig. 8. 1-D model parameter, γ variation versus rotation angle. The rotation angle is incremented from -30 degrees to +30 degrees with 0.5 degree steps.

2.3.3 Mean of the standard deviations along the X-Axis

The mean of the standard deviations along the x axis can be expressed as follows;

$$I = \frac{\sum_{i=1}^{M} \sigma_{x_i}}{M} \qquad \Phi_x = \frac{I}{\Psi} \tag{14}$$

where I is the mean of standard deviation along the x axis of the texture image. We divide I to Ψ, which is the mean of the gray level image pixels, in order to eliminate the ambient illumination effects of the environment. σ_{x_i} is the standard deviations of the ith row, and M, represents the width of the image. The variation of the mean of standard deviations along the x-axis versus rotation is shown in Fig. 9.

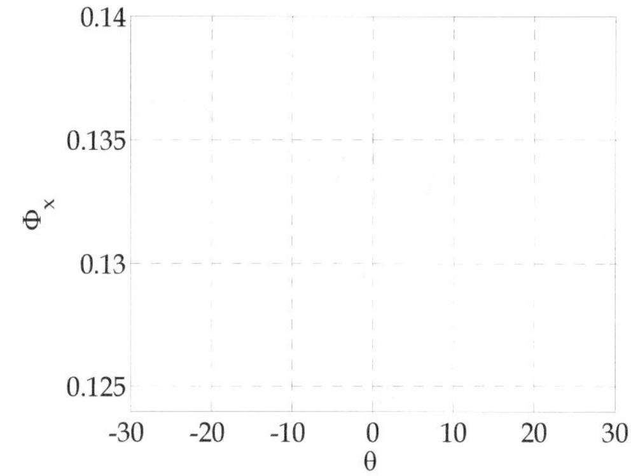

Fig. 9. Mean of standard deviations along *x*-axis versus rotation angle.

2.3.4 Mean of the standard deviations along Y axis

Similarly, the mean of the standard deviations along the y axis can be expressed as follows;

$$J = \frac{\sum_{i=1}^{N} \sigma_{y_i}}{M} \qquad \Phi_y = \frac{J}{\Psi} \tag{15}$$

J is the mean of standard deviations along the y-axis of the texture image. Similarly, it is divided to mean intensity level of the image to find Φ_y. The standard deviation of the ith column is denoted by σ_{y_i}, and N represents the height of the image. The variation of Φ_y with the rotation angle is shown in Fig. 10.

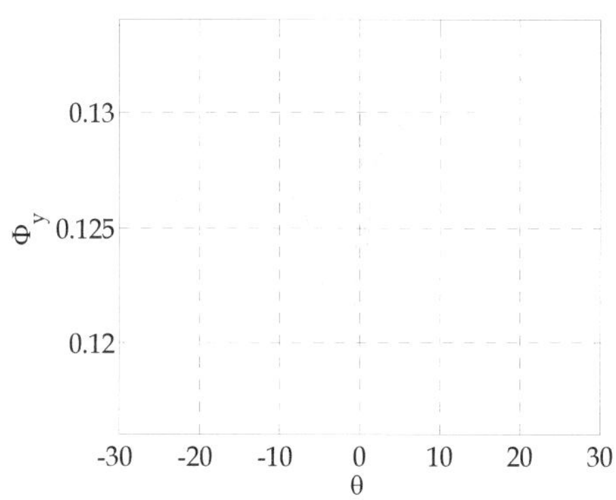

Fig. 10. Mean of standard deviations along *y*-axis.

2.3.5 Statistical feature based on mean of the standard deviations along diagonal axes

$$K_1 = \frac{\sum_{i=1}^{D} \sigma_{d_{ii}}}{D} \qquad K_2 = \frac{\sum_{i=1}^{D} \sigma_{d_{i(D-i)}}}{D} \qquad \Phi_d = \frac{K_1 + K_2}{2\Psi} \tag{16}$$

K_1 and K_2 are the means of standard deviations on the diagonal and off-diagonal axes of the texture image. D stands for the number of the diagonal elements. Actually, for all study, *M*, *N*, and *D* are the same size since we work with the square images. Fig. 11 shows the variation of this new parameter with respect to the rotation angle.

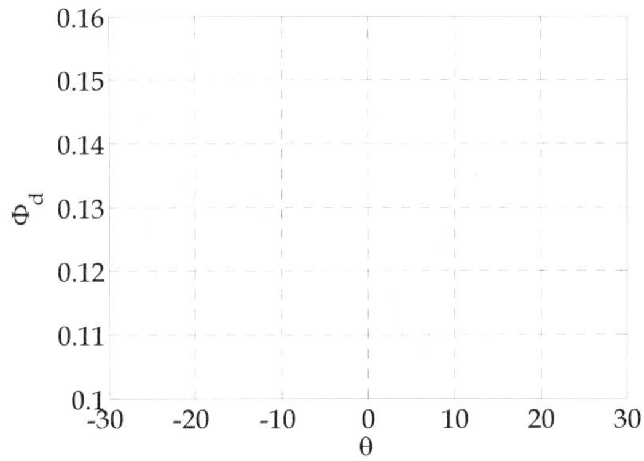

Fig. 11. Mean of standard deviations along diagonal axes.

2.4 Rotation angle estimation by using statistical features and model parameters

The main problem in rotation angle estimation is to find the useful statistical parameters which change significantly and linearly with rotation. To be able to find the best candidates, variations of each parameter are drawn as in the previous section and looked for the linear and important changes. Then, we decide if these parameters might be used or not. After determining the statistical features, a look up table is generated by rotating the reference image in a range and its statistical parameters are stored. In the estimation process, for the rotated image, the statistical features are computed, and the parameters are searched through the look up table by using a nearest neighborhood search (NNS) method. The closest parameter combination is accepted as rotation estimation. However, this idea best works by assuming that the features change linearly with the rotation, and all of them have the same importance. In fact, this is not true for many cases. Because, neither the parameters change linearly nor have the same importance. For this reason, we append the artificial neural networks to the proposed method to overcome this problem.

Another issue is to use sufficient number of statistical features. To show the importance of the number of parameters, the experiments are divided into two parts. In the first section, only 2-D model parameter is chosen as feature parameters, which are the most effective ones. In the second section, all statistical parameters are exploited to show the improvements.

2.4.1 Using only 2D model parameters

In this subsection, the 2-D model parameter is used as the statistical feature. A look-up table is generated by rotating the reference image in a desired region and calculating the 2D model parameters for each rotation.

$$\Gamma(\theta_i) = [\alpha_i \ \beta_i] \qquad 1 \le i \le M \tag{17}$$

where, θ, denotes the amount of rotation, and i is the index of each rotation starting from the first to last rotation, M. After the look-up table is built, the system performance is tested in the same region with higher resolution. For example, one can generate the look-up table in the region of -30 degrees to +30 degrees with 0.5 degree steps, and then the method is tested in the same region with 0.1 degree steps.

Another important point is the problem of choosing distance parameters, d_x and d_y. These parameters have to be chosen properly because they significantly affect the linearity of the variations. To decide which distance values are suitable for the reference texture image, we followed two ways. The first one is to draw the parameter-rotation graph for each, d_x and d_y combinations and look at the linearity and amount of change. The second and more professional one is to calculate the sum of least square errors between actual and estimated rotations for d_x and d_y combinations and accept the combinations which give the smallest error.

After we decide the distance parameters, d_x and d_y, the measured model parameters, α and β, are searched through the look-up table and the closest variations are used for rotation estimation. In general case, if we have foreknowledge about the weights of the parameters, we can use the weighted nearest neighborhood search as,

$$e = \sum_{i=1}^{N} w^T \{\Gamma(\theta_i) = [\alpha_i \;\; \beta_i]\} \tag{18}$$

where $\Gamma(\theta_i)$ is the ith row of the look-up table if the parameters are put on the columns. The weights are represented as w which emphasizes some statistical parameters over others. In this section, w is chosen as unity vector.

2.4.2 Using all statistical parameters

To get better results in rotation estimation algorithm, all statistical features explained in Section 2.3 are used. The idea is very similar to the two model rotation estimation algorithm; the first difference is that we generate the look-up table with all these statistical parameters. The other difference is the computational cost. In this method, the processing and searching time increases due to the number of the parameters that are used. The look-up table has the following structure;

$$\Gamma(\theta_i) = [\alpha_i \;\; \beta_i \;\; \gamma_i \;\; \Phi_{x_i} \;\; \Phi_{y_i} \;\; \Phi_{d_i}] \qquad 1 \le i \le M \tag{19}$$

2.4.3 Drawbacks of nearest neighborhood search (NNS)

The NNS is used to solve one norm distance problem. However, the NNS algorithm may fail if the parameters do not have the equal weights and linear changes with the rotation. We observe that the statistical parameters, explained in Section 2.3, are neither exactly linear nor have same importance. The second problem with the NNS is the searching time. If the look-up table is generated with high resolution, the table becomes so large and takes long time to estimate the amount of rotation. The method can be accelerated from *O(N)* to *O(logN)* by using kd-tree space quantization methods (Bently, 1980). Another and better solution is to use artificial neural networks in order to find the weighting factor and speed up the algorithm significantly. In neural networked based solution, the method becomes much faster since it does not need any searching method, and it is sufficient to train look up table.

2.5 Neural network improvement to statistical parameters based approach

Statistical parameters do not have equal importance on estimating the rotation angle; therefore, artificial neural networks are used to choose optimum weights for these parameters. To do this, Fletcher-Reeves updates etc., as shown in Fig. 12, we formed a global user interface in Matlab to test the performance of the various neural network training methods such as Levenberg-Marquardt (LM), BFGS quasi-Newton back propagation, conjugate gradient back propagation with Fletcher-Reeves updates etc.

Statistical features based method is computationally very attractive. Experimental results show quite affirmative results: An accuracy of less than 0.2 degree can easily be achieved with relatively little computational effort. In our tests, we observed that LM training method provides better performance over others.

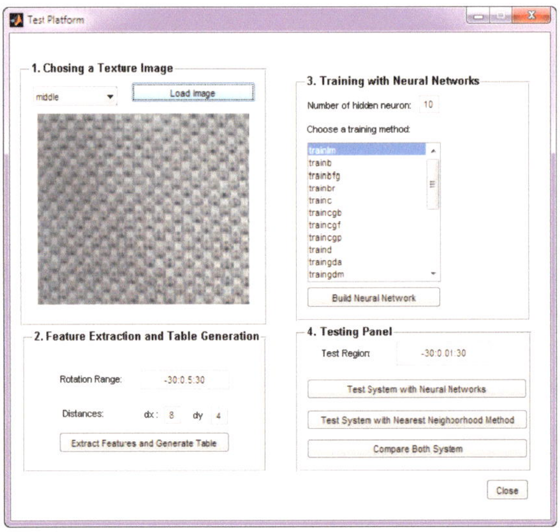

Fig. 12. A Test Platform formed in Matlab to test Neural Network with various training methods and compare the results with Nearest Neighborhood method.

3. Experimental results

3.1 Experimental setup

Tattile Smart Camera has a StrongARM SA-1100 processor based board, and Lattice iM4A3 programmable logic device. The StrongARM SA-1100 board is a low power embedded system running a Linux port. The architecture of the board is heavily based on the LART project done at Delft University (LART project). Tattile Smart Camera also has a progressive CCD sensor, and image acquisition is done by the high speed logic circuit implemented on the Lattice iM4A3 programmable logic device. In principle, it is possible to use a desktop PC, cross compile the code for StrongARM and then upload (ftp) it to the board via network connection. However, one still has to work with cross compilers, and simulators for prototyping experiments.

On the other hand, the FU-SmartCam embedded system which does not use that low power, but has more processing power, memory, and flash storage. It has Vortex86 processor (Vortex system) in it, and runs the Intel x86 port of Linux, i.e. the Linux port which runs on regular desktop PCs. Because of this, there is no need for cross compilers, and simulators. Complete prototyping experiments can be performed on the desktop PC, and the generated code will run without any modification on the target board, which is slow as the Vortex86 but not as powerful as a Pentium 4. The image acquisition was performed using a low cost interlaced scan camera with a Conextant Bt848 chip. The FU-SmartCam is shown in Fig. 13.

The FU-SmartCam has VGA, Keyboard, Ethernet, and RS-232 connections, is extremely flexible, and easily reconfigurable. Currently, we are also developing a relay board with a small 8-bit microcontroller interfaced to the Vortex over RS-232. This relay board will enable direct connection of the FU-SmartCam to several industrial equipment.

Fig. 13. FU-SmartCam : On the top, we have a low cost interlaced scan camera, in the middle we have the Vortex system, and in the bottom we have a dual output power supply. The Vortex system itself consists of two boards. The overall system is quite small.

3.2 FGT constellation performance analysis

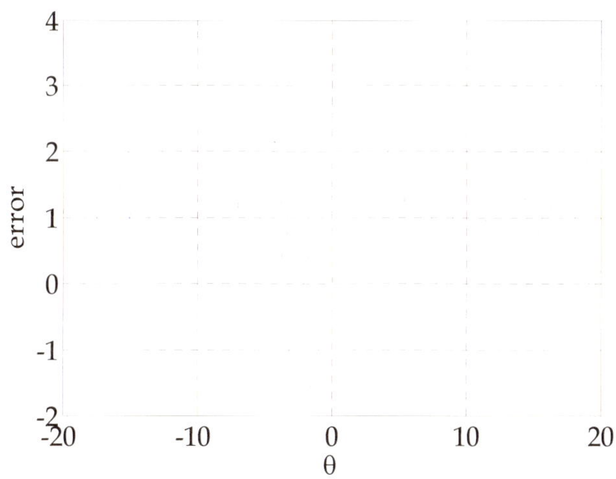

Fig. 14. Error versus rotation in the range of -20 degrees to 20 degrees with 0.1 degree steps.

In this part, the performance of the FGT constellation approach is investigated, as seen from the Fig. 14, the maximum absolute error is about 3 degree and the average of error is about 1 degree. Due to periodicity, although we expect to see algorithm to work in the range of -45 to +45, we observe that the method works fine in the range of -30 to +30 (see Fig. 15.) However, for weft-strengthening problem, this range can be acceptable for rotation angle estimation.

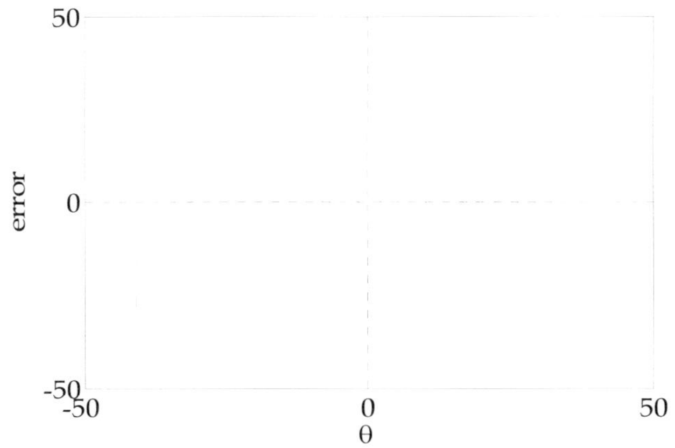

Fig. 15. The drawback of the FGT constellation based approach. The algorithm cannot estimate rotation angle if there is a 30 degrees rotation in both directions.

3.3 Using only 2-D model parameters

In order to test the performance of the statistical parameters based approach, we plot the errors with respect to rotations. The look-up table is generated in the region of -30 degrees to 30 degrees with 0.5 degree steps and it is tested in the same region with 0.1 degree resolution.

First of all, to show the effect of the distance parameters, the distance values are taken randomly as $d_x = 1$ and $d_y = 1$ and the error variations versus rotation is given in Fig.16. From this figure it can be seen that, the error becomes considerably high in some rotations and cannot be acceptable as acceptable.

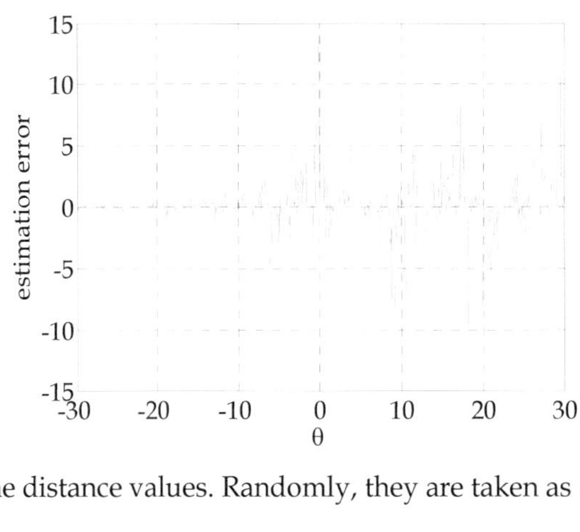

Fig. 16. The effect of the distance values. Randomly, they are taken as $d_x = 1$ and $d_y = 1$.

However, as explained in Section 2.4, if the proper distance parameters are chosen as $d_x = 9$ and $d_y = 8$, the error variation versus rotation is shown in Fig. 17, and can be considered as acceptable. The average absolute error of the estimation is less than 0.5 degree.

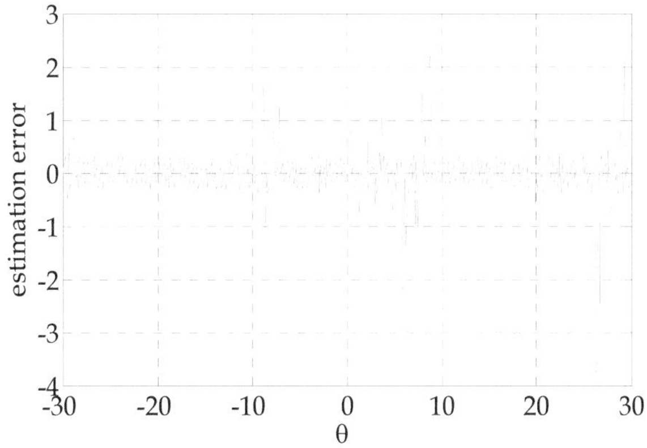

Fig. 17. The error variation versus rotation for the optimum distance, $d_x = 9$ and $d_y = 8$.

3.4 Using all statistical parameters with nearest neighborhood method

In this section, all statistical feature parameters are used for rotation angle estimation. The testing region is again from -30 degrees to +30 degrees with 0.1 degree steps. In this case the results are very attractive and the absolute error is about 0.2 degree. The distance parameter were chosen as $d_x = 9$ and $d_y = 8$. The variation of estimation error versus rotation angle is shown in Fig. 18.

3.5 Using all statistical parameters with artificial neural networks

To compare the results of neural networks and nearest neighborhood based methods, we used the same texture image and same optimum distance values in the same testing region. In Fig. 19, the error variation for only 2-D model parameters is shown.

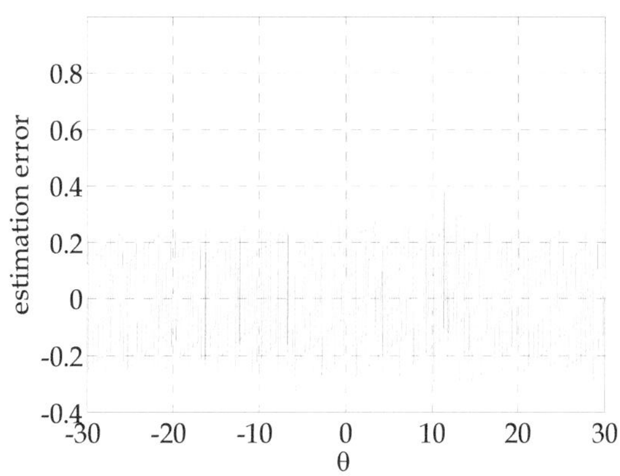

Fig. 18. Rotation estimation error versus rotation angle. The estimation error is about 0.2 degree.

In order to show the power of using all statistical parameters with neural networks, we increased the testing resolution from 0.1 degree to 0.01 degree and plot the result in Fig. 20. The estimation error is decreased up to 0.1 degree, which was about 0.2 degree in nearest neighborhood method. Therefore, the performance is increased almost two times. If we compare the computation time, neural network is much faster than nearest neighborhood based method.

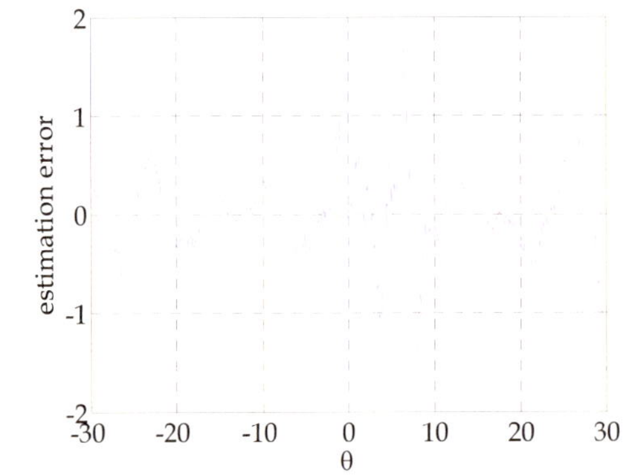

Fig. 19. Rotation estimation error variation versus rotation angle. Here only 2-D model parameters are trained in neural networks.

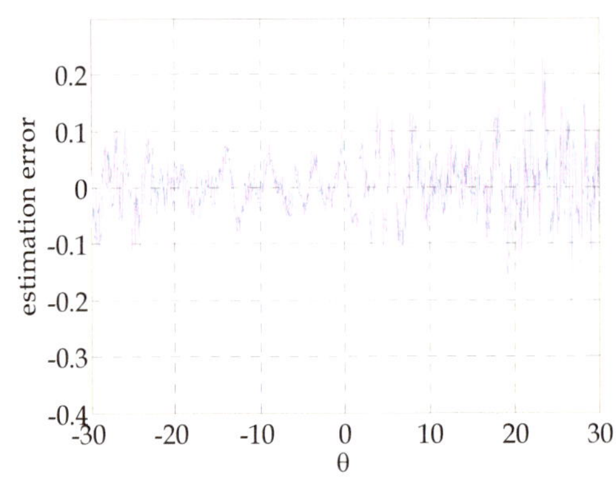

Fig. 20. Rotation estimation error variation versus rotation angle. Here only all model parameters are trained in neural networks. The testing resolution is increased to 0.01 degree to show robustness of the algorithm.

4. Conclusion

In this chapter, the weft-straightening problem encountered in the textile industry is described. As a solution of rotation angle estimation which is the fundamental part of the weft-straightening problem, three different algorithms are introduced. The first algorithm is based on the Polar Transform which is applied to auto-correlated images; therefore, the translation in the θ direction gives the rotation angle. The second one is based on FGT constellation approach, and it depends on the autocorrelation of the thresholded image. FGT constellation consists of regularly distributed bright spots and the rotation angle is estimated by finding the brightest point in the first quadrant of the coordinate axis. The third algorithm is based on statistical parameters. These parameters, firstly, are computed for the reference image and a look up table is generated for its artificially rotated images. The statistical parameters of the input image are searched in the look up table and the closest one is found as rotation angle. Finally, in order to improve the statistical parameters based approach, neural networks are used to choose optimum weight factors since not all parameters have the same importance. Various neural network training methods are tested to find the best performance. The results show that the proposed methods can be successfully implemented in the real-time systems.

5. Acknowledgements

This work is supported by the Scientific Research Fund of Fatih University under the project number P50061001_2.

6. References

Araiza, R., M. G. Averill, G. R. Keller, S. A. Starks, and C. Bajaj ,"3-D Image Registration Using Fast Fourier Transform, with Potential Applications to Geoinformatics and Bioinformatics," Proceedings of the International Conference on Information Processing and Management of Uncertainty in Knowledge-Based Systems IPMU'06, Paris, France, July 2-7, 2006, pp. 817-824.

Bentley, L. Multidimisional divide and conquer, Communications of the ACM, vol. 22, no. 4, pp. 214–229, 1980.

Josso, B., D. R. Burton, M. J. Lalor, "Texture orientation and anisotropy calculation by Fourier transform and Principal Component Analysis", Mechanical Systems and Signal Processing 19 (2005) 1152–1161.

Kim, W.Y., Kim, Y.S. (1999). Robust Rotation Angle Estimator, *IEEE Transactions on Pattern Analysis and Machine Intelligence*, Vol. 21, No. 8, pp. 768-773.

Lefebvre, A. Corpetti, T. Hubert-Moy, L. (2011). Estimation of the orientation of textured patterns via wavelet analysis. *Pattern Recognition Letters,* Vol. 32, pp. 190-196.

Li, B.; Xu, Y. & Choi, J. (1996). Applying Machine Learning Techniques, *Proceedings of ASME 2010 4th International Conference on Energy Sustainability*, pp. 14-17, ISBN 842-6508-23-3, Phoenix, Arizona, USA, May 17-22, 2010

Li, S. Z., Wang, H., and Soh, W.Y.C. (1998). Robust Estimation of Rotation Angles from Image Sequences Using the Annealing M-Estimator, *Journal of Mathematical Imaging and Vision* Vol. 8, pp. 181–192.

Loh, A. M., A. Zisserman, "Estimating the affine transformation between textures," Proceedings of the Digital Imaging Computing: Techniques and Applications (DICTA 2005).

Tuceryan, M., and A. K. Jain (1998), "Texture Analysis," The Handbook of Pattern Recognition and Computer Vision 2nd Edition), by C. H. Chen, L. F. Pau, P. S. P. Wang (eds.), pp. 207-248, World Scientific Publishing Co.

Ulas, C., Demir, S., Toker, O., Fidanboylu, K. (2007). Rotation Angle Estimation Algorithms for Textures and Their Real-Time Implementation on the FU-SmartCam, *Proceedings of the 5th International Symposium on image and Signal Processing and Analysis,* pp. 469-475, Istanbul, Turkey.

LART project page at Delft University

http://www.lartmaker.nl/

Vortex system
http://www.icop.com.tw/products_detail.asp?ProductID=119

Seiren Electronics' DENSIMATIC-SVW II
http://textileinfo.com/en/it/seiren-e/p01-01.html

Erhart+Leimer's ElStraight
http://www.erhardt-leimer.it/prodotti/raddrizzatrama_eng.html

Tattile Smart Camera
http://www.tattile.com/new.site/start.htm

6

Characterization of the Surface Finish of Machined Parts Using Artificial Vision and Hough Transform

Alberto Rosales Silva, Angel Xeque-Morales,
L.A. Morales-Hernandez and Francisco Gallegos Funes
National Polytechnic Institute of Mexico and Autonomous University of Queretaro
Mexico

1. Introduction

The surface finish of machined parts is of the utmost importance in determining their quality. This is not only for aesthetic purposes. Since in several industrial applications the machined parts have to be in contact with other parts, surface finish is also a determining factor in defining the capacity of wear, lubrication, and resistance to fatigue (i.e. service life). To determine the quality of machined parts, the roughness **is analyzed to be a representation** of the surface texture. Therefore, mathematical techniques have been developed to measure this criterion; such as the roughness meter, X-ray diffraction, ultrasound, electrical resistance, and image analysis (Alabi et al., 2008; Xie, 2008; Bradley & Wong, 2001).

The surface finish is the factor that determines whether the edge is sharp or not because this presents linear and continuous segments when the tools are sharpened and discontinuous segments when the tool begins to dull. In this chapter, the surface finish is analyzed utilizing images of machined parts. The texture of the surface of these parts is characterized by lines representing the valleys and ridges formed by the machining process. The continuity of the scratched surface is segmented by applying the standard modified Hough transform, and the quality of the surface is assessed by analyzing the continuity of the scratch.

The Hough Transform has been studied by many different authors (Leavers, 1993; Illingworth & Kittler, 1988), through which various techniques have been developed. The principal differences between these techniques are the parameters employed for the study of the space generated. It has been mentioned that the Hough transform is a case of the Radon Transform.

The stages of the standard Hough transform, developed by Duda and Hart (Duda & Hart, 1972) are the following: 1. Determination of θ and ρ; 2. Accumulator registration; 3. Maximum location in the accumulator; and 4. Image reconstruction. This methodology has been used to analyze uniform machined parts in milling machines (Mannan, Mian & Kassim, 2004). **In this chapter, machined parts will be analyzed on Computer Numerical Control (CNC) lathes, and will be improved by the methodology mentioned**. An intermediate stage has been added between stages 1 and 2 to discriminate possible pixels that are not part of the lines to be segmented. Another contribution has been introduced in stage 4, where the reconstruction is undertaken using the straight lines that form the image without projection, up to the edges of the image. These modifications were implemented to optimize the developed application.

The Hough transform is a mathematical method that uses an edge detector to locate points that could form a perceptible edge. The method determines if the points are specific components of a parameterized curve. This was developed in 1962 by Paúl Hough (Davies, 1990) in order to find, in nuclear physics, the straight paths of high energy particles in a bubble chamber; **but not until 1969 was the Rosenfeld (Illingworth & Kittler, 1988) proposed algorithm introduced for use in the image processing area**.

2. The hough transform

The Hough transform is a method used for line segmentation. It is based on the transformation of a straight line in an *x-y* plane (eq.(1)) into a point in an *m-b* plane (eq. (2)). The line equation (eq. (1)) defines each one of the lines in the *x-y* plane by means of a slope (*m*) and a *y*-axis intersection (*b*).

The points that form the line in the *x-y* plane are represented by one point in the new *m-b* plane, as shown in Fig. 1. In this, the line of the *x-y* plane has a slope value equal to the unit and an intersection equal to two. One can see that this line is represented by only one point in the *m-b* plane, whose coordinates correspond to the parameter values (*m*,*b*) that define the straight line in the *x-y* plane (Leavers, 1993).

$$y=mx+b, \tag{1}$$

$$b=y-mx. \tag{2}$$

a) *x-y* plane b) *m-b* plane

Fig. 1. The Hough transform. a) Representation of the straight line in the *x-y* plane. b) Representation of the same line in the *m-b* plane.

When using the Hough transform to solve the problem of locating straight lines and then searching for points with greater frequency in the *m-b* plane, the following routines and elements need to be created and evaluated:

1. **Acummulator**: A table whose cells are initialized to zero. The number of rows and columns in the accumulator are determined by the maximum values of the parameters

(m,b) of the lines to be segmented. For example: a line in the x-y plane is defined by two points and the points that are found between these two points belong to a straight line, as the points shown in Fig. 2a. The expected slope of the straight line formed by these three points is in the interval $0 \leq m \leq 3$, while its intersection with the y axis is in the interval $0 \leq b \leq 5$. Thus, the accumulator has to be constructed with four columns and six rows to accommodate the expected values of m and b, for the line defined by these three collinear points.

2. **Evaluation of points and registration in the accumulator**. Eq. (2) is solved by maintaining constant the values of the coordinates for each point (x,y), and varying the value of m in its expected interval. Each point describes a straight line in the m-b plane, as shown in Fig. 2b. In the cells of the accumulator, their values are increased by one unit according to the cells coinciding with the projection of the points in the expected values for (m,b). This can be seen in Fig. 2c, in those cells with values other than zero.
3. **Search for the greatest number of intersections in the accumulator:** All the collinear points belonging to a given line intersect (or are counted) in the same coordinate (m,b) within the Hough space. The intersections are registered in the accumulator cells (frequency *fr*). The cell with the highest frequency of intersections defines the parameters (m,b) of the straight line in the x-y plane.
4. **Construction of the straight line:** Equation (1) is evaluated by maintaining constant the values of m and b obtained in the accumulator, and by varying the value of x in an interval determined from the amplitude of the x axis in the x-y plane. Fig. 2d shows the construction of the line depicted in Fig. 2a. The limitation of employing the m-b parameters manifests itself when lines perpendicular to the x axis are to be segmented. In such a case, m tends to infinity. To solve this indefinition, Duda and Hart (Duda & Hart, 1972) propose to use the parameters (θ, ρ) of a vector starting at the origin and oriented perpendicular to the line to be segmented. In this parameterization, θ is the angle sustained by the vector and the x axis, and ρ is the distance measured from the origin to the intersection between the vector and the line (see Fig. 3). This parameterization is described by eq. (3); it is known as the Standard Hough Transform (SHT).

$$\rho = x_i \cos\theta + y_i sin\theta \text{, where } -\frac{\pi}{2} \leq \theta < \frac{\pi}{2}. \tag{3}$$

In this new space, each point with parameters (θ,ρ) is mapped into a sinusoid when its coordinates x_i and y_i are kept constant (see eq. (3)), and the value of θ is varied within the specified interval (Duda & Hart, 1972).

The process used to identify straight lines with the SHT is presented in Fig. 3, where we can see four points (Fig. 3a). Three of the describe the straight line with greater length. Each one of these points is to be projected into a different sinusoid in the θ-ρ space (see Fig. 3b). These projections are registered in the accumulator (Fig. 3c), whose dimensions are obtained according to the expected intervals for θ and ρ. The interval for ρ is defined as (Duda & Hart, 1972):

$$-2\sqrt{\left(a^2+h^2\right)} \leq \rho \leq 2\sqrt{\left(a^2+h^2\right)}, \tag{4}$$

where a is the width and h is the height of the x-y plane, respectively.

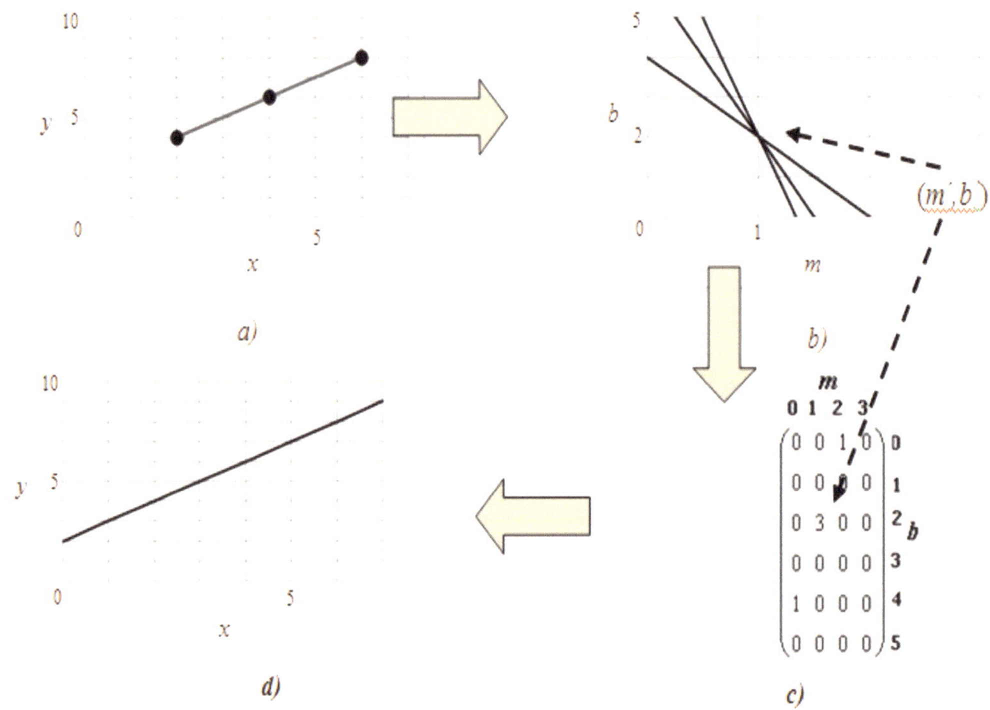

Fig. 2. The Hough transform: a) Collinear points in the *x-y* plane; b) Mapping and intersection of in the points in the *m-b* plane; c) Accumulator registration; d) Reconstruction of the straight line in the *x-y* plane.

Expression 4 is used to obtain the possible lengths of the parameterization or radius vector, from its origin to the line. When the points are collinear the cells with greater values provide the parameters of the lines to be identified. In Fig. 3c we can see the maximum value of three, this corresponds to the greatest number of collinear points that define the line of Fig. 3a having the greatest length. By plotting the values of the intersections in the accumulator on a coordinate system with three dimensions (θ, ρ, fr) (where fr indicates the number of registered intersections in each cell), one can discriminate, in the accumulator, the cells with a high frequency of intersection, as shown in Fig. 3c. In the third axis, corresponding to the frequency of intersection, the value of the highest frequency occurs only once; that is, there is only one maximum point. Accordingly, there is only one set of points with the highest possible collinearity (see Fig. 3a), and therefore a line of greater length in the *x-y* plane of the image.

The reconstruction of the lines located with the standard Hough Transform is carried out by evaluating the values of x in eq. (5) Said values for x are from zero to the value of the width of the *x-y* plane. The evaluation of Eq. (5) for x, is performed by maintaining constant the values of the θ and ρ parameters obtained from the rows and columns' values of the cells having maximum values in the accumulator.

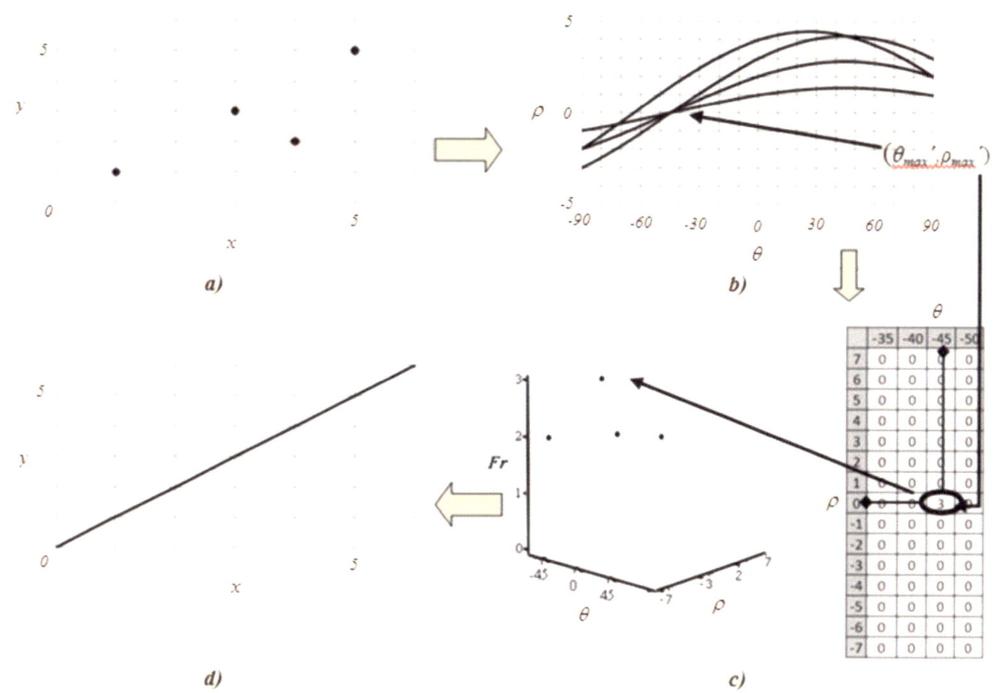

Fig. 3. The standard Hough transform: a) Points in the x-y plane, b) Projection of the points in the θ-ρ plane, c) Accumulator and its graphical representation of the intersections in the cells, d) Straight line identified in the x-y plane.

The straight line located from the four points processed is shown in Fig. 3d; it is used to evaluate the points and parameters in (Duda & Hart, 1972):

$$y_i = \frac{\rho - x_i \cos\theta}{\sin\theta}. \quad (5)$$

The Hough transform is an efficient method in detecting lines, circles, and ellipses, but with a high computational cost. With the objective of improving its efficiency, several proposals have been done (Xie, 2008; Illingworth & Kittler, 1988). The methods to improve the performance of the Hough transform are basically in the lines segmentation, and these are classified into two groups (Genswein & Yang, 1999): *Non Probabilistics Hough Transformations (NPHT)* (Lee, Yoo & Jeong, 2006) - where one finds the Fast Hough Transform, the Adaptive Hough Transform, the Combinatorial Hough Transform, and the Hierarchical Hough Transform; and *Probabilistics Hough Transformations (PHT)* (Xie, 2008) - where the Hough Transform of Dynamic Combination, Probabilistic Hough Transform, and Random Hough Transform are found.

The *NPHTs* propose methods to optimize the discretization of the accumulator as well as the identification of the cells with the larger number of intersections or records. The *PHTs* conduct, in an iterative manner, the random selection of small sets of points until a cell in the accumulator exceeds a defined limit in line identification. The points that define the line in the image are deleted and the selection of new points starts again, this process is done in an iterative manner until the processing of all points in the image is accomplished.

Yun-Seok et. al. (Lee, Yoo & Jeong, 2006), stated that the developed techniques in the non-probabilistic Hough transformations decrease the complexity of finding the maximum point in the accumulator. However, in these methods, all the possible orientations of θ must be considered. Meanwhile, with the probabilistic Hough transformations, the number of points to be processed in order to avoid incorrect results must be carefully selected.

3. Edge detection and binarization of the image

The creation of a binary version of the original image is carried out as a pre-processing stage; this provides a significant decrease in the number of points to be processed. Through this preprocessing stage, only the points that form part of the objects of interest in the image are considered for further analysis, while those that belong to the bottom are discarded.

3.1 Edge detection

Edge detection is based on the abrupt variation of pixel intensity between the borders and the background, where the borders define the edges of the objects that form the image. This permits definition of false edges due to noise in the image. Techniques to identify edges based on discontinuities in the pixels' intensity are based on the gradient obtained in the image *f(x,y)* at (x,y), where the point *(x,y)* is defined as a vector perpendicular to the edge (Szeliski, 2008):

$$G\left[f(x,y)\right]=\left[G_x \quad G_y\right]=\left[\frac{\partial f}{\partial x}\ \frac{\partial f}{\partial y}\right], \tag{6}$$

where G is the maximum variation in the intensity of *f* (which corresponds to the intensity and position of each pixel in the image) in the *(x,y)* point per unit of distance, with magnitude and direction given by (Szeliski, 2008):

$$|G|=\sqrt{G_x^2+G_y^2} \quad \propto(x,y)=\arctan\left(\frac{G_y}{G_x}\right). \tag{7}$$

Several gradient operators have been divided into two groups, the first-derivative operators and the second-derivative operators. In the first group are the Sobel, Prewitt, Roberts, Kirsch, Robinson, and Feid-Chen operators. In the second group are the Laplacian and Gaussian Laplacian operators. In this work, we focus on the Sobel operator to detect edges and smooth the image, thereby minimizing the noise due to false edge resulting from the noise enhancement produced by the derivative operators (Gonzalez & Woods, 2008).

The Sobel operator computes the intensity values change approximation at a point when it is considered to be a neighborhood of a 3×3 size, taking the point as the center. The Sobel mask is shown in Fig. 4.

$$G_x = \begin{bmatrix} -1 & 0 & 1 \\ -2 & 0 & 2 \\ -1 & 0 & 1 \end{bmatrix} \qquad G_y = \begin{bmatrix} -1 & -2 & -1 \\ 0 & 0 & 2 \\ 1 & 2 & 1 \end{bmatrix}$$

a) b)

Fig. 4. Sobel masks. a) Mask used to obtain G_x . b) Mask used to obtain G_y .

Edge detection using the Sobel operator implies the computation of the sum of the coefficient of the gray scale levels of the pixels contained in the region enclosed by the mask:

$$G_x = \left[f\left(x_{i+1}, y_{j-1}\right) - f\left(x_{i-1}, y_{j-1}\right) \right] + 2\left[f\left(x_{i+1}, y_j\right) - f\left(x_{i-1}, y_j\right) \right] + \left[f\left(x_{i+1}, y_{j+1}\right) - f\left(x_{i-1}, y_{j+1}\right) \right], \tag{8}$$

$$G_y = \left[f\left(x_{i+1}, y_{j+1}\right) - f\left(x_{i-1}, y_{j-1}\right) \right] + 2\left[f\left(x_i, y_{j+1}\right) - f\left(x_i, y_{j-1}\right) \right] + \left[f\left(x_{i-1}, y_{j+1}\right) - f\left(x_{i-1}, y_{j-1}\right) \right], \tag{9}$$

An example of the resulting image applying the Sobel methodology is shown in Fig. 5b.

A threshold is applied for the construction of the binary image. This threshold is used in the gradient image to identify and separate the pixels which belong to the edges of the image from those that form the background:

$$g(x,y) = \begin{cases} 1 & if\ G\left[f(x,y)\right] > T \\ 0 & if\ G\left[f(x,y)\right] \le T \end{cases}. \tag{10}$$

a) b) c)

Fig. 5. Binary image construction from a threshold of the gradient image: a) 8-bit gray scale image. b) Gradient image. c) Binary image.

The binary image shown in Fig. 5c is obtained after the application of the threshold (*T*=*32*) to the gradient image in Fig. 5b. In the binary image (Fig. 5c) one notices the presence of discontinuous edges. A process that significantly reduces the discontinuous edges is the algorithm developed by Canny (Canny, 1986); which is implemented in the edge extraction stage.

The Canny algorithm detects the edges applying error criteria, location, and response; these conditions can be broken down into three modules. In the first module, the image is processed using a Gaussian filter, with the purpose of smoothing both the image and the existing noise (Fig. 6):

σ = 0.625 pixels

1	2	3	2	1
2	7	11	7	2
3	11	17	11	3
2	7	11	7	2
1	2	3	2	1

Fig. 6. Typical Gaussian mask.

After the image is smoothed, the Sobel operator is applied to obtain the magnitude and direction of the gradient. These values are employed as criteria in the second module in order to construct a new image, whose edges must have a width of one pixel. In the third module, the false modules are determined by applying two thresholds (*T1* and *T2*, *T1* being less than *T2)* to the last image obtained. These values are derived from the intensity of the pixels, by which it is expected that the edges of objects in the image will be found. The intensities of the pixels that are greater than *T2* form part of the edges of the binary image, as do the pixels whose intensities are greater than *T1,* and which also have at least one neighbor with intensity greater than *T2*. Figure 7 shows the binary image of Fig. 5a obtained with the Canny edge detector (to be compared with Fig. 5).

Fig. 7. The binary image of Fig. 5a, this time obtained using the Canny algorithm.

3.2 Histogram modification

Image enhancement using intensity distribution is now undertaken. To do this, a histogram equalization is proposed. The procedure used to conduct the histogram equalization is as follows. The histogram of an image is a discrete function defined as:

$$P(g) = \frac{N(g)}{M}, \tag{11}$$

where $P(g)$ is the probability that a given gray value (intensity) occurs in the image, M is the total number of pixels in the image, and $N(g)$ is the number of pixels with intensity g. Redistribution or transformation of intensities during the histogram equalization is expressed as (Shanmugavadivu & Balasubramanian, 2010):

$$V_{ki} = \frac{(L-1)(S_k - S_{k\min})}{S_{k\max} - S_{k\min}}, \tag{12}$$

where V_{ki} is the new intensity value for the i-th pixel in the equalized image, L is the number of gray levels in the image, S_k is the number of accumulated pixels with a determined value of the i-th intensity, $S_{k\ \min}$ is the smallest number of frequencies accumulated that are greater than zero, and $S_{k\ \max}$ is the largest number of frequencies accumulated from pixels.

For detection of straight lines in the surface of a polished-finish cutting tool using the Hough transform, a binary image of the tool's surface must be obtained from the original grayscale image. To preserve the characteristics of the straight lines, the histogram equalization defined by Eq. (12) is used.

In the proposed framework, instead of predicting the values for the θ parameter in the interval $\left[-\frac{\pi}{2}, \frac{\pi}{2}\right]$, the binary image is processed with the Sobel mask, in order to find the direction of the straight lines and to determine the value of θ. Knowing the directions of the straight lines reduces the size of the accumulator. Also, the number of iterations in the SHT to straight-line segmentation is decreased, compared with previously published methods (Illingworth & Kittler, 1988; Leavers, 1993). Next, the pixels that retain the zero value in the binary image are selected. The selected points are processed with the SHT. The (θ,ρ) cell values with the largest number of intersections in the accumulator define the parameters that describe the number of straight lines on the tool's surface, as well as the width of each one. This allows for quantifying the straight lines on the machined surface. The width measurements of the straight lines on the tool's surface can be related to its quality.

4. Proposed framework phases

4.1 Cutting tool

A cutting tool has two characteristics to be taken into account, the material and the geometry of the tool. The second characteristic may lead to defective machining due to gradual wear and even loss of the tool's shape by, for example, tearing at the radius of the nose, as shown in Fig. 8. This damage requires a reworking of the surface or possibly discardation of the piece. The damage can be avoided if the cutting tool is changed before a catastrophic failure of the edge or cutting edge happens.

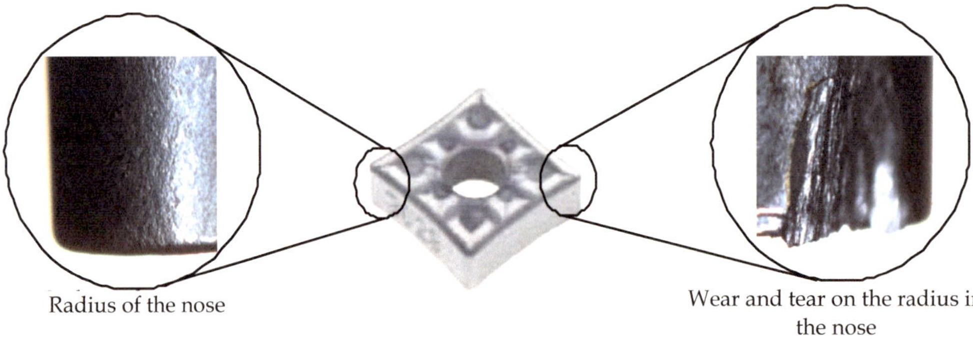

Fig. 8. Defective cutting tool.

The images processed to assess the quality of the finish of the studied surfaces are obtained from a round bar of a 6061 T6 aluminum with a diameter of 38.1mm (1 ½ inches). This bar was machined by a process of hammering without coolant, utilizing a QNMG 432 TF cutting tool with a depth of cut at 0.5mm, a cutting speed of 300 mm/min, and a spindle speed of 700 RPM. The process was carried out on a CNC lathe, with a Model DMC 1820 Galil ® driver (Fig. 9).

Fig. 9. CNC lathe use to produce the test cylinders.

4.2 Framework

The proposed stages to identify the straight line characteristics of the machined surface's finish by applying the SHT line segmentation procedure, is shown in Fig. 10:

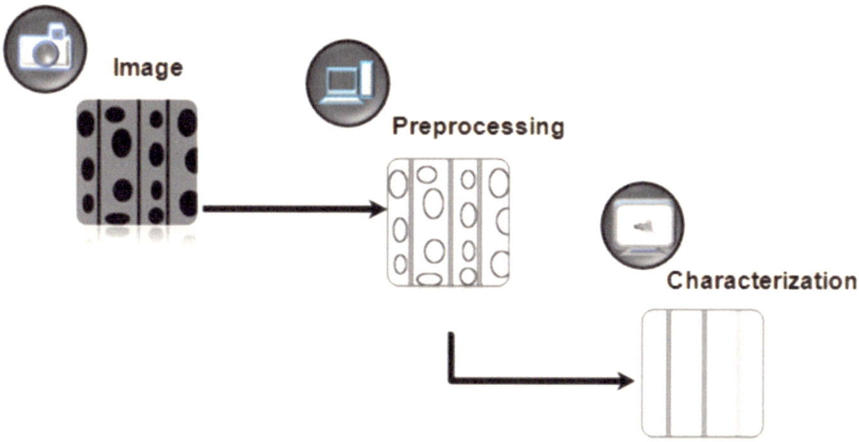

Fig. 10. Stages of the proposed method.

- **Image:** Capture of the surface finish in 8-bit grayscale image with 640×480 pixel resolution, in BMP format.
- **Preprocessing:** Creation of a binary image from the captured 8-bit grayscale image for latter processing in the characterization stage.
- **Characterization:** Description of the surface finishes straight lines via the parameters of the SHT.

4.3 Image capture

The image acquisition of the surface of the pieces machined on the CNC lathe was carried out using a 16X optical microscope (LEICA, Model EZ4D with an integrated 3 megapixel digital camera), in BMP format. Pixel intensity was recorded in gray scale with an 8 bits-per-pixel resolution.

4.4 Preprocessing

The preprocessing stage consists of two processes: histogram equalization and gradient angle computation. During the equalization of the histogram, the binary image of the surface is obtained. At the gradient angle computation stage, the θ information is obtained.

4.5 Binary image construction

This process begins with the computation of the frequency and cumulative histograms of the original grayscale image. To accomplish this, the process described in Fig. 11.

Upon obtaining the records in the cumulative histogram, the cells with the highest and lowest registrations other than zero are sought, assigning them values of $S_{K\max}$ and $S_{K\min}$, respectively. The new intensity values are found from in the image pixels by evaluating each register of the table S_K of Fig. 11 in eq. (12).

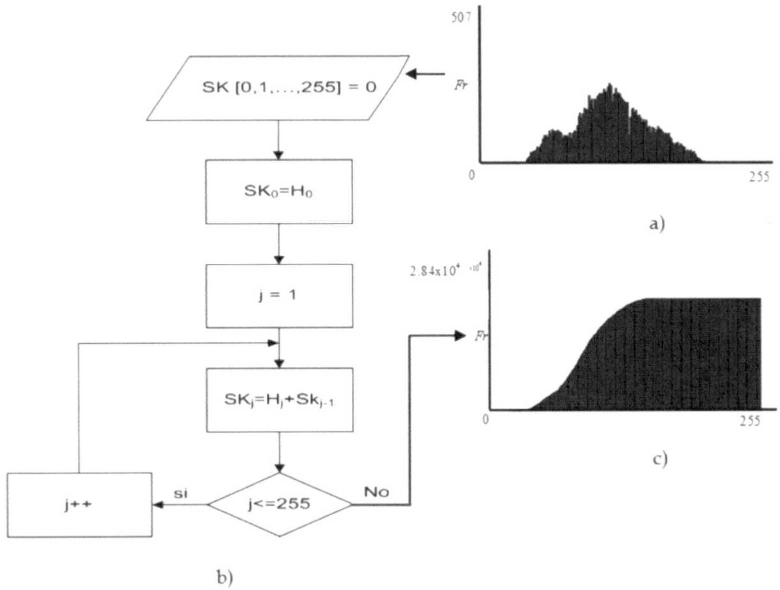

Fig. 11. Construction of the frequency and cumulative histograms of the original image. a) Frequency histogram; b) Flow diagram of the process; c) Cumulative histogram.

Having established the new intensity values V_k, the shift in intensity of each pixel is made as follows:

- The intensity of the pixel is obtained.
- This intensity points to the entry in the table of the V_ks, whose value indicates the new intensity to be assigned to the pixel.

The binary image shown in Fig. 12 was obtained, using a threshold value of 128. The pixels with intensity values that are less than or equal to an established value are assigned a zero for its intensity value, while those pixels with an intensity greater than the threshold value are assigned a 255 magnitude value.

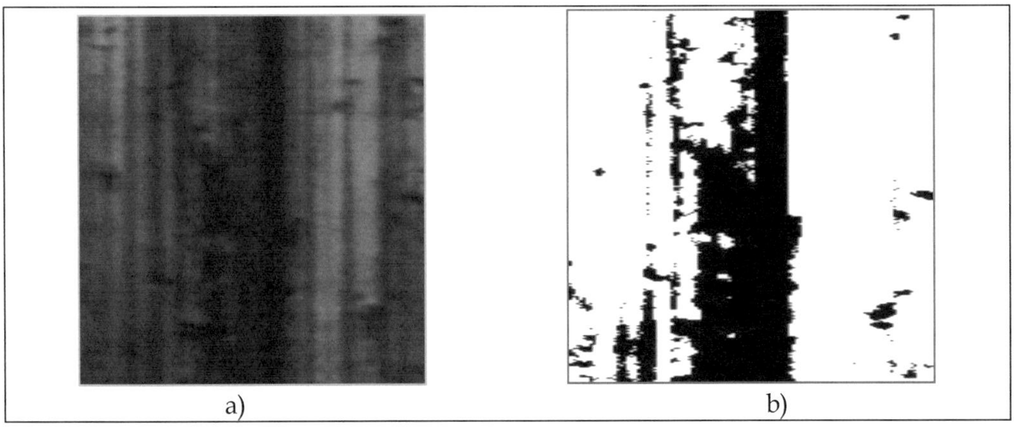

Fig. 12. a) Original grayscale image; b) Binary image.

4.6 The gradient angle

To define the values of the θ angle, the information from the gradient is employed during the processing of the original grayscale image. In this chapter, the gradient information is used to obtain the angle of the edges having more frequency than in the binary image obtained.

Equation (7) describes the magnitude and gradient angle. These values are gathered using the Sobel mask on the binary image obtained by means of the histogram equalization of the original image, evaluating each point that does not form part of the background.

In Fig. 13a, the binary image is observed and in Fig. 13b, the edges are observed. Here, the pixels' intensity within the segment of the specified edge is taken as a reference for the angle gradient computation described below.

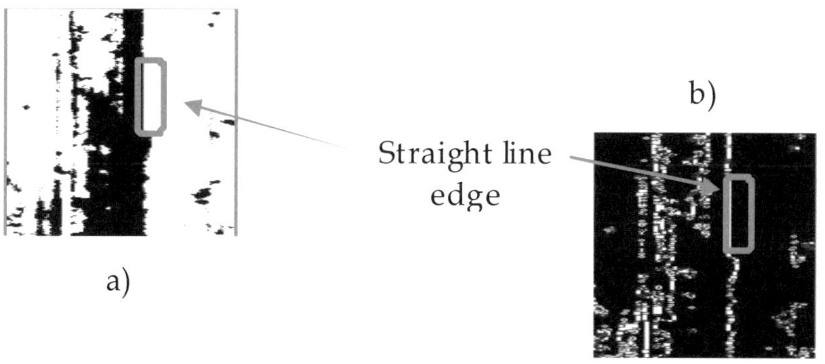

Fig. 13. Gradient magnitude computation. a) Binary image; b) Enhanced edges.

With edges highlighted by the gradient magnitude, the following steps are undertaken to determine the angle magnitude, obtained by the process described as follows:

- A table is constructed, consisting of a single column and 361 rows for each of the expected angles; that is, 0, 1, 2,…,360. Within these cells, the registration corresponding to the angle value is noted. Negative entries correspond to not-yet-evaluated angles.
- The edge angle is estimated and registered taking only those points that comply with the condition that its value in the binary image is zero, and the magnitude of the gradient is other than zero.
- In the algorithm, the angles registered are entered into a table, where the times that the angles are given are accumulated to obtain the global information of the edge direction.

Based on the angles registrations, the interval of the θ parameter of the Hough transform is determined. This interval is obtained from the cumulative frequency indicated in Fig. 14, in the range proposed (0, 1-45, 46-89, 90, 271-315, 316-359), which corresponds to all of the possible angle values of the gradient of the image in such a way that, in the interval where the greatest frequency is registered, the interval for θ is determined.

As an example, the result of this angle grouping scheme is shown in Fig. 14, where the first column determines the upper boundary of the θ interval, while the second column shows the corresponding cumulative frequency (*fr*).

$$\begin{array}{c|c} \theta & fr \\ 0 & 776 \\ 1 & 562 \\ 46 & 398 \\ 90 & 188 \\ 271 & 411 \\ 316 & 276 \end{array}$$

Fig. 14. Angle grouping process used to determine the direction of the straight lines from the greatest registration value.

The definition of all the angles of the detected edges is shown in Table 1. One can observe that the intervals go from a minimum of 1 to a maximum of 44 values, using the information obtained from the gradient angle.

Initial value	Final value	Interval width	Edge orientation
0	0	1	90, 270
1	45	44	91-135, 315-359
46	89	43	136-179, 271-314
90	90	1	0, 180
271	315	44	1-45, 181-225
316	359	43	46-89, 226-269

Table 1. Gradient angle intervals and edge directions in the binary image.

To conclude, the preprocessing stage is conducted by filtering the points in the binary image with the gradient magnitude, decreasing the width of the straight lines due to the histogram equalization process. The filtering process is carried out as follows: only the pixels having a value of zero are selected. For each one of these pixels, its intensity in the table of gradient values is checked to ascertain that it is equal to or greater than 255. If this is the case, a value of 255 is assigned to the pixel. This way, we have a binary image with fewer points without affecting the straight lines of the surface finish during the filtering process. This helps to improve the performance of the Hough transform.

5. Characterization

The general process concludes in the characterization stage of the surface finish with the maximum picks in the parameter space of the Hough transform. The parameter space is obtained starting from the values of θ-ρ in the image; these values describe the length of the straight line of the surface finish. The stages of this module are: estimation of the values of the θ and ρ parameters, construction and evaluation of the accumulator for each one of the points in the straight line, and location of the maximum value registration of the intersections in the parameters space.

5.1 Parameter interval definition

For the ρ parameter, the expected values are defined by the diagonal of the image obtained from the width a and height b values of the image by utilizing eq. (4). Values for the θ interval are obtained from the proposed stage in the angle gradient computation, whose processes are quantified from the angle frequencies observed in Fig. 14. In this registration, the highest value must be located in order to define the values to be assigned to the θ parameter. These values can be observed in the first row (indicated by the second column value), by which the θ parameter will take the zero value, as described in Table 1.

5.2 Accumulator construction and points' evaluation in the straight line normal equation

A table for the accumulator is constructed whereby the columns and rows are determined, respectively, by the values defined for the (θ, ρ) parameters. Within those cells, the values in the expected interval for ρ are recorded. Each one of the values that are not part of the binary image background of the surface finish are evaluated in eq. (13), in the interval defined by θ.

$$\rho = x_i \cos\theta + y_i \sin\theta \quad (13)$$

With the binary image of the surface finish, the pixels that do not correspond to the background are evaluated using eq. (13), where the θ value is equal to zero (the value assigned in the definition stage of the parameters' interval). If the value of ρ is found within the interval defined, it is registered in the accumulator. Fig. 15 shows the registers in the accumulator that correspond to the binary image of Fig. 13b. The θ-ρ parameters are located in the x-y axis; in the z axis, the frequency of the points that are collinear are shown (those registering more than a single occurrence within the same cell of the accumulator). In this way, the straight lines of the surface finish are identified by the higher frequencies in the graphic. In the graphic of Fig. 15, only a single peak is observed; this describes the straight line observed in the surface finish of Fig. 12a.

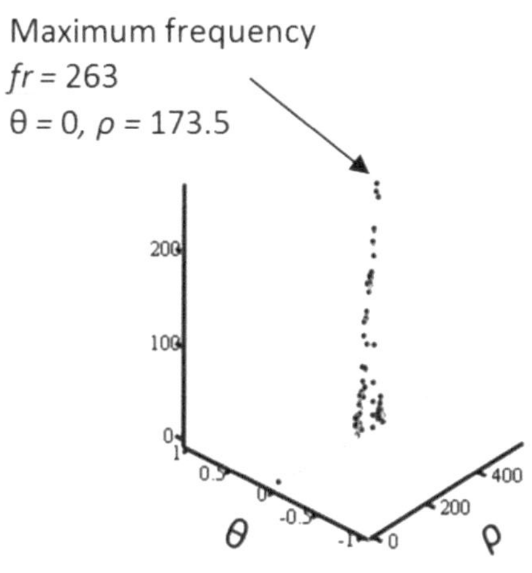

Fig. 15. Peak of the straight line in the surface finish of Fig. 12a.

5.3 Location of the maximum in the Hough space

The accumulator table is now explored in order to locate the cell or cells that have the highest registration. This is used to obtain the θ and ρ values, which allows for the positioning and orientation of the straight lines of the surface finish.

In Fig. 15, corresponding to the space frequency of the parameters, we can see only one point at the top. That point in the Hough transform represents the straight line that is observed in Fig. 16. The values shown in the θ-ρ figure of this cell describe the orientation and position thereof.

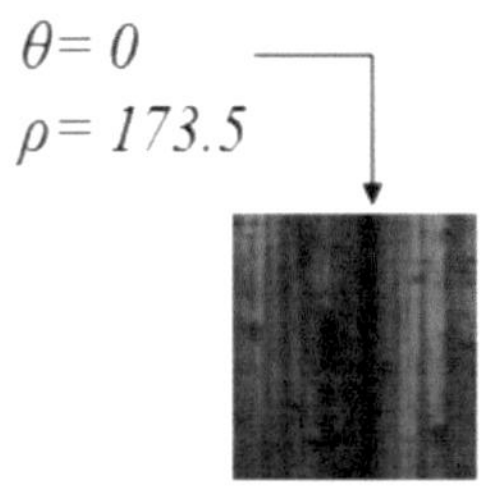

Fig. 16. θ-ρ parameters illustrating maximum frequency.

6. Results

6.1 Image acquisition

The captured image of the surface finish of a piece machined on a CNC lathe is presented in BMP format, with 640×480 pixels in gray scale with 8 bits per pixel. From this, a sub-image of 256×256 pixels was used to segment the straight lines of the surface finish. This image is used to segment the eight straight lines observed, which are characteristic of the surface finish of machined pieces using a sharp tool. The purpose of this stage is to achieve the construction of the binary image and to compare the obtained results with the algorithm based on the Sobel mask and the Canny edge detector.

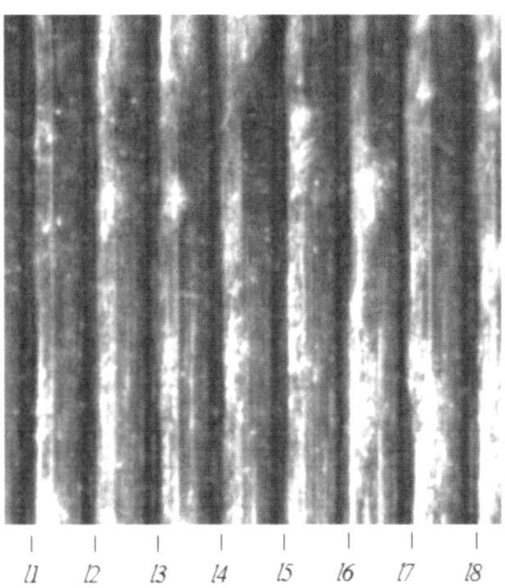

Fig. 17. Straight lines typical of surface finish (1 pixel = 0.0052 mm, scale 505:1).

6.2 Preprocessing

6.2.1 Construction of the binary image

To build the binary image of the surface finish using the Sobel mask (see Fig. 18b) and the Canny algorithm (see Fig. 18c), it is observed that both images present fewer points to be processed, but both lose information. This is because the binary image constructed with the Sobel mask presents discontinuous straight lines of the surface finish, wherein none of these present continuity from the beginning to the end of the straight lines observed in the image of the surface finish (Fig. 18a). The image constructed with the Canny algorithm preserves the continuity in the majority of the straight lines, but they are seen to be distorted. The image constructed with the proposed method (Fig. 18d) does not lose information since all of the straight lines of the surface finish **preserve their continuity and definition**.

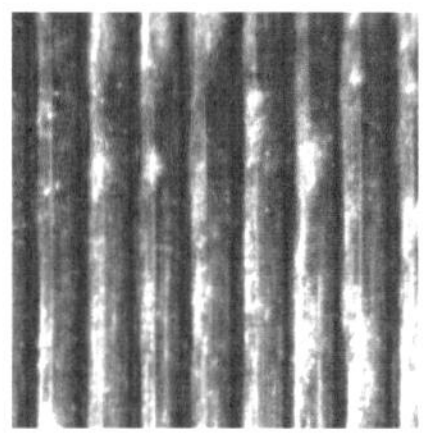

Number of points to be processed = 65, 536

(100 %)

a)

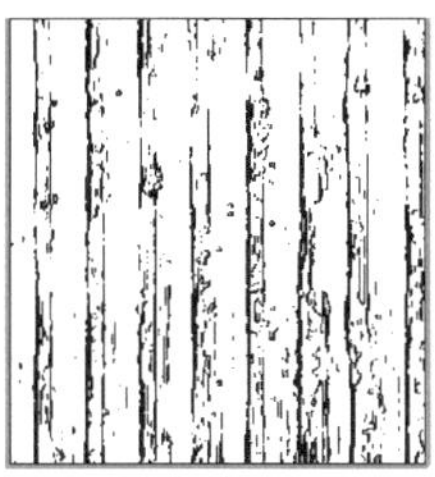

Number of points to be processed = 9,896

(15.1 % of points in the original image)

b)

Number of points to be processed = 7,179

(9.5 %of points in the original image)

c)

Number of points to be processed = 38, 720

(59 % of points in the original image)

d)

Fig. 18. Construction of the binary image. a) Original grayscale image; b) Sobel mask construction; c) Canny algorithm construction; d) Histogram equalization construction.

6.3 The θ y ρ parameters interval definition

The method to obtain the orientation description of the edges is applied on the binary images constructed with the gradient by applying the Sobel mask and the Canny algorithm. The tables with the computed angle frequencies are presented in Fig. 19. Here one can see that the tables of the frequencies that describe the straight line orientation of the surface finish, starting from the interval with the greatest frequency are the binary images constructed with the Sobel mask and the proposed equalization method. Therefore, the θ parameter can be defined in both images using only one value.

On the contrary, the binary image constructed with the Canny algorithm does not permit the description of straight line orientation with the information from the gradient angle, because the interval presenting the highest frequency corresponds to the orientations in the interval [0,45]. For this reason, the θ parameter must be defined in the interval indicated to make the correct segmentation of the straight lines observed on the surface finish.

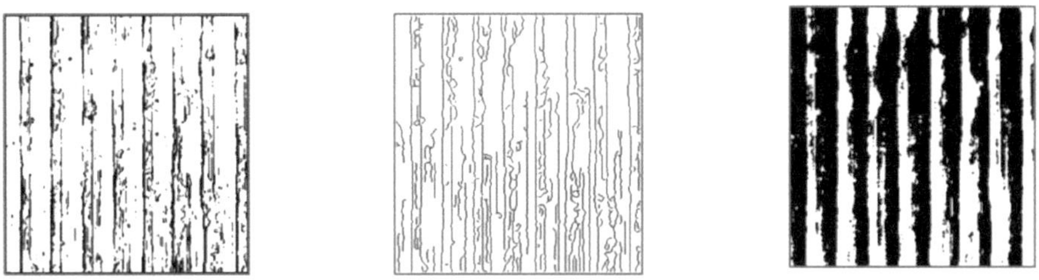

Sobel | Canny | Equalization

Sobel:

Gradient angle values		Fr
0	0	3 780
1	45	1 631
46	89	818
90	90	716
271	315	746
316	359	857

Definition of the optimum parameters with $\theta = 0$

Canny:

Gradient angle values		Fr
0	0	132
1	45	925
46	89	924
90	90	213
271	315	853
316	359	6

Definition of the optimum parameters with $-\frac{\pi}{2} \leq \theta \leq \frac{\pi}{2}$

Equalization:

Gradient angle values		Fr
0	0	3 134
1	45	1 716
46	89	1 033
90	90	567
271	315	992
316	359	804

Definition of the optimum parameters with $\theta = 0$

Fig. 19. Edge orientation frequency from the gradient information for the correct definition of the θ parameter.

The values expected for the ρ parameter are defined in the literature for the interval $-2\sqrt{(a^2+h^2)} \leq \rho \leq 2\sqrt{(a^2+h^2)}$, which has been reduced by applying the proposed method of grouping frequencies in the edges' orientation on the intervals utilizing the information obtained from the gradient angle. Since the angle identification of the predominant edges corresponds to zero, the straight lines to be segmented are perpendicular. Therefore, the magnitude of the radius vector is found in the width dimensions of the image. From this conclusion, the values expected for the ρ parameter are reduced to 82%.

6.4 Processed points in the equalized image

The image constructed using the proposed equalization method renders a greater percentage of points than the 50% found in the original image. This value is reduced by eliminating the isolated points, while the distance is increased between the straight line edge segments that are not part of the straight lines on the surface finish. The aim of this is to reduce false segmentations due to points that are collinear, but that are not necessarily part of straight lines on the surface finish.

In Fig. 20b, a fragment of the binary image constructed is shown. In this image, a couple of isolated points can be observed, as well as a fragment of the straight edge that does not form part of the actual straight lines on the finish. The intensities of these pixels and the magnitude of the gradient are shown in the tables presented in Fig. 20c and Fig. 20d, respectively. An example worthy of consideration is the point located in Column 217, Row 102. Its intensity value is equal to zero, but the magnitude of the gradient is other than zero, so the pixels are considered to be part of the background image, changing its intensity to 255, as can be seen in Fig. 20e. Accordingly, this point is not seen in the image fragment shown in Fig. 20f. This criterion is applied to all of the pixels in the image and a new image is constructed, as shown in Fig. 20g, which contains 48% of the points in the original image.

Once the point-filtering process is applied, we can see, in Fig. 20g, how isolated points are eliminated and the distances are incremented between segments of the edges that do not belong to the straight lines on the surface finish, without affecting the definition and continuity of them.

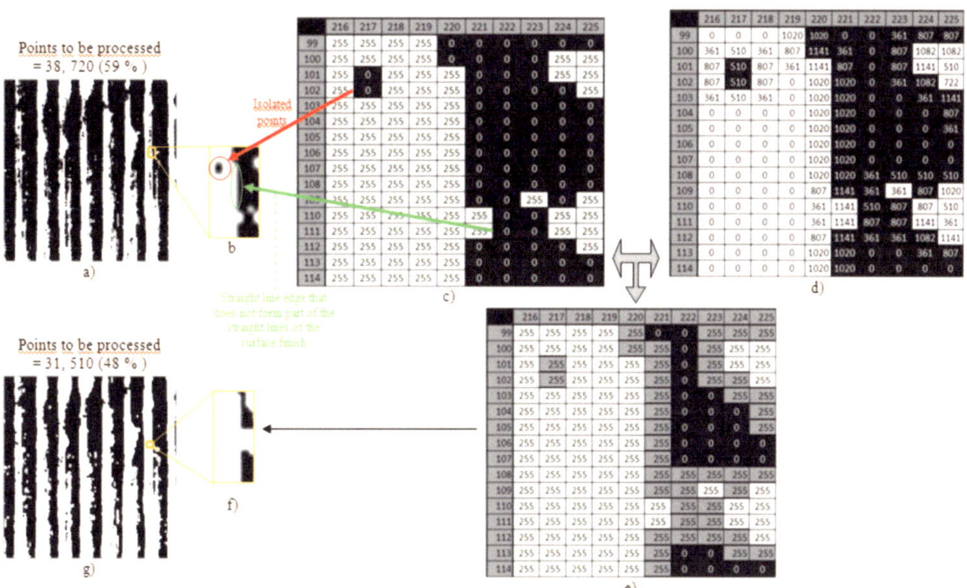

Fig. 20. Points filtered: a) Binary image; b) Binary image zoomed; c) Table of intensity from the fragment in the binary image; d) Gradient magnitude of the binary image zoomed; e) Filtered pixels intensity of the binary image zoomed; f) Filtered binary image zoomed; g) Filtered binary image.

6.5 Hough characterization of the straight lines of the surface finish

6.5.1 Construction of the accumulator

The accumulator tables of the binary images constructed are shown in Fig. 21. Here we can observe that, for the binary image constructed with the Sobel mask and the proposed method of equalization, the gradient information can be employed. This decreases the accumulator dimensions by 99% with respect to the accumulator constructed for the binary image formed by the Canny algorithm. The Canny accumulator also does not allow for defining the orientations of the straight lines of the surface finish by the frequency of the angles obtained.

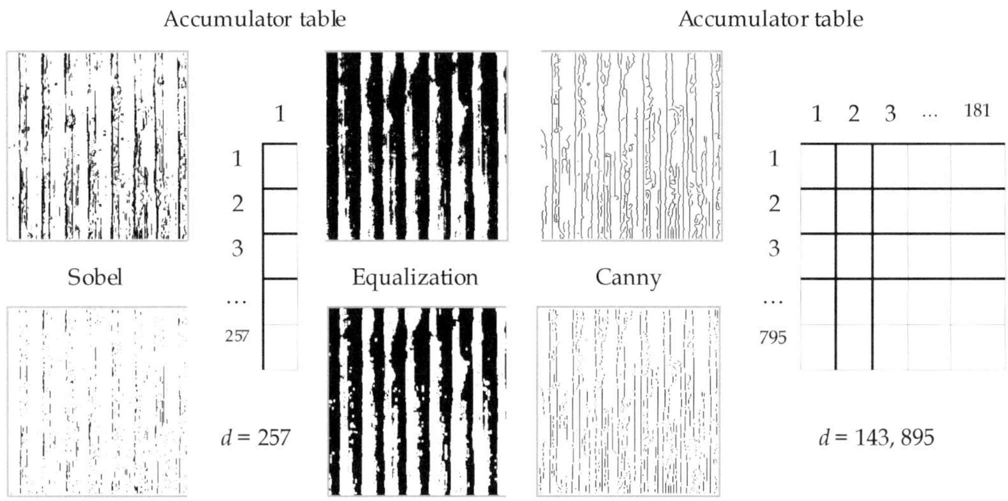

Fig. 21. Accumulator dimensions for the binary images constructed.

6.5.2 Evaluation of points into the equation of the straight line and its accumulator registration

Fig. 22 presents the registration in the accumulator of the images constructed in order to filter the points with the proposed algorithm employing the gradient magnitude and the pixels' intensity from the binary images. This decreases the number of iterations through the reduction of the number of processed points. Note that the greatest percentage in the reduction of iterations in the constructed image, with respect to iterations realized in the images without filtering, is with the Sobel mask (70%), followed by the Canny algorithm (29%), and finally the creation of the image by means of the equalization algorithm (18%). In addition to reducing the iterations in the processing of binary images filtered with the Hough transform, the filtering process provides the reduction of collinear points that are not necessarily part of a continuous straight line. This factor is reflected in the location stage of the maximum in the parameters space, as described below.

The time employed to process the image of the surface finish by means of the algorithm developed in Mathcad version 14 (on a computer with a 1.66 GHz processor and 1GB RAM, running Windows XP) was 15.23 s. without point filtering and 9.17 s. with point filtering. This processing time was spent to obtain the segmented straight lines on the surface finish image observed in Fig. 22.

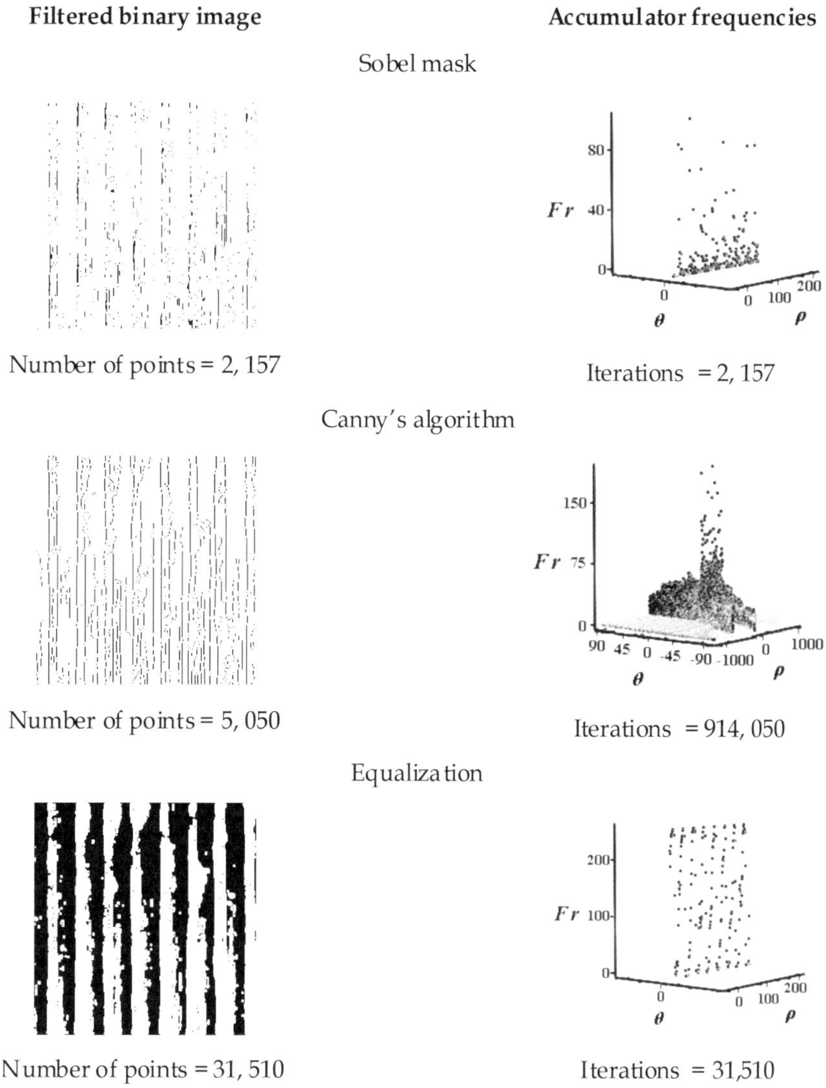

Fig. 22. Accumulator frequency registration for the filtered binary image.

6.5.3 Maximum frequencies location in the parameters space

In Fig. 23, the maximum-frequency points located in the accumulators are shown. One can notice that for the points of the binary images constructed with the Sobel mask, the maximum value is 244, while for the Canny algorithm the value is 176; presenting a singular point in the upper part of their respective graphics. Although the values of the cells describe the parameters within the expected values, they offer only the parameters of one of the eight straight lines in the image. The threshold value can be decreased to obtain the parameters of more straight lines. This would implicate an additional process to define an optimum threshold value. On the contrary, with the method proposed here to obtain the image and accumulator, the maximum frequency value is high enough to allow for the segmentation all of the straight lines, as can be seen in the graphic corresponding to the binary image constructed with the equalization method.

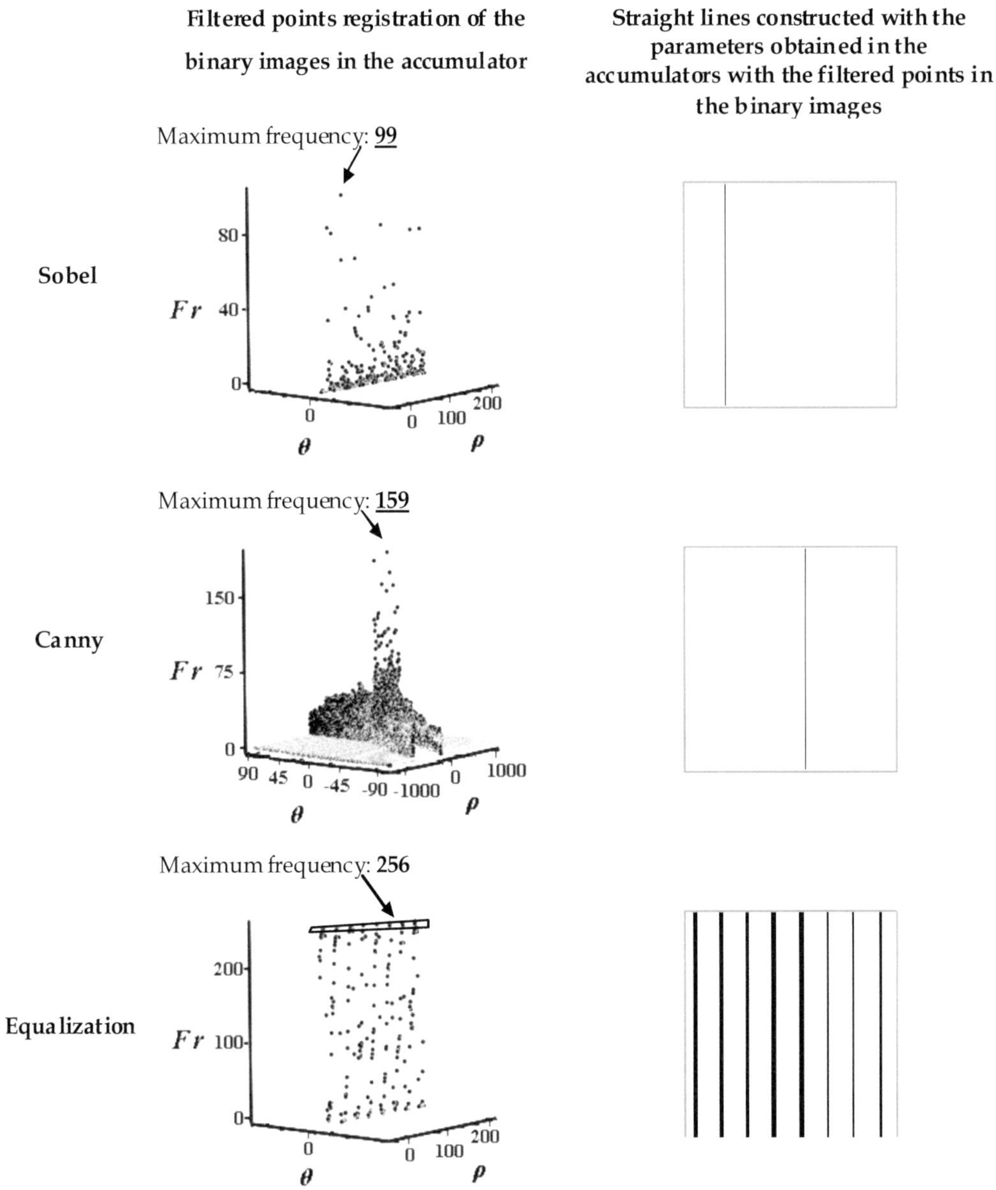

Fig. 23. Maximum frequencies in the accumulators of the constructed binary images.

The discontinuity observed in the segmented straight line widths in the binary image obtained by means of the equalization method is due to the surface finish dimension not being constant along each one of the straight lines' widths. This characteristic can be observed in the parameters space of the Hough transform. These values are presented in groups in Fig. 24, where the straight line describes the obtained parameters.

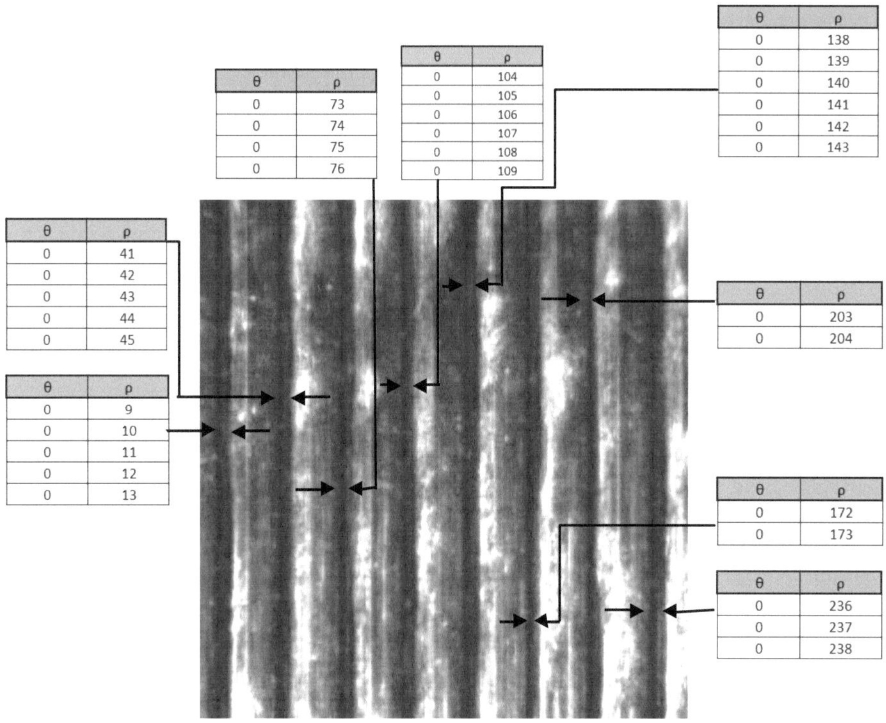

Fig. 24. Widths indentified with the Hough transform in each one of the straight lines of the surface finish (scale 562:1).

Is important to underline that the intervals quantity in which the cells are grouped with the highest frequency correspond to the number of straight lines in the images. The information for each one of them is obtained from the parameters space, using the maximum frequency taken from the accumulator.

7. Conclusion

By equalizing the histogram of the surface finish image, a binary image is constructed without losing the characteristics of the straight lines in the image.

The proposed method defines the θ and ρ parameters and the selection of the pixels to be processed. Gradient-information processing reduces the number of iterations in the standard Hough transform.

With the proposed method, the necessary data can be determined to describe the length, width, and straight line numbers of the surface finish from the corresponding parameter values of the cells having the highest frequency in the accumulator table of the Hough transform without an additional algorithm being needed.

The number of pixels to be processed is directly related to the number of iterations used to segment the lines with the standard Hough transform. Therefore, larger images require a higher number of iterations, increasing the processing time for the segmentation of straight lines on machined surfaces.

8. Acknowledgment

The authors would like to thank the National Polytechnic Institute of Mexico and Autonomous University of Queretaro for their support in this study. This work was funded by the government agency CONACyT (134481) and the PROMEP program, Mexico.

9. References

Alabi, B.; Salau, T.; Oke, S. (2008). Surface finish quality characterization of machined workpieces using fractal analysis; *Mechanika*, Vol. 64, No. 2, pp 65-71, ISSN 1392-1207.

Bradley, C. & Wong, Y. (2001). Surface Texture Indicators of Tool Wear - A Machine Vision Approach; *International Journal of Advanced Manufactory Technology*, Springer, Vol. 17, No. 6, pp 435 - 443, ISSN 0268-3768.

Canny, J. (1986). A computational approach to edge detection; *IEEE transactions on pattern analysis and machine intelligence*, Vol. 8, No. 6, pp: 679-698.

Davies, E. (1990). *Machine Vision: Theory, Algorithms, Practicalities*; Morgan Kaufmann, ISBN 0122060938.

Duda, R. & Hart, P. (1972). Use of the Hough transform to detect lines and curves in pictures; *Commun ACM*, Vol. 15, No. 1, pp: 11-15.

Genswein, B. & Yang, Y. (1999). A Fast Rule-Based Parameter Free Discrete Hough Transform; *International Journal of Pattern Recognition and Artificial Intelligence*, Vol. 13, No. 5, pp: 615 - 64.

Gonzalez, R. & Woods, R. (2008). *Digital image processing*; Prentice Hall, ISBN number 9780131687288.

Illingworth, J. & Kittler, J. (1988). A survey of the Hough Transform; *Computer Vision, Graphics, and Image Procesing*, Vol. 44, No. 1, pp. 87-116, ISSN: 0734189X.

Leavers, V. (1993). Which Hough Transform?; *CVGIP: Image Understanding*, Vol. 58, pp. 250-264.

Lee, Y.; Yoo, S. & Jeong C. (2006). Modified Hough Transform for Images containing many textured regions; *Lecture Notes in Computer Science*, Springer Berlin/Heidelberg, Vol. 4259, pp: 824 - 833.

Mannan, M; Mian, Z. & Kassim, A. (2004). Tool wear monitor using a fast Hough transform of images of machined surfaces; *Machine Vision and Applications*, Vol. 15, No. 3, pp 150-163, ISSN 1432-1769.

Shanmugavadivu, P. & Balasubramanian, K. (2010). Image inversion and Bi Level Histogram equalization for contrast enhancement; *International Journal of computer applications*, Vol. 1, No 15, pp: 61- 65.

Szeliski, R. (2008). Computer Vision: Algorithms and Applications; Springer, *Texts in Computer Science*, ISBN 978-1-84882-934-3.

Xie, X. (2008). A Review of Recent Advances in surface defect detection using Texture Analysis Techniques; *Electronic Letters on Computer Vision and Image Analysis*, Vol. 7, No. 3, pp 1-22 2008, ISSN:1577-5097.

7

Methods for Ellipse Detection from Edge Maps of Real Images

Dilip K. Prasad[1] and Maylor K.H. Leung[2]
[1]Nanyang Technological University
[2]Universiti Tunku Abdul Rahman (Kampar)
[1]Singapore
[2]Malaysia

1. Introduction

Detecting geometric shapes like ellipses from real images have many potential applications. Some examples include pupil tracking, detecting spherical or ellipsoidal objects like fruits, pebbles, golf balls, etc. from a scene for robotic applications, detecting ellipsoidal objects in underwater images, detecting fetal heads, cells, or nuclei in biological and biomedical images, identifying the eddy currents and zones using oceanic images, forming structural descriptors for objects and faces, traffic sign interpretation[1], etc.

Given the very wide scope of applications, it is important that the ellipses can be detected from real images with high reliability. Specifically, it is important that the structures that are elliptical are indeed detected and the structures that are non-elliptical are not detected as

(a) an example of real image

(b) its edge map (after histogram equalization and Canny edge detection)

Fig. 1. An example of a real image and the problems in detection of ellipses in real images

---

[1] (Antonaros and Petrou, 2001; Belaroussi et al., 2005; Bell et al., 2006; Burrill et al., 1996; Chia et al., 2009; Dijkers et al., 2005; Fernandes, 2009; Feyaerts et al., 2001; Foresti, 2002; Foresti et al., 2005; Fu and Huang, 1995; He et al., 2009; Hua et al., 2007; Hwang et al., 2006; Iles et al., 2007; Ji et al., 1999; Kayikcioglu et al., 2000; Kumar et al., 2009; Kuno et al., 1991; Liu et al., 2007; Lu et al., 2005; Matson et al., 1970; O'Leary et al., 2005; Prasad 2011c; Rosin and West, 1992; Salas et al., 2006; Shen et al., 2009; Shih et al., 2008; Smereka and Glab, 2006; Soetedjo and Yamada, 2005; Sood et al., 2005; Takimoto et al., 2004; Tang et al., 2000; Wang et al., 2006; Wu and Wang, 1993; Yuasa et al., 2004; Zaim et al., 2006; Zhang et al., 2003; Zhou and Shen, 2003; Zhou et al., 2009)

ellipses. However, the problem of detecting ellipses in real images is very challenging due to many reasons. Consider the example given in **Fig. 1**.

Besides the obvious and ubiquitous presence of the problem of digitization in any digital image, there are other problems as seen in **Fig. 1**. The various challenges are summarized below succinctly:

1. **Digitization:** In digital images, the elliptic curves get digitized. As a consequence of digitization, the ellipses cannot be retrieved correctly and only an estimate of the parameters of the ellipses can be retrieved (Prasad and Leung, 2010b).
2. **Presence of incomplete ellipses due to overlap and occlusion:** One issue that can be easily envisaged in real images is that objects are usually present in overlap with each other, see the labels in **Fig. 1**(b) for examples. If the overlapping object is transparent or translucent, the boundaries of overlapped objects might still be available in the edge map as small disconnected edges (obtained after edge detection), see Fig. 2(a,c). However, if the overlapping object is opaque, the overlapped object is occluded and its incomplete boundary will appear in the edge map, see Fig. 2(b,d). If an image is cluttered by various objects of such nature, the problem gets very complicated to handle as such scenario results in various incomplete small edges in the image.

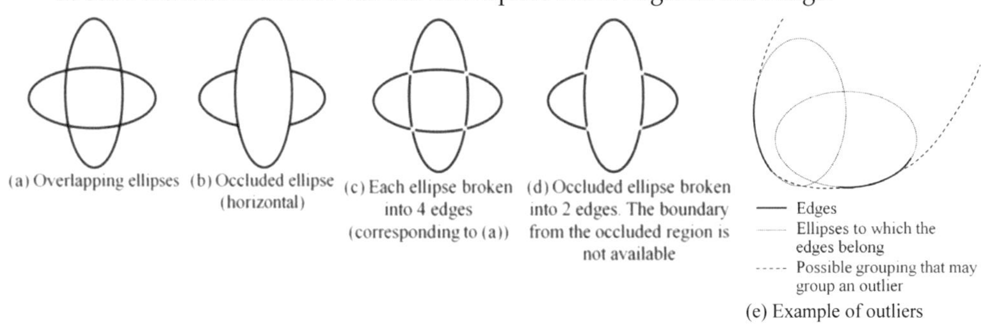

Fig. 2. Illustration of the presence of overlapping ellipses, occluded ellipses, and outliers

3. **Presence of outliers:** In cluttered background, some curved edges that are non-elliptic, may appear as if they are part of some elliptic edge. An example is shown in Fig. 2(e). The presence of such edges, referred to as the outliers, often results in false ellipse detections and degrades the performance of ellipse detection algorithms. Due to this reason, the incorrectly detected ellipses need to be filtered out at the end of the ellipse detection method.
4. **Corruption in the quality of edge:** Another problem that is encountered in real images is the deterioration of the boundary of the edge map due to the light and shadow conditions and the perspective of the object. Under different lighting conditions, boundaries in some region may be emphasized and appear sharp and clear, while boundaries in other regions might blur and deteriorate the edge in that region. Shadow effect may blur the region in which the boundaries of two objects overlap. Due to this, the boundaries of two objects may merge and appear to be smooth. Further, noise can appear in image due to imperfect imaging instruments and other external conditions like fog, glare, etc. Noise corrupts the quality of edge, rendering it to be non-smooth over small section of edge, and abrupt breaks in the boundaries. One such example can be found in Fig. 3.

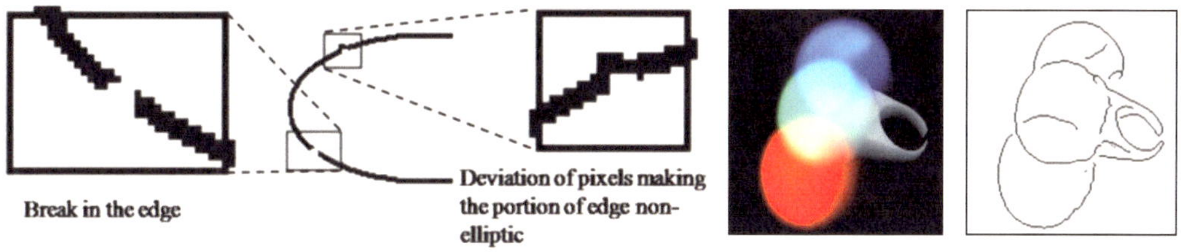

(a) Corruption of edges due to noise

(b) Elliptic shapes look non-elliptic in edgemap due to light and shadow effect

Fig. 3. Illustration of the corruption in the quality of edges

5. **Lack of a priori information:** Another aspect of the considered problem is that no a priori information is available. It might have helped if the expected number of elliptic shapes, expected size of ellipses, or expected regions of focus were known a priori. However, real images may vary greatly in the scale and content, and in general such a priori information cannot be generated reliably.

2. Contemporary ellipse detection methods

Extraction of elliptic shapes from images has captured the interest of researchers for a long time. For two decades, numerous researchers are working on this problem. Many methods have been proposed for ellipse extraction in real images. The methods used for ellipse detection can be primarily categorized into four categories, viz., Hough transform (HT) based methods, least squares based methods, genetic algorithms based methods, and hybrid ellipse detection methods.

Hough Transform (HT) and its adaptations are widely used in many ellipse detection methods (Lu and Tan, 2008; McLaughlin, 1998; McLaughlin and Alder, 1998; Yuen et al., 1989). The key advantage of HT is that it does not require perfect connectivity of all the edge pixels belong to an ellipse and which makes it useful for ellipse detection in noisy and cluttered images compared to edge following based ellipse detection method. However, since the ellipse detection problem involves 5-dimensional parametric space in HT (Illingworth and Kittler, 1988). HT turns out to be a computation intensive method, requiring huge computation time and memory. In the last two decades, the researchers using HT to detect elliptical and circular segments have focused on providing computationally more efficient and faster HT adaptations. Two popular approaches used by researchers are dimensionality reduction in parametric space (Aguado et al., 1995; Aguado et al., 1996; Goneid et al., 1997; Ser and Siu, 1995; Yip et al., 1992; Yip et al., 1995; Yuen et al., 1989) and piecewise linear approximation of curved segments (Yip et al., 1992; Yip et al., 1995).

Least squares based methods usually cast the ellipse fitting problem into a constrained matrix equation in which the solution should give least squares error. From the mathematical perspective, important work for ellipse detection has been done by (Fitzgibbon et al., 1999; Rosin, 1993b; Rosin and West, 1995; Tsuji and Matsumoto, 1978). In terms of application, some interesting works include (Ellis et al., 1992; Kim et al., 2002; Meer et al., 1991). Fitzgibbon (Fitzgibbon et al., 1999) proposed a very robust and efficient least squares method for ellipse detection. This method is invariant to affine transformation and is computationally efficient. While Fitzgibbon (Fitzgibbon et al., 1999) is arguably the most popular method used in the image processing community, some further work in this direction has been done (Harker et al., 2008; Maini, 2006; O'Leary et al., 2005). We also note the family of work (Ahn and Rauh, 1999; Ahn et al., 2002; Ahn et al., 2001) which uses geometric properties within the least squares method. However, they still require non-linear constrained optimization and are not specific to ellipses. We emphasize here that though some work has been done for fitting ellipses to a cluster of data, they may not be specifically useful in detecting ellipses without suitable modifications.

Genetic algorithms are good at dealing with non-linear optimization problems in which there are many local minima. However, these algorithms are generally computation intensive and require a lot of time for convergence. The stochastic nature of such algorithms make them time consuming. Further, the possibility of premature saturation cannot be fully excluded in most cases and the algorithms have to be carefully designed for each problem. Some interesting adaptations of genetic algorithm for the ellipse detection (Kasemir and Betzler, 2003; Kawaguchi and Nagata, 1998a; Kawaguchi and Nagata, 1998b; Procter and Illingworth, 1994).

The final category is the hybrid ellipse detection methods (Chia et al., 2011; Chia et al., 2008; Kim et al., 2002; Mai et al., 2008; Prasad and Leung, 2010a; Prasad and Leung, 2010c). The hybrid ellipse detection methods use one or more of the above approaches as just an intermediate step in the ellipse detection algorithm. Other steps like sophisticated digital curve pre-processing techniques (Bhowmick and Bhattacharya, 2007; Carmona-Poyato et al., 2010; Masood, 2008; Prasad and Leung, 2010d; Prasad et al., 2011b; Prasad et al., 2012), curvature estimation and correction techniques (Anderson and Bezdek, 1984; Cazals and Pouget, 2005; Heikkila, 1998; Matas et al., 1995; Prasad, 2011; Prasad et al., 2011a; Worring and Smeulders, 1993; Zhong et al., 2009), partial retrieval of ellipses' parameters using some geometric properties of ellipses (Guil and Zapata, 1997; Ho and Chen, 1995; Yuen et al., 1989; Zhang and Liu, 2005) are usually added before the actual ellipse detection method. Further, most hybrid methods include some or other form of grouping mechanism to group the edges that possibly belong to the same ellipse (Chia et al., 2011; Hahn et al., 2008; Kawaguchi and Nagata, 1998b; Kim et al., 2002; Mai et al., 2008). Finally, there are some essential ellipse refinement and selection steps to deal with the outliers and reduce false positives (Basca et al., 2005; Cheng, 2006; Ji and Haralick, 2001; Princen et al., 1994; Qiao and Ong, 2007; Wang et al., 2007; Prasad et al., 2010e)).

3. Least squares based high selectivity ellipse detection method

The least squares methods currently in use are based on fundamental work by Rosin (Rosin, 1993a; Rosin, 1993b; Rosin, 1996a; Rosin, 1996b) and Fitzgibbon (Fitzgibbon et al., 1999), in which the algebraic equation of general conics is used for defining the minimization problem and additional numeric constraints are introduced in order to restrict the solutions to elliptic curves. The non-linear optimization problem is then solved for finding the parameters of the ellipses. These methods do not explicitly use the geometric properties of ellipse and as a consequence give high false positive and false negative rates.

We propose an elliptic geometry based least squares method that does not require constrained optimization and is highly selective of ellipses. Since it uses a set of unconventional variables which are related to the actual parameters of ellipses in a non-linear manner. The constraints are directly incorporated in the definition of the new variables and the need of non-linear optimization is avoided. The main idea behind the proposed method is that since this method has to be applied on the digitized images (pixels), we can incorporate the effect of digitization in the development of least squares formulation. The main concept is that, rather than designing a least squares formulation using a general quadratic equation and satisfying certain constriants, we can use the geometric model of ellipse as the basic model and the distance of the pixels from the fitted ellipse as the criteria for designing the least squares formulation.

3.1 Proposed method

Consider the simplest form of ellipse, whose equation is as below:

$$\frac{x^2}{a^2}+\frac{y^2}{b^2}=1 \tag{1}$$

For a point $P(x_0,y_0)$ on the ellipse, the equation of the tangent at the point is given as $y=-(b/a)\cot\theta_0 x+b\csc\theta_0$, where we have used the parametric notation for the points on ellipses: $x_0=a\cos\theta_0$; $y_0=b\sin\theta_0$. If the ellipse (1) has to be fitted on a sequence of pixels $P_i(x_i,y_i)$; $x_i,y_i\in\mathbb{Z}$, we have to find a and b such that the distance of the pixels from the ellipse, or alternatively from the nearest tangents on the ellipse, is minimum. Thus, we want to minimize the residue:

$$\min(\text{w.r.t } a,b):\left|y_i+\frac{b}{a}x_i\cot\theta_0-b\csc\theta_0\right|=\left|y_i+\frac{b^2}{a^2}\frac{x_0}{y_0}x_i-\frac{b^2}{y_0}\right| \tag{2}$$

where $|\bullet|$ denotes the absolute value in the case of scalars and Euclidean norm in the case of vectors and the point $P(x_0,y_0)$ on the ellipse nearest to a pixel $P_i(x_i,y_i)$ satisfies $x'=a\cos\theta_0+\Delta x$, $y'=b\sin\theta_0+\Delta y$. Considering that the pixels in images are digitized form of the actual ellipse (rounding to the nearest integer), and assuming no other form of noise is present, it can be concluded that $|\Delta x|,|\Delta y|\le 0.5$. In other words, the point $P(x_0,y_0)$ on the ellipse which is nearest to the pixel $P_i(x_i,y_i)$ is within one pixel region of $P_i(x_i,y_i)$. Thus, using $\Delta x,\Delta y\le 0.5$, the intended maximum distance between $P_i(x_i,y_i)$ and $P(x_0,y_0)$ is $1/\sqrt{2}$. Thus, we can write the upper limit of (2) as $\left|y_i+(b^2/a^2)(x_0/y_0)x_i-(b^2/y_0)\right|\le 1/\sqrt{2}$. The above indicates that there is a definite upper bound (which itself is very small) to the expression to be minimized. This indicates that the minimization problem (2) should be easily solvable. Since $P(x_0,y_0)$ is not known, but $|\Delta x|,|\Delta y|\le 0.5$ are small, we can safely replace the values of (x_0,y_0) with (x_i,y_i) in (2) and rewrite (2) as follows:

$$\min(\text{w.r.t } a,b):\left|y_i+\frac{b^2}{a^2}\frac{x_i^2}{y_i}-\frac{b^2}{y_i}\right| \tag{3}$$

Thus, using the above minimization goal, we formulate a matrix equation as follows:

$$\begin{bmatrix} \vdots & \vdots \\ x_i^2/y_i & -1/y_i \\ \vdots & \vdots \end{bmatrix} \begin{bmatrix} b^2/a^2 \\ b^2 \end{bmatrix} = \begin{bmatrix} \vdots \\ -y_i \\ \vdots \end{bmatrix} \tag{4}$$

However, for numerical stability, (4) can be modified as:

$$\begin{bmatrix} \vdots & \vdots \\ x_i^2 & -1 \\ \vdots & \vdots \end{bmatrix} \begin{bmatrix} b^2/a^2 \\ b^2 \end{bmatrix} = \begin{bmatrix} \vdots \\ -y_i^2 \\ \vdots \end{bmatrix} \tag{5}$$

The above model is used to compute values of b^2 and b^2/a^2 using least squares inversion (Weisstein), which can be used subsequently to calculate the values of a and b. Following the same logic, we can formulate the least squares problem for a general ellipse, with centre at $O(\tilde{x},\tilde{y})$, semi-major and semi-minor axes a and b, and the angle of orientation α. For brevity, we present the model without incorporating the actual details of derivation:

$$\begin{bmatrix} \vdots & \vdots & \vdots & \vdots & \vdots \\ x_i^2 & 2x_iy_i & -2x_i & -2y_i & -1 \\ \vdots & \vdots & \vdots & \vdots & \vdots \end{bmatrix} \begin{bmatrix} \phi_1 & \phi_2 & \phi_3 & \phi_4 & \phi_5 \end{bmatrix}^{\mathrm{T}} = \begin{bmatrix} \vdots \\ -y_i^2 \\ \vdots \end{bmatrix} \tag{6}$$

and the parameters of the ellipse can be found using:

$$\begin{gathered} \tilde{x} = (\phi_3 - \phi_4\phi_2)\big/\left(\phi_1 - \phi_2^2\right); \; \tilde{y} = (\phi_1\phi_4 - \phi_3\phi_2)\big/\left(\phi_1 - \phi_2^2\right); \; \alpha = 0.5\tan^{-1}\left(2\phi_2/(\phi_1 - 1)\right) \\ a = \sqrt{2\left(\phi_5 + \tilde{y}^2 + \tilde{x}^2\phi_1 + 2\phi_2\right)\Big/\left((1+\phi_1) - \sqrt{(1-\phi_1)^2 + 4\phi_2^2}\right)} \\ b = \sqrt{2\left(\phi_5 + \tilde{y}^2 + \tilde{x}^2\phi_1 + 2\phi_2\right)\Big/\left((1+\phi_1) + \sqrt{(1-\phi_1)^2 + 4\phi_2^2}\right)} \end{gathered} \tag{7}$$

3.2 Numerical examples

In this section, we consider various numerical examples and compare the performance of the proposed method with Fitzgibbon's method (Fitzgibbon et al., 1999). We consider three main categories of curves: elliptic, non-elliptic conical, and non-conical curves. Using the results, we demonstrate that the proposed method is more robust and generates less false positives than Fitzgibbon (Fitzgibbon et al., 1999).

3.2.1 Elliptic curves

For elliptic curves, we consider three experiments. The experiments are performed on a family of digital ellipses, whose minimum semi-minor distance is 10 pixels, and maximum semi-major distance is 2000 pixels. Thus, the eccentricities of the ellipses may vary between 0 and 0.999987. The centre of the ellipses lie within a square region of size 2000 pixels and centred at the origin. The angle of orientation lies in the range $[-90°, 90°]$. For comparing the actual ellipses with the detected ellipses, we use the following measures:

$$\delta\tilde{x} = |\tilde{x} - \tilde{x}_{est}|;\ \delta\tilde{y} = |\tilde{y} - \tilde{y}_{est}|;\ \delta a = |a - a_{est}|/a;\ \delta b = |b - b_{est}|/b;$$
$$\delta\alpha = |\alpha - \alpha_{est}|\sqrt{1 - b^2/a^2} \text{ (in degrees)} \quad (8)$$

where the subscript *'est'* is used to denote the values estimated by the ellipse detection method.

In practice, the complete elliptic curve is not available due to occlusion or lighting conditions. Thus, it is important that an ellipse detection method is capable of detecting ellipses from partial curves as well. This is the motivation of this example. We consider the curves from $[(\theta_0 + \alpha), (\theta_0 + \alpha + \Delta\theta)]$, where we vary $\Delta\theta$ from 90° to 360° at a step of 10°, thus the experiment contains 36 cases. For each case, we generate 10,000 digital elliptic curves. For each of the curve, the parameters $\tilde{x}$, $\tilde{y}$, a, b, α, and θ_0 are generated randomly. The range for θ_0 is $[0, 360°]$.

The measures in (8) are computed for each curve and then averaged for all the 10,000 curves corresponding to a particular value of $\Delta\theta$. The residues and the measures (8) are presented in Fig. 4.

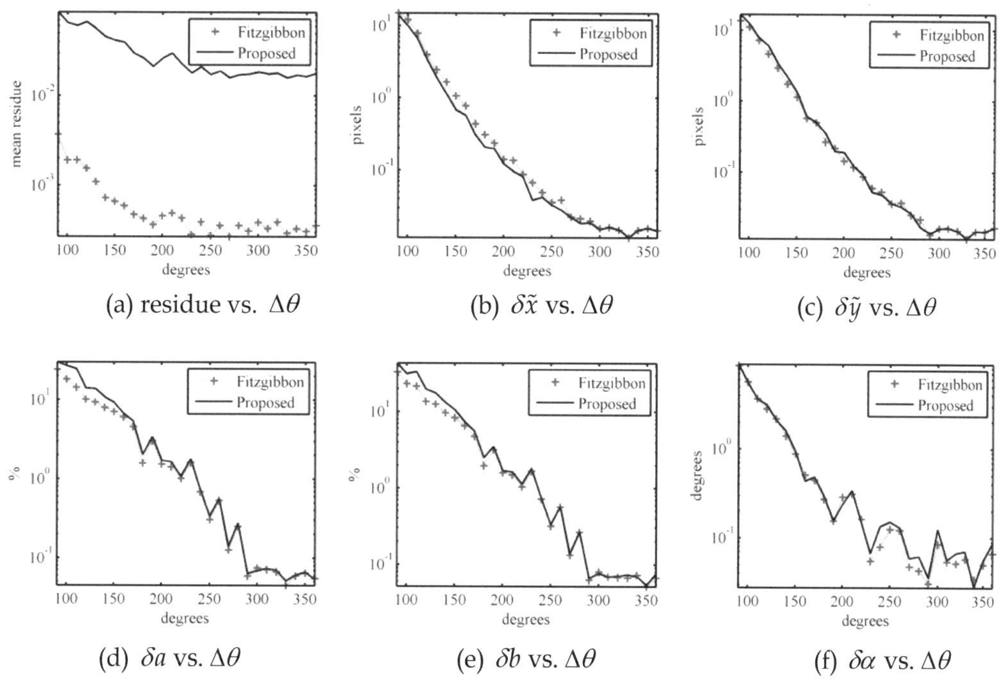

(a) residue vs. $\Delta\theta$ (b) $\delta\tilde{x}$ vs. $\Delta\theta$ (c) $\delta\tilde{y}$ vs. $\Delta\theta$

(d) δa vs. $\Delta\theta$ (e) δb vs. $\Delta\theta$ (f) $\delta\alpha$ vs. $\Delta\theta$

Fig. 4. Residue and error measures for experiment in section 3.2.1.

The errors are very small when we have the entire curve available, i.e., $\Delta\theta = 360°$. However, even for other values $\Delta\theta > 90°$, the errors are in practically useful range. We note that Fitzgibbon (Fitzgibbon et al., 1999) generated invalid conics for about 56% of the all the curves, thus indicating a false negative rate of 56%. We also note that the smaller the curve, the higher is the error for all the parameters, which is expected, since lesser and lesser curvature is available. Further, δa and δb are very large for small curves, indicating that larger curve is essential for detecting the ellipses accurately. It is seen that though Fitzgibbon (Fitzgibbon et al., 1999) has lower residue, the proposed method has lower or comparable error for almost all the parameters.

3.2.2 Non-elliptic conical curves

We intend to test if Fitzgibbon (Fitzgibbon et al., 1999) and the proposed method generate false positives, i.e., detect non-elliptic curves as ellipses. For this, we consider a family of conics given by:

$$x = l\cos\theta/(1-e\cos\theta); y = l\sin\theta/(1-e\cos\theta) \tag{9}$$

where $l \in [20,200]$ and eccentricity $e \in [1,2]$. We generate portions of this curve corresponding to $\theta \in [180-\Delta\theta/2, 180+\Delta\theta/2]$, where $\Delta\theta$ is stepped from 45° to 180° in steps of $5°$. As before, for each value of $\Delta\theta$, 10,000 random curves using (9) are generated. The family of curves corresponding to $\Delta\theta = 180°$ is shown in Fig. 5(a). It is seen that though the residue is small for the proposed method (Fig. 5(b)), the proposed method identifies all the curves as non-elliptic (i.e., generates imaginary value for at least one of the parameters a and b of the fitted ellipse). On the other hand, Fitzgibbon (Fitzgibbon et al., 1999) fits some real valued ellipses on the most non-elliptic curves as well (see Fig. 5(c)). When the curves are digitized, though the performance of the proposed method gets poor, for small values of $\Delta\theta$, in general, it still outperforms Fitzgibbon (Fitzgibbon et al., 1999) in its selectivity for the elliptic curves (see Fig. 5(d)).

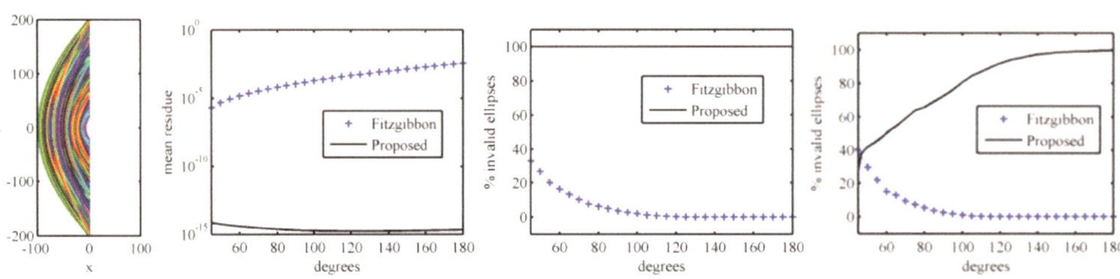

(a) family of conics (b) residue vs. $\Delta\theta$ (c) % of true negative detections (d) % of true negative detection (in the presence of digitization)

Fig. 5. Residue and percentage of invalid ellipses for experiment 3.2.2.

3.2.3 Non-conical fourth order curves

Now, we consider a family of curves given by the equation:

$$\frac{x^4}{a^4} + \frac{y^4}{b^4} = 1 \tag{10}$$

where $a,b \in [20,200]$, $x = a\sqrt{\cos\theta}$, $x = b\sqrt{\sin\theta}$, and $\theta \in [0,\Delta\theta]$. For this family, we step $\Delta\theta$ from 45° to 90° at steps of $5°$. As before, for each value of $\Delta\theta$, we generate 10,000 randomly chosen curves. The family of curves for $\Delta\theta = 90°$ is plotted in Fig. 6(a). In the absence of digitization, Fitzgibbon (Fitzgibbon et al., 1999) is not at all selective about elliptic curves, while the proposed method rejects all the curves as non-elliptic (see Fig. 6(b)). In the presence of digitization, though the selectivity of the proposed method decreases for small values of $\Delta\theta$, Fitzgibbon (Fitzgibbon et al., 1999) continues to perform poorly (see Fig. 6(c)).

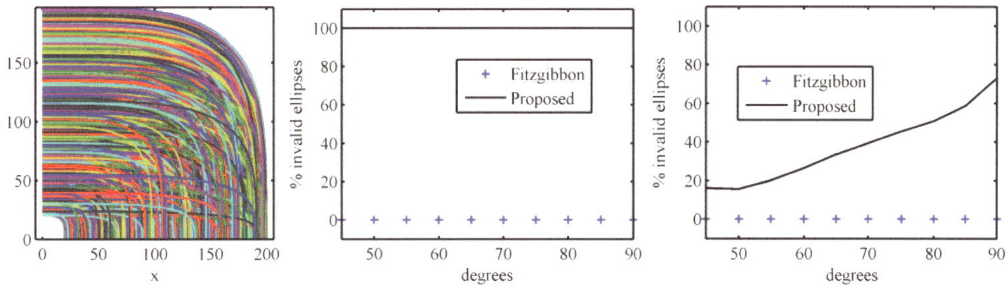

(a) family of curves (b) % of true negative detections (c) % of true negative detection (in the presence of digitization)

Fig. 6. Percentage of invalid ellipses for experiment 3.2.3.

4. Hybrid ellipse detection method for real images

We propose a hybrid ellipse detection method for real images. The proposed method consists of three major stages, viz., edge preprocessing, ellipse detection, and salient ellipse selection. Various steps in these methods address to various problems in the problem of ellipse detection. We succinctly highlight the steps which serve an important purpose in the algorithm in the following list:

- The dominant point detection (step 3 of section 4.1) is helpful in reducing the effect of digitization and corruption in the quality of edges due to noise.
- The curvature correction (step 4 of section 4.1) is useful for dealing with the corruption in the quality of image due to light and shadow conditions. Specifically, it is useful for dealing with the merging of edges of two objects, with one or both possibly being elliptic.
- The determination of the search region (step 1 of section 4.2) is useful in making the method more selective of ellipses by targeting two problems, looking for the presence of incomplete ellipses and simultaneously reducing the possibility of obvious outliers (which will definitely not present in the search region of an edge).
- The associated convexity (step 2 of section 4.2) further increases the method's selectivity for ellipses by weeding out other edges that may be outliers.
- The geometric center finding (step 3 of section 4.2) is the core step that enables the grouping of edges possibly belonging to the same ellipse. In conjunction with steps 1 and 2 of section 4.2, this step enables the grouping of the edges that possibly belong to the same ellipse with much greater reliability than most contemporary methods.
- The relationship score (step 4 of section 4.2) is a technique of quantifying the reliability of relevance of an edge in the candidate ellipse. This technique enables the ellipse detection in the next step with very good reliability because the next step can further remove some edges with lower relationship score (thus identifying them as outliers).

- Detecting similar ellipses (step 1 of section 4.3) is a way of reducing multiple detections for one elliptic object in the image. This step chooses the best representative ellipse for an object, thus improving the selectivity of the method.
- The saliency parameters (steps 2-5 of sections 4.3) quantify the quality of detected ellipses (precision as well as reliability) which are used in step 6 of section 4.3 for further reducing the false positives. Step 6 of section 4.3 is a technique of non-heuristically choosing the detected ellipses with above average quality.

4.1 Edge preprocessing

The various steps of this stage are listed below:

1. **Extraction of edge map**: The real image is converted to the gray image and Canny edge detector is used to obtain the edge map.
2. **Obtaining edges**: Then, sequences of continuous edge pixels are identified, and called edges. The edges begin and end with either 1-connected or >2connected edge pixels. There are many approaches in vogue for implementing this step. We use the codes by (Kovesi) in our method. But, other methods may be used as well.
3. **Dominant point detection (polygonal approximation of edges)**: After this, we approximate the edges with a polygonal approximation (or dominant point detection method). We note that we have used a recently proposed parameter independent line fitting method which provides robust approximation for edges of any length and curvature and noise (Prasad et al., 2011b). In general, if other polygonal approximation methods are used, the performance depends upon the control parameters very strongly. Also, if these methods are used, we note that one value of control parameter may be suitable for one image, but may result in poor performance for another image. Such issues are absent in (Prasad et al., 2011b) and it provides good performance for all the images used for testing the performance of the algorithm.
4. **Curvature correction**: The sequence of dominant points obtained for an edge is used to perform curvature correction. For this step, it is useful to define the sequence of chords of the polygonal approximation of an edge e. Suppose $\{l_1, l_2, \ldots, l_N\}$ is the sequence of chords formed by the dominants point of the edge e. Let the angles between all the pairs of consecutive line segments be denoted as $\{\theta_1, \theta_2, \ldots, \theta_{N-1}\}$, where $\theta_i \in [-\pi, \pi]$ is the anticlockwise angle from l_{i+1} to l_i (see Fig. 7(a) for illustration).

a. **Removal of sharp turns:** In the sequence of the angles, $\{\theta_1, \theta_2, \ldots, \theta_{N-1}\}$, if any angle θ_i is large, i.e., equal to or greater than a chosen threshold θ_0 (say $\pi/2$, empirically determined), then the change in the curvature of the edge at such points P_{i+1} (the intersection point of line segments l_i and l_{i+1}) is considered to be large (or sharp) , and the edge is split at P_{i+1} to ensure smooth curvature.

b. **Removal of inflexion points:** From the above definition of the angles $\{\theta_1, \theta_2, \ldots, \theta_{N-1}\}$, the change in direction of curvature occurs in the change of the sign of the angles (negative or positive). Thus, we can create a Boolean sequence $\{b_1, b_2, \ldots, b_{N-1}\}$, where b_i is '0' if the sign of θ_i and θ_1 is the same. This Boolean sequence can be used to identify the inflexion points and decide the exact places where the edge contour should be split. The three possibilities of occurrence of inflexion points are shown in Fig. 7(b). The points where the edge needs to be split are also shown in the same figure.

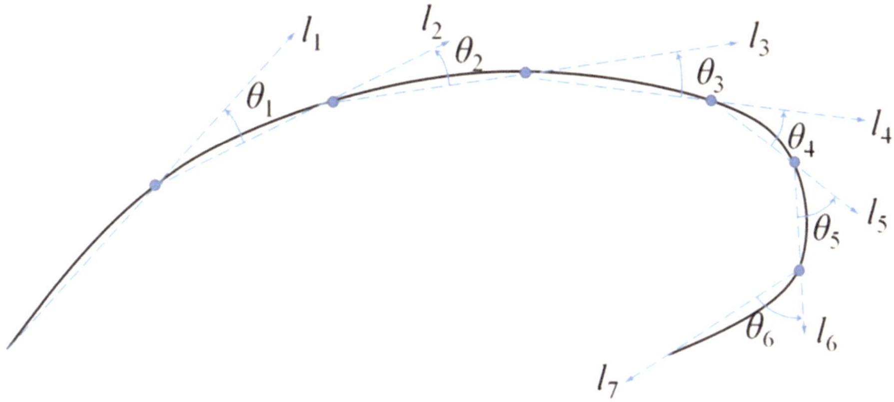

(a) Illustration of the chords formed by the dominant points (polygonal approximation) of an edge

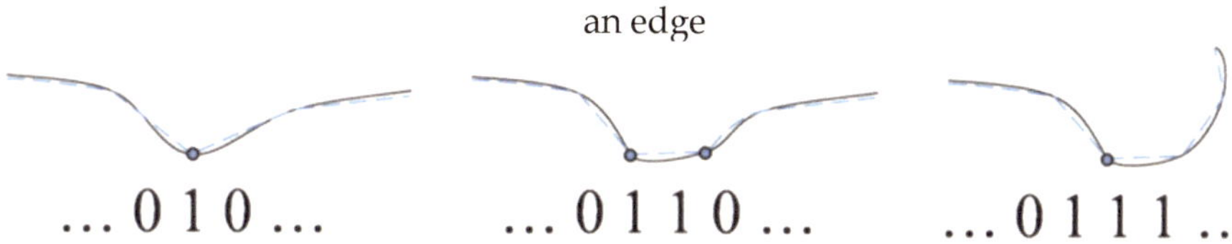

(b) Illustration of varieties of inflexion points and the dominant points to be deleted

Fig. 7. Illustrations for the curvature correction techniques used in step 4 of section 4.1.

4.2 Ellipse detection

The various steps of this stage are discussed below:

1. **Determining search region of an edge:** We define the search region R as following. For a given edge e, let the tangents to the edge at its end points P_1 and P_2 be denoted by l_1 and l_2, and the line segment connecting the end points P_1 and P_2 be denoted by l_3. The two tangents l_1 and l_2 and the line l_3 divide the image into two regions, R_1 and R_2, as shown in Fig. 8(a). Then the search region R is the region that does not contain P_{mid}, where P_{mid} is the middle pixel of the edge e. Mathematically, the search region R is defined as follows:

$$R = \begin{cases} R_2 & P_{mid} \notin R_2 \\ R_1 & \text{otherwise} \end{cases} \tag{11}$$

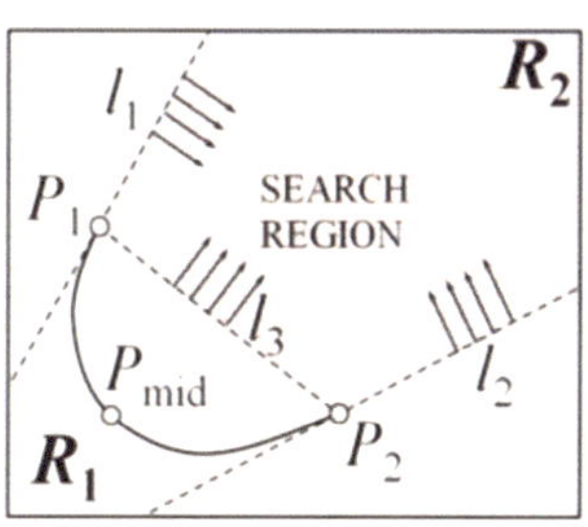

(a) Search region

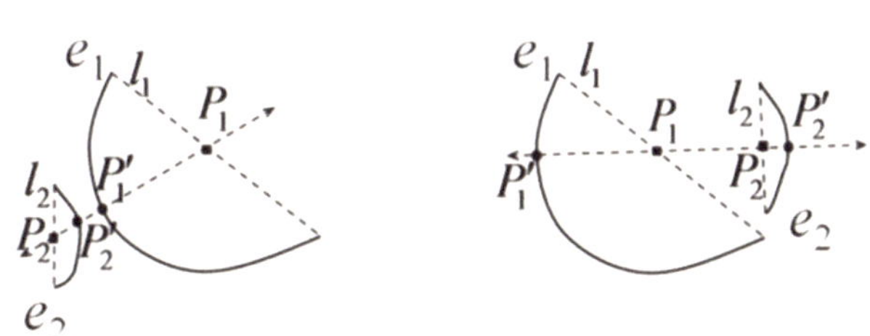

(b) demonstration of associated convexity

Fig. 8. Illustrations of search region and associated convexity (steps 1 and 2 of section 4.2.)

2. **Computing associated convexity:** Let us consider the line segments l_1 and l_2 formed by joining the end points of e_1 and e_2, respectively, as shown in Fig. 8. Let P_1 and P_2 be the midpoints of the line segments l_1 and l_2. Let l_3 be a line passing through P_1 and P_2, such that it intersects the edges e_1 and e_2 at P_1' and P_2' respectively. The pair of edges e_1 and e_2 are suitable for grouping if and only if:

$$P_1'P_2' \approx P_1P_1' + P_1P_2 + P_2P_2' . \tag{12}$$

3. **Detection of the geometric center of the ellipse corresponding to an edge:** In this step, we use the method proposed in (Yuen et al., 1989) for estimating the geometric center of the ellipse to which the edge may belong. This method uses the tangents estimated at three points on the edge. It was shown in (Prasad and Leung, 2010b) that Yuen's method is sensitive to the error in tangent estimation. Thus, we use a recently proposed upper-bounded tangent estimator for digital curves (Prasad et al., 2011a) using $R = 4$, which gives better performance than other contemporary tangent estimation methods. We split an edge into three sub-edges and choose points randomly from each sub-edge to form several (upto 200) sets of three points. Then for each set, a geometric center is found.
4. **Determination of relationship score:** The image space is divided into equal square bins where the size of one bin is given using the Rayleigh distribution presented in (Prasad and Leung, 2010b). For an edge and a bin, the relationship score proposed in (Prasad and Leung, 2010a) is given by:

$$r_e^b = S_e^b \, r_1 \, r_2 \tag{13}$$

where $r_1 = \left(\frac{S_e^b}{S_e}\right)\exp\left(\frac{S_e^b}{S_e} - 1\right)$, $r_2 = \left(\frac{S_e}{S}\right)\exp\left(2\left(\frac{S_e}{S} - 1\right)\right)$, $S = 200$ is the total number of sets of three pixels, S_e is the number of sets for which no geometric exception occurred, and S_e^b is the number of sets that generated geometric centers inside the square bin b. More details can be found in (Prasad and Leung, 2010a).

5. **Grouping of edges and ellipse detection:** All the edges having a common bin b may initially be considered as a group. However, every edge in the group should also fall in the search region of every other edge and satisfy the condition of associated convexity. The edges that do not fulfill these conditions are removed from the group. The edges in a group are ranked in the descending order of their bin-edge relationship scores r_e^b. The edge pixels of the edges in a group are appended and least squares fitting technique (Fitzgibbon et al., 1999)[2] is used on the group to find all the parameters of the ellipse. Now, the quality of the group is evaluated using the two criteria listed below:

- Criterion 1 (C1): Error of least squares fitting $\leq \varepsilon_{ls}$, a chosen threshold error value.
- Criterion 2 (C2): The bin b of the group is inside the detected elliptic hypothesis.

the threshold used for C1 is $\varepsilon_{ls} = 0.01$. If both C1 and C2 are satisfied, then the parameters of the ellipses computed using the least squares fitting are passed to the next stage. If

[2] We recommend that the least squares method presented in section 3 is used. However, the results for section 4 using the method in section 3 were not available at the time of writing this chapter.

anyone of the two criterions is not fulfilled, the weakest edge (with the lowest relationship score r_e^b) is removed from the group and the above process is repeated till either the above criteria are satisfied or the group becomes empty.

4.3 Salient ellipse selection

The various steps of this stage are discussed below:

1. **Detecting similar ellipses:** For two ellipses E_1 and E_2, we obtain the Boolean matrices I_1 and I_2 of the same size as the actual image, such that the pixels outside the ellipse are assigned Boolean '0' amd the remaining are assigned Boolean '1'. Then, the similarity measure D is computed using:

$$D = 1 - \frac{\text{count}(XOR(I_1, I_2))}{\text{count}(OR(I_1, I_2))} \tag{14}$$

where $\text{count}(A)$ gives the number of Boolean '1' elements in the matrix A. For a given ellipse, all the ellipses that have overlap ratio $D > 0.9$ are clustered together. Among the ellipses in a cluster, the choice of representative candidate is done by choosing the ellipse that was formed by maximum amount of data. Thus, we have used percentage circumference of an ellipse (15) for choosing the representative.

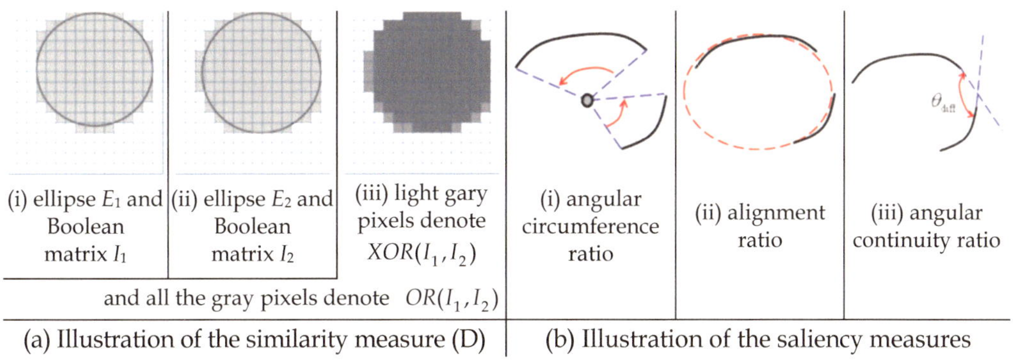

Fig. 9. The illustration of the concept of similarity measures and the saliency measures

2. **Computing circumference ratio:** Suppose an ellipse E was fitted to a group G, then we define circumference ratio $c(E,G)$ as below:

$$c(E,G) = \sum_{\forall e \in G} \alpha(E,e)/2\pi \tag{15}$$

where $\alpha(E,e)$ is the angle subtended by the ends of the edge e at the centre of the ellipse E. A higher value of $c(E,G)$ implies a larger support of E on G.

3. **Computing alignment ratio:** We consider all the pixels the pixels $\{P_i; i = 1 \text{ to } N_G\}$ in the group of edges that generated an elliptic hypothesis, and compute their Euclidean distance d_i from the elliptic hypothesis. This is used to compute a function $s(E, P_i)$ as follows:

$$s(E,P_i) = \begin{cases} 1 & if\ d_i < d_0 \\ 0 & \text{otherwise} \end{cases} \tag{16}$$

The value of threshold chosen here is $d_0 = 2$. Then the alignment ratio is calculated as:

$$a(E,G) = \sum_{i=1}^{N_G} s(E,P_i)/N_G \tag{17}$$

4. **Computing angular continuity ratio**: The angular continuity ratio is computed as:

$$\phi(E,G) = \begin{cases} 1 & \text{if } N = 1 \\ \dfrac{1}{N-1} \displaystyle\sum_{i=1}^{N-1} \dfrac{\theta_{\text{diff}}(e_i, e_{i+1})}{\pi} & \text{if } N > 1 \end{cases} \tag{18}$$

where N is the number of edge curves in the group G.

5. **Computing net saliency score**: The net saliency score is computed using circumference ratio, alignment ratio, and angular continuity ratio, as follows:

$$\sigma_{\text{add}}(E,G) = \frac{a(E,G) + c(E,G) + \phi(E,G)}{3}. \tag{19}$$

6. **Final selection of ellipses**: In order to make the selection of the elliptic hypotheses non-heuristic, the decision of selecting the elliptic hypothesis E is made using the expression below:

$$\text{AND}\begin{pmatrix} a(E,G) \geq avg\{a(E,G)\}, \\ c(E,G) \geq avg\{c(E,G)\}, \\ \phi(E,G) \geq avg\{\phi(E,G)\}, \\ \sigma_{add}(E,G) \geq avg\{\sigma_{add}(E,G)\} \end{pmatrix} \tag{20}$$

Here, $avg\{a(E,G)\}$ is the average value of the alignment ratios calculated for all the elliptic hypotheses remaining after the similar ellipses identification. The same applies for the other expressions in (20).

5. Experimental results for section 4

5.1 Synthetic dataset: Overlapping and occluded ellipses

To generate the synthetic images, we consider an image size of 300×300 and generate $\alpha \in \{4,8,12,16,20,24\}$ ellipses randomly within the region of image. The parameters of the ellipses are generated randomly: center points of the ellipses are arbitrarily located within the image, lengths of semi-major and semi-minor axes are assigned values randomly from the range $\left[10, 300/\sqrt{2}\right]$, and the orientations of the ellipses are also chosen randomly. The only constraint applied is that each ellipse must be completely contained in the image and overlap with at least one ellipse. For each value of α, we generate 100 images containing occluded ellipses and 100 other images containing overlapping ellipses. In the occluded ellipses, the edges of the overlapped regions are not available, while in the overlapping images all the edge contours of the ellipses are available. Thus, in total there are 600 images with occluded ellipses and 600 images with overlapping ellipses.

The true positive elliptic hypotheses are identified as the elliptic hypotheses that have high overlap with the ellipses in the ground truth. We use precision, recall, and F-measure (Baeza-Yates and Ribeiro-Neto, 1999) for measuring the performance of the proposed method. The performance of method is compared with the results of Mai (Mai et al., 2008), Kim (Kim et al., 2002), simplified Hough transform (McLaughlin, 1998), and randomized Hough transform (McLaughlin, 1998). The comparative results are presented in Table 1 for synthetic images with occluded ellipses and in Table 2 for synthetic images with overlapping ellipses.

The proposed method gives the best performance among all the methods considered for both occluded and overlapping ellipses. Some example images are also provided in Fig. 10.

5.2 Synthetic dataset: Overlapping and occluded ellipses

Now, we compare the performance of various methods and the proposed method (scheme 3) for the real image dataset. The average value of the performance metrics for the 400 real

	Recall			
α	Hybrid method (section 4)	Chia	Mai	Kim
4	0.99	0.90	0.39	0.48
8	0.98	0.80	0.20	0.20
12	0.95	0.70	0.13	0.09
16	0.92	0.65	0.05	0.05
20	0.90	0.62	0.04	0.04
24	0.86	0.55	0.01	0.02
	Precision			
4	0.93	0.9	0.62	0.75
8	0.91	0.82	0.42	0.59
12	0.89	0.73	0.34	0.36
16	0.87	0.70	0.18	0.30
20	0.85	0.68	0.12	0.20
24	0.82	0.60	0.04	0.18
	F-measure			
4	0.96	0.90	0.48	0.58
8	0.94	0.80	0.28	0.3
12	0.91	0.71	0.19	0.15
16	0.90	0.69	0.09	0.10
20	0.87	0.65	0.05	0.06
24	0.84	0.58	0.01	0.05

Table 1. Result of the hybrid ellipse detection method for synthetic images with occluded ellipses.

images is shown in Table 3. The total time taken by each method for the complete dataset is also shown. In terms of the performance, the proposed method not only outperforms the other methods, it also shows practically acceptable level of performance.

Though the time taken by Mai (Mai et al., 2008) is small, the proposed method takes lesser time than the remaining methods. Further, the superior performance of the proposed method as compared to Mai (Mai et al., 2008) clearly outweighs the longer time taken by the proposed method. The proposed method is easily parallelizable and can give real time performance when optimized for specific applications.

We also give some examples of real images and our results in Fig. 11 to Fig. 14.

		Recall		
α	Hybrid method (section 4)	Chia	Mai	Kim
4	1.00	1.00	0.78	0.80
8	0.99	0.96	0.62	0.55
12	0.98	0.96	0.53	0.35
16	0.95	0.87	0.45	0.20
20	0.91	0.79	0.35	0.10
24	0.89	0.69	0.32	0.09
	Precision			
4	1.00	1.00	0.81	0.90
8	0.98	0.96	0.70	0.81
12	0.95	0.96	0.65	0.75
16	0.94	0.91	0.58	0.64
20	0.92	0.89	0.48	0.44
24	0.90	0.82	0.46	0.44
	F-measure			
4	1.00	1.00	0.79	0.85
8	0.98	0.96	0.66	0.66
12	0.97	0.96	0.48	0.59
16	0.94	0.90	0.30	0.50
20	0.91	0.82	0.15	0.41
24	0.89	0.75	0.14	0.39

Table 2. Result of the hybrid ellipse detection method for synthetic images with overlapping ellipses.

Fig. 10. Example synthetic image and ellipses detected on the images.

	Hybrid method (section 4)	Mai	Kim
Average Precision	0.8748	0.2862	0.1831
Average Recall	0.7162	0.1632	0.1493
Average F-measure	0.7548	0.1831	0.1591
Average time taken (seconds)	38.68	11.41	60.87

Table 3. Performance metrics for the proposed method (Section 4), Mai (Mai et al., 2008), and Kim (Kim et al., 2002) for real dataset (400 real images (Griffin et al.)).

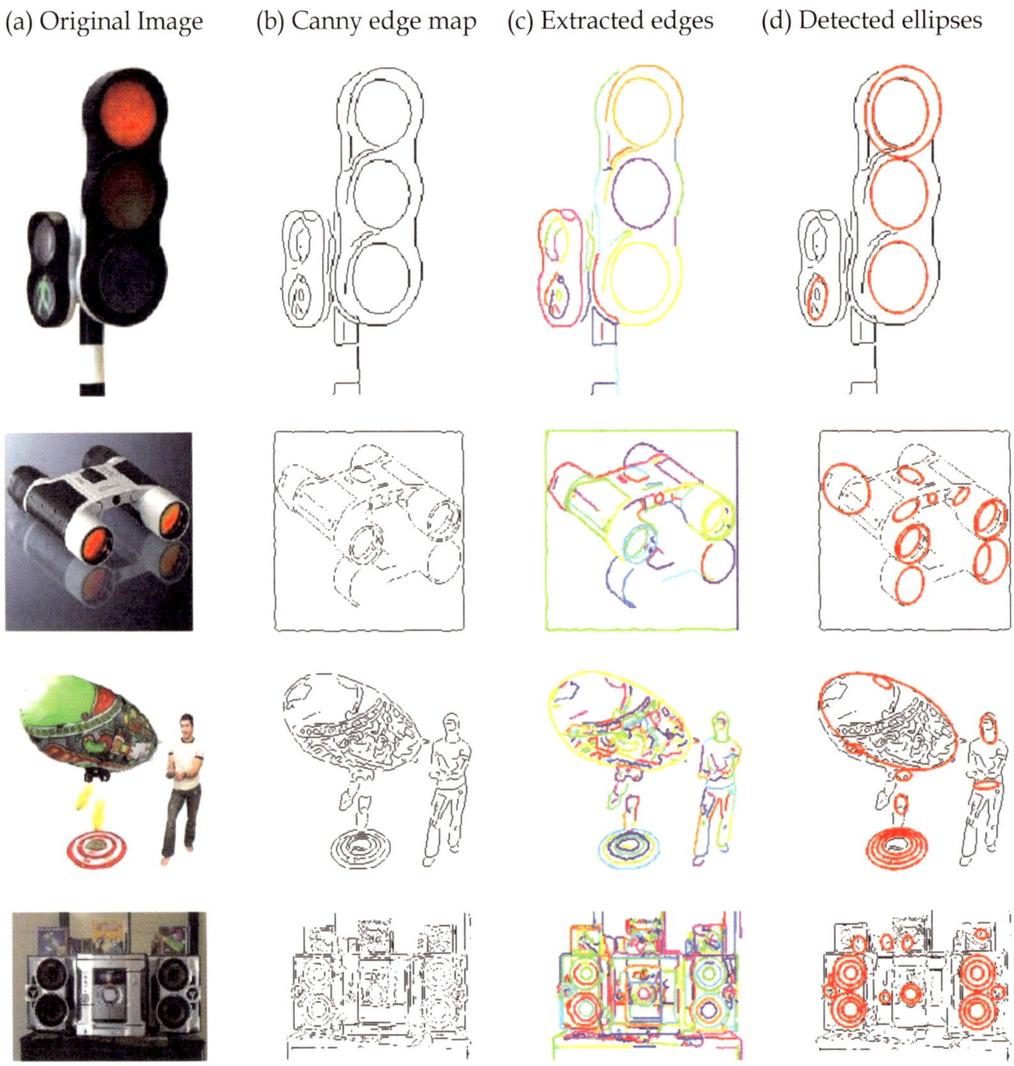

Fig. 11. Examples of real images and ellipse detection using the proposed method.

Fig. 12. Examples of real images and ellipse detection using the proposed method (continued).

Fig. 13. Examples of real images and ellipse detection using the proposed method (continued).

(a) Original Image (b) Canny edge map (c) Extracted edges (d) Detected ellipses

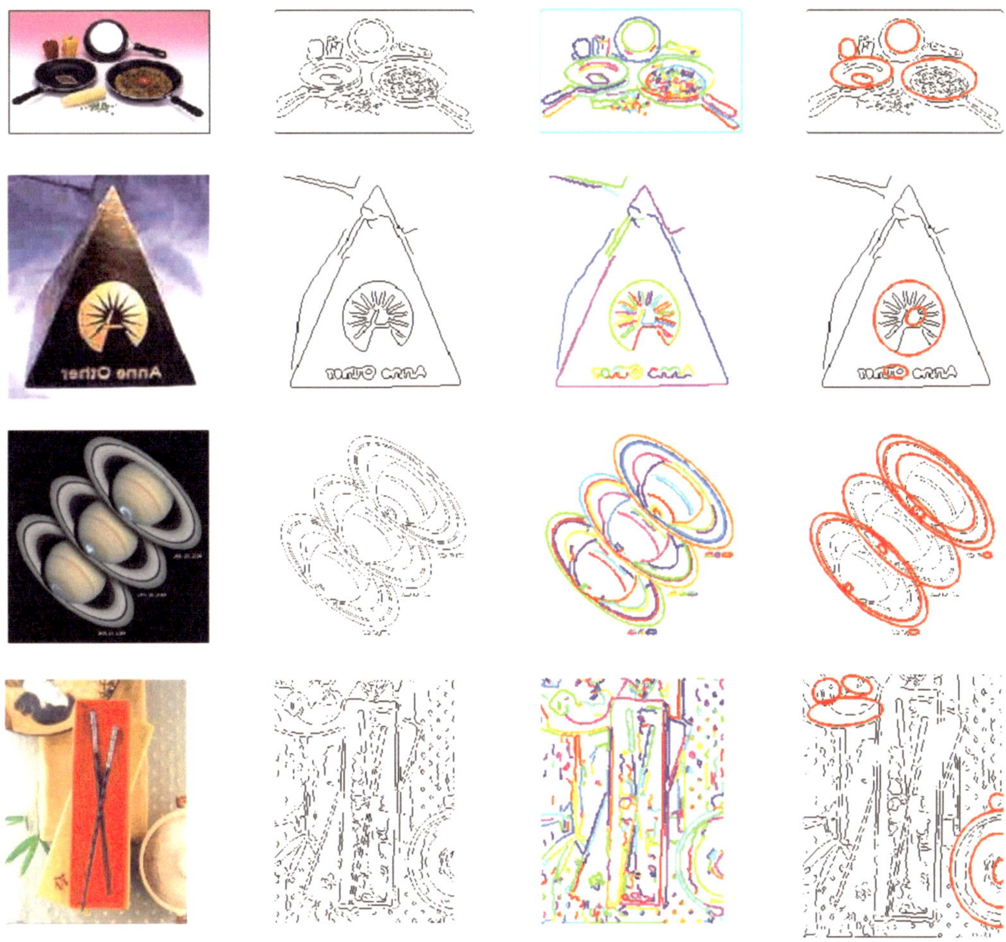

Fig. 14. Examples of real images and ellipse detection using the proposed method (continued).

6. Conclusion

The challenges in the problem of ellipse detection in real images are clearly identified. The need to use more selective ellipse detection methods, that can detect elliptic shapes with greater accuracy and reduce the false detections, is highlighted. A new unconstrained linearly computable least squares method is proposed. This method uses a linear matrix formulation and transfers the non-linearity of the ellipse detection problem to a new set of variables that are linked non-linearly to the geometric parameters of the ellipses. This method shows significantly better performance than the widely used non-linear constrained least squares formulation of Fitzgibbon (Fitzgibbon et al., 1999). The proposed method has lower false positive as well as false negative rates.

A hybrid ellipse detection method is also presented. This method uses various steps to deal with the various challenges of the problem of ellipse detection. Due to these sophisticated pre-processing, grouping, ellipse detection, and ellipse selection steps, the proposed method is highly selective and detects ellipses in various challenging scenario.

7. References

Aguado, A.S., Montiel, M.E. and Nixon, M.S., 1995. Ellipse detection via gradient direction in the Hough transform, Proceedings of the International Conference on Image Processing and its Applications, pp. 375-378.

Aguado, A.S., Montiel, M.E. and Nixon, M.S., 1996. Improving parameter space decomposition for the generalised Hough transform, Proceedings of the IEEE International Conference on Image Processing, pp. 627-630.

Ahn, S.J. and Rauh, W., 1999. Geometric least squares fitting of circle and ellipse. International Journal of Pattern Recognition and Artificial Intelligence, 13(7): 987-996.

Ahn, S.J., Rauh, W., Cho, H.S. and Warnecke, H.J., 2002. Orthogonal distance fitting of implicit curves and surfaces. IEEE Transactions on Pattern Analysis and Machine Intelligence, 24(5): 620-638.

Ahn, S.J., Rauh, W. and Warnecke, H.J., 2001. Least-squares orthogonal distances fitting of circle, sphere, ellipse, hyperbola, and parabola. Pattern Recognition, 34(12): 2283-2303.

Anderson, I.M. and Bezdek, J.C., 1984. Curvature and tangential deflection of discrete arcs: a theory based on the commutator of scatter matrix pairs and its application to vertex detection in planar shape data. IEEE Transactions on Pattern Analysis and Machine Intelligence, PAMI-6(1): 27-40.

Antonaros, G.I. and Petrou, L.P., 2001. Real time map building by means of an ellipse spatial criterion and sensor-based localization for mobile robot. Journal of Intelligent and Robotic Systems: Theory and Applications, 30(4): 331-358.

Baeza-Yates, R. and Ribeiro-Neto, B., 1999. Modern Information Retrieval. ACM Press, Addison-Wesley, New York.

Basca, C.A., Talos, M. and Brad, R., 2005. Randomized hough transform for ellipse detection with result clustering, Proceedings of the International Conference on Computer as a Tool, Belgrade, SERBIA MONTENEG, pp. 1397-1400.

Belaroussi, R., Prevost, L. and Milgram, M., 2005. Classifier combination for face localization in color images, Lecture Notes in Computer Science, pp. 1043-1050.

Bell, A.A., Herberich, G., Meyer-Ebrecht, D., Bocking, A. and Aach, T., 2006. Segmentation and detection of nuclei in silver stained cell specimens for early cancer diagnosis, Proceedings of the International Conference on Image Processing, San Antonio, TX, pp.VI-49 - VI-52.

Bhowmick, P. and Bhattacharya, B.B., 2007. Fast polygonal approximation of digital curves using relaxed straightness properties. IEEE Transactions on Pattern Analysis and Machine Intelligence, 29(9): 1590-1602.

Burrill, J.R., Wang, S.X., Barrow, A., Friedman, M. and Soffen, M., 1996. Model-based matching using elliptical features, Proceedings of SPIE - The International Society for Optical Engineering, pp. 87-97.

Carmona-Poyato, A., Madrid-Cuevas, F.J., Medina-Carnicer, R. and Muñoz-Salinas, R., 2010. Polygonal approximation of digital planar curves through break point suppression. Pattern Recognition, 43(1): 14-25.

Cazals, F. and Pouget, M., 2005. Estimating differential quantities using polynomial fitting of osculating jets. Computer Aided Geometric Design, 22(2): 121-146.

Cheng, Y.C., 2006. The distinctiveness of a curve in a parameterized neighborhood: Extraction and applications. IEEE Transactions on Pattern Analysis and Machine Intelligence, 28(8): 1215-1222.

Chia, A.Y.-S., Rahardja, S., Rajan, D. and Leung, M.K.H., 2011. A Split and Merge Based Ellipse Detector with Self-Correcting Capability. IEEE Transactions on Image Processing, 20(7): 1991 - 2006.

Chia, A.Y.S., Rahardja, S., Rajan, D. and Leung, M.K.H., 2009. Structural descriptors for category level object detection. IEEE Transactions on Multimedia, 11(8): 1407-1421.

Chia, A.Y.S., Rajan, D., Leung, M.K.H. and Rahardja, S., 2008. A split and merge based ellipse detector, Proceedings of the IEEE International Conference on Image Processing, San Diego, CA, pp. 3212-3215.

Dijkers, J.J. et al., 2005. Segmentation and size measurement of polyps in CT colonography, Lecture Notes in Computer Science, pp. 712-719.

Ellis, T., Abbood, A. and Brillault, B., 1992. Ellipse detection and matching with uncertainty. Image and Vision Computing, 10(5): 271-276.

Fernandes, A.M., 2009. Study on the Automatic Recognition of Oceanic Eddies in Satellite Images by Ellipse Center Detection-The Iberian Coast Case. IEEE Transactions on Geoscience and Remote Sensing, 47(8): 2478-2491.

Feyaerts, F., Vanroose, P., Fransens, R. and Van Gool, L.J., 2001. Using shape to correct for observed non-uniform color in automated egg grading, Proceedings of SPIE - The International Society for Optical Engineering, pp. 93-101.

Fitzgibbon, A., Pilu, M. and Fisher, R.B., 1999. Direct least square fitting of ellipses. IEEE Transactions on Pattern Analysis and Machine Intelligence, 21(5): 476-480.

Foresti, G.L., 2002. Detecting elliptical structures in underwater images. Journal of Electronic Imaging, 11(1): 59-68.

Foresti, G.L., Micheloni, C. and Piciarelli, C., 2005. Detecting moving people in video streams. Pattern Recognition Letters, 26(14): 2232-2243.

Fu, G. and Huang, S., 1995. Computer-aided control in nonround process, Proceedings of SPIE - The International Society for Optical Engineering, pp. 184-189.

Goneid, A., ElGindi, S. and Sewisy, A., 1997. A method for the Hough Transform detection of circles and ellipses using a 1-dimensional array, Proceedings of the IEEE International Conference on Systems, Man, and Cybernetics, Orlando, Fl, pp. 3154-3157.

Griffin, G., Holub, A. and Perona, P., Caltech-256 object category database. California Institute of Technology, pp. http://authors.library.caltech.edu/7694.

Guil, N. and Zapata, E.L., 1997. Lower order circle and ellipse Hough transform. Pattern Recognition, 30(10): 1729-1744.

Hahn, K., Jung, S., Han, Y. and Hahn, H., 2008. A new algorithm for ellipse detection by curve segments. Pattern Recognition Letters, 29(13): 1836-1841.

Harker, M., O'Leary, P. and Zsombor-Murray, P., 2008. Direct type-specific conic fitting and eigenvalue bias correction. Image and Vision Computing, 26(3): 372-381.

He, Z., Tan, T., Sun, Z. and Qiu, X., 2009. Toward accurate and fast iris segmentation for iris biometrics. IEEE Transactions on Pattern Analysis and Machine Intelligence, 31(9): 1670-1684.

Heikkila, J., 1998. Moment and curvature preserving technique for accurate ellipse boundary detection. In: A.K. Jain, S. Venkatesh and B.C. Lovell (Editors), Proceedings of the International Conference on Pattern Recognition, Brisbane, Australia, pp. 734-737.

Ho, C.T. and Chen, L.H., 1995. A fast ellipse/circle detector using geometric symmetry. Pattern Recognition, 28(1): 117-124.

Hua, C., Wu, H., Chen, Q. and Wada, T., 2007. Object tracking with target and background samples. IEICE Transactions on Information and Systems, E90-D(4): 766-774.

Hwang, S., Oh, J., Tavanapong, W., Wong, J. and De Groen, P.C., 2006. Polyp detection in colonoscopy video using elliptical shape feature, Proceedings of the International Conference on Image Processing, San Antonio, TX, pp. II-465 - II–468.

Iles, P.J., Brodland, G.W., Clausi, D.A. and Puddister, S.M., 2007. Estimation of cellular fabric in embryonic epithelia. Computer methods in biomechanics and biomedical engineering, 10(1): 75-84.

Illingworth, J. and Kittler, J., 1988. A survey of the hough transform. Computer Vision, Graphics and Image Processing, 44(1): 87-116.

Ji, G.-R., Lu, B.-L., Chen, X. and Wang, J., 1999. Object searching in scale-space, Proceedings of the IEEE International Conference on Systems, Man and Cybernetics, Tokyo, Japan, pp. I-565 - I-570.

Ji, Q. and Haralick, R.M., 2001. Error propagation for the Hough transform. Pattern Recognition Letters, 22(6-7): 813-823.

Kasemir, K.U. and Betzler, K., 2003. Detecting ellipses of limited eccentricity in images with high noise levels. Image and Vision Computing, 21(2): 221-227.

Kawaguchi, T. and Nagata, R., 1998a. Ellipse detection using a genetic algorithm. In: A.K. Jain, S. Venkatesh and B.C. Lovell (Editors), Proceedings of the International Conference on Pattern Recognition, Brisbane, Australia, pp. 141-145.

Kawaguchi, T. and Nagata, R.I., 1998b. Ellipse detection using grouping of edgels into line-support regions, Proceedings of the IEEE International Conference on Image Processing, pp. 70-74.

Kayikcioglu, T., Gangal, A. and Ozer, M., 2000. Reconstructing ellipsoids from three projection contours. Pattern Recognition Letters, 21(11): 959-968.

Kim, E., Haseyama, M. and Kitajima, H., 2002. Fast and Robust Ellipse Extraction from Complicated Images, Proceedings of the International Conference on Information Technology and Applications, pp. 357-362.

Kovesi, P.D., MATLAB and Octave Functions for Computer Vision and Image Processing, pp. http://www.csse.uwa.edu.au/~pk/Research/MatlabFns/index.html.

Kumar, N., Kohlbecher, S. and Schneider, E., 2009. A novel approach to video-based pupil tracking. 2009 IEEE International Conference on Systems, Man and Cybernetics, SMC 2009, San Antonio, TX, pp. 1255-1262.

Kuno, Y., Okamoto, Y. and Okada, S., 1991. Robot vision using a feature search strategy generated from a 3-D object model. IEEE Transactions on Pattern Analysis and Machine Intelligence, 13(10): 1085-1097.

Liu, Y., Ikenaga, T. and Goto, S., 2007. Geometrical, physical and text/symbol analysis based approach of traffic sign detection system. IEICE Transactions on Information and Systems, E90-D(1): 208-216.

Lu, W., Tan, J. and Floyd, R.C., 2005. Fetal head detection and measurement in ultrasound images by a direct inverse randomized hough transform, Proceedings of the SPIE on Progress in Biomedical Optics and Imaging, pp. 715-722.

Lu, W. and Tan, J.L., 2008. Detection of incomplete ellipse in images with strong noise by iterative randomized Hough transform (IRHT). Pattern Recognition, 41(4): 1268-1279.

Mai, F., Hung, Y.S., Zhong, H. and Sze, W.F., 2008. A hierarchical approach for fast and robust ellipse extraction. Pattern Recognition, 41(8): 2512-2524.

Maini, E.S., 2006. Enhanced direct least square fitting of ellipses. International Journal of Pattern Recognition and Artificial Intelligence, 20(6): 939-953.

Masood, A., 2008. Dominant point detection by reverse polygonization of digital curves. Image and Vision Computing, 26(5): 702-715.

Matas, J., Shao, Z. and Kittler, J., 1995. Estimation of curvature and tangent direction by median filtered differencing. Lecture Notes in Computer Science, 974: 83-88.

Matson, W.L., McKinstry, H.A., Johnson Jr, G.G., White, E.W. and McMillan, R.E., 1970. Computer processing of SEM images by contour analyses. Pattern Recognition, 2(4).

McLaughlin, R.A., 1998. Randomized Hough transform: Improved ellipse detection with comparison. Pattern Recognition Letters, 19(3-4): 299-305.

McLaughlin, R.A. and Alder, M.D., 1998. The hough transform versus the upwrite. IEEE Transactions on Pattern Analysis and Machine Intelligence, 20(4): 396-400.

Meer, P., Mintz, D., Rosenfeld, A. and Kim, D.Y., 1991. Robust regression methods for computer vision: A review. International Journal of Computer Vision, 6(1): 59-70.

O'Leary, P., Harker, M. and Zsombor-Murray, P., 2005. Direct and least square fitting of coupled geometric objects for metric vision, pp. 687-694.

Prasad, D.K. and Leung, M.K.H., 2012. Polygonal Representation of Digital Curves, Digital Image Processing, Stefan G. Stanciu (Editor), InTech, pp. 71-90.

Prasad, D.K., 2011. A geometric technique for tangent estimation for digital curves. ForseLab-11-001, Nanyang Technological University, Singapore.

Prasad, D.K., Gupta, R.K. and Leung, M.K.H., 2011a. An Error Bounded Tangent Estimator for Digitized Elliptic Curves, Lecture Notes in Computer Science. Lecture Notes in Computer Science. Springer Berlin / Heidelberg, pp. 272-283.

Prasad, D.K. and Leung, M.K.H., 2010a. An ellipse detection method for real images, 25th International Conference of Image and Vision Computing New Zealand (IVCNZ 2010), Queenstown, New Zealand.

Prasad, D.K. and Leung, M.K.H., 2010b. Error analysis of geometric ellipse detection methods due to quantization, Fourth Pacific-Rim Symposium on Image and Video Technology (PSIVT 2010), Singapore, pp. 58 - 63.

Prasad, D.K. and Leung, M.K.H., 2010c. A hybrid approach for ellipse detection in real images. In: K. Jusoff and Y. Xie (Editors), 2nd International Conference on Digital Image Processing. SPIE, Singapore, pp. 75460I-6.

Prasad, D.K. and Leung, M.K.H., 2010d. Reliability/Precision Uncertainty in Shape Fitting Problems, IEEE International Conference on Image Processing, Hong Kong, pp. 4277-4280.

Prasad, D.K. and Leung, M.K.H., 2010e. Clustering of Ellipses based on their Distinctiveness: An aid to Ellipse Detection Algorithms, 3rd IEEE International Conference on Computer Science and Information Technology (ICCSIT 2010), Chengdu, China, pp. 292 - 297.

Prasad, D.K., Leung, M.K.H., Cho, S.Y. and Quek, C., 2011b. A parameter independent line fitting method, 1st IAPR Asian Conference on Pattern Recognition (ACPR 2011), Beijing, China, pp. 441-445.

Prasad, D.K. 2011c. Adaptive traffic signal control system with cloud computing based online learning, Eighth International Conference on Information, Communications, and Signal Processing (ICICS 2011), Singapore.

Princen, J., Illingworth, J. and Kittler, J., 1994. Hypothesis testing - a framework for analyzing and optimizing Hough transform performance. IEEE Transactions on Pattern Analysis and Machine Intelligence, 16(4): 329-341.

Procter, S. and Illingworth, J., 1994. A comparison of the randomised Hough transform and a genetic algorithm for ellipse extraction, Pattern Recognition in Practice IV. Elsevier, Amsterdam, pp. 449-460.

Qiao, Y. and Ong, S.H., 2007. Arc-based evaluation and detection of ellipses. Pattern Recognition, 40(7): 1990-2003.

Rosin, P.L., 1993a. Ellipse fitting by accumulating five-point fits. Pattern Recognition Letters, 14(8): 661-669.

Rosin, P.L., 1993b. A note on the least squares fitting of ellipses. Pattern Recognition Letters, 14(10): 799-808.

Rosin, P.L., 1996a. Analysing error of fit functions for ellipses. Pattern Recognition Letters, 17(14): 1461-1470.

Rosin, P.L., 1996b. Assessing error of fit functions for ellipses. Graphical Models and Image Processing, 58(5): 494-502.

Rosin, P.L. and West, G.A.W., 1992. Detection and verification of surfaces of revolution by perceptual grouping. Pattern Recognition Letters, 13(6): 453-461.

Rosin, P.L. and West, G.A.W., 1995. Nonparametric segmentation of curves into various representations. IEEE Transactions on Pattern Analysis and Machine Intelligence, 17(12): 1140-1153.

Salas, J., Avalos, W., CastaÃ±eda, R. and Maya, M., 2006. A machine-vision system to measure the parameters describing the performance of a Foucault pendulum. Machine Vision and Applications, 17(2): 133-138.

Ser, P.K. and Siu, W.C., 1995. Novel detection of conics using 2-D Hough planes, Proceedings of the IEE Vision, Image and Signal Processing, pp. 262-270.

Shen, Y., Yu, J. and Wang, Y., 2009. Fetal skull analysis in ultrasound images based on iterative randomized hough transform, Proceedings of the SPIE on Progress in Biomedical Optics and Imaging.

Shih, F.Y., Cheng, S., Chuang, C.F. and Wang, P.S.P., 2008. Extracting faces and facial features from color images. International Journal of Pattern Recognition and Artificial Intelligence, 22(3): 515-534.

Smereka, M. and Glab, G., 2006. Detection of pathological cells in phase contrast cytological images, Lecture Notes in Computer Science, pp. 821-832.

Soetedjo, A. and Yamada, K., 2005. Fast and robust traffic sign detection, Proceedings of the IEEE International Conference on Systems, Man and Cybernetics, pp. 1341-1346.

Sood, V., John, B., Balasubramanian, R. and Tandon, A., 2005. Segmentation and tracking of mesoscale eddies in numeric ocean models, Proceedings of the International Conference on Image Processing, pp. 469-472.

Takimoto, H., Mitsukura, Y. and Akamatsu, N., 2004. Face Identification Based on Ellipse Parameter Independent of Varying Facial Pose and Lighting Condition, Lecture Notes in Computer Science, pp. 874-880.

Tang, C.Y., Chen, Z. and Hung, Y.P., 2000. Automatic detection and tracking of human heads using an active stereo vision system. International Journal of Pattern Recognition and Artificial Intelligence, 14(2): 137-166.

Tsuji, H. and Matsumoto, F., 1978. Detection of ellipses by a modified Hough transformation. IEEE Transactions on Computers, C-27(8): 777-781.

Wang, C., Newman, T.S. and Cao, C., 2007. New hypothesis distinctiveness measure for better ellipse extraction, Lecture Notes in Computer Science, pp. 176-186.

Wang, M., Yang, J. and Liu, W., 2006. Extraction of line and rounded objects from underwater images. Journal of Harbin Institute of Technology (New Series), 13(5): 613-620.

Weisstein, E.W., CRC concise encyclopedia of mathematics, 2. CRC Press, Florida.

Worring, M. and Smeulders, A.W.M., 1993. Digital Curvature Estimation. Computer Vision and Image Understanding, 58(3): 366-382.

Wu, W.Y. and Wang, M.J.J., 1993. Elliptical object detection by using its geometric properties. Pattern Recognition, 26(10): 1499-1509.

Yip, R.K.K., Tam, P.K.S. and Leung, D.N.K., 1992. Modification of hough transform for circles and ellipses detection using a 2-dimensional array. Pattern Recognition, 25(9): 1007-1022.

Yip, R.K.K., Tam, P.K.S. and Leung, D.N.K., 1995. Modification of hough transform for object recognition using a 2-dimensional array. Pattern Recognition, 28(11): 1733-1744.

Yuasa, M., Yamaguchi, O. and Fukui, K., 2004. Precise Pupil Contour Detection Based on Minimizing the Energy of Pattern and Edge. IEICE Transactions on Information and Systems, E87-D(1): 105-112.

Yuen, H.K., Illingworth, J. and Kittler, J., 1989. Detecting partially occluded ellipses using the Hough transform. Image and Vision Computing, 7(1): 31-37.

Zaim, A., Quweider, M., Scargle, J., Iglesias, J. and Tang, R., 2006. A robust and accurate segmentation of iris images using optimal partitioning, Proceedings of the International Conference on Pattern Recognition, pp. 578-581.

Zhang, J.Z., Wu, Q.M.J. and Gruver, W.A., 2003. Active head tracking based on chromatic shape fitting. International Journal of Pattern Recognition and Artificial Intelligence, 17(4): 529-544.

Zhang, S.C. and Liu, Z.Q., 2005. A robust, real-time ellipse detector. Pattern Recognition, 38(2): 273-287.

Zhong, B., Ma, K.K. and Liao, W., 2009. Scale-space behavior of planar-curve corners. IEEE Transactions on Pattern Analysis and Machine Intelligence, 31(8): 1517-1524.

Zhou, J. and Shen, J., 2003. Ellipse detection and phase demodulation for wood grain orientation measurement based on the tracheid effect. Optics and Lasers in Engineering, 39(1): 73-89.

Zhou, Q.C., Liu, H.S., Kondrashkov, V.V., Li, G.D. and Lin, Y.H., 2009. Ellipse evolving common reflection point velocity analysis and its application to oil and gas detection. Journal of Geophysics and Engineering, 6(1): 53-60.

8

Detection and Pose Estimation of Piled Objects Using Ensemble of Tree Classifiers

Masakazu Matsugu, Katsuhiko Mori, Yusuke Mitarai and Hiroto Yoshii
Canon Inc.
Japan

1. Introduction

Detection and pose estimation of 3D objects is a fundamental machine vision task. Machine vision for bin-picking system (Figure 1 (a)), especially for piles of objects, is a classical robot vision task.

To date, however, there has been only limited success in this longstanding problem (e.g., picking piled objects), and, to the best of our knowledge, existing algorithms (e.g., Drost, et al., 2010; Ulrich et al., 2009; Hinterstoisser et al., 2007) fall short of practical use in automatic assembly of electronic products composed of various parts with differing optical and surface properties as well as with differing shapes and sizes. We found that even the state-of-the-art, commercially available machine vision software cannot be practically used for picking such piles of parts with unknown pose and occlusion. Specifically, as exemplified in Figure 1 (b), for black (or white)-colored and untextured parts with some degree of complexity in shape, conventional methods turned out to be of little use.

In this chapter, we present a potential solution to this classical, unsolved problem (i.e., Detection and pose estimation of each of piled objects) with an efficient and robust algorithm for object detection together with 3D pose estimation for practical use in robot vision systems. We consider the detection and 3D pose estimation as a classification problem (Lepetit & Fua, 2006) which constitutes a preprocessing stage of subsequent model fitting for further precise estimation (Tateno et al., 2010), and explore the use of ensemble of classifiers in a form of Random Forests (Lepetit & Fua, 2006; Gall & Lempitsky, 2009; Shotton et al., 2011) or Ferns (Bosch et al., 2007; Oshin et al., 2009; Özuysal et al., 2010) that can handle multi-categories.

Based upon sliding window approach, we formulate the problem as classifying a set of patches of input image (local regions) into a sufficient number of conjunct pose and location categories which are supported by distributed representation of leaf nodes in trees.

Main contributions of this paper are 1) spatially restricted and masked sampling (SRMS) scheme, 2) voting through local region-based evidence accumulation for pose categories, 3) cancellation mechanism for fictitious votes (CMFV) suggesting ill-conditioned and degenerated sampling queries for pose estimation, altogether leading to robust detection and 3D pose estimation of piles of objects.

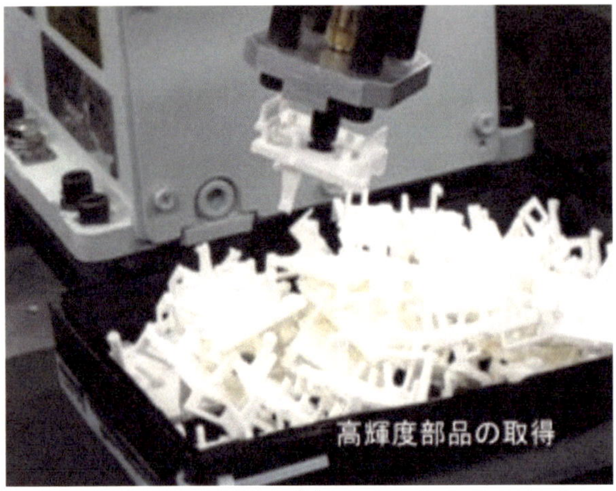

(a)

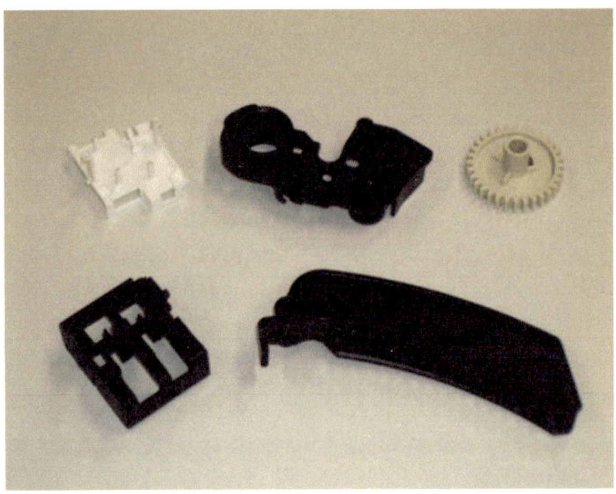

(b)

Fig. 1. (a) Picking system for piles of parts. (b) Parts with various shape and surface properties.

2. Basic formulation

Detection task of piled objects composed of the same category parts as shown in Figure 1 (b), inherently requires following three properties. 1) Robustness to occlusion, 2) Robustness to high background clutters (i.e., noisy clutters are by themselves some other objects of the same class in the neighborhood of a specific object to be detected), 3) Robustness to drastic variability of object appearance due to varying pose and illumination change, especially for objects with higher specularity.

In this section, we show details about basic formulation of the proposed algorithm. We show here a new patch based method for object localization as well as pose estimation, which is a class of generalized Hough transforms and similar in spirit with Hough Forests (Gall & Lempitsky, 2009), and the basic strategy for the improvement is given in the next section.

Through construction of trees in the training, pose classes of an object are defined in a form of tree-like structure and described by a set of local cues inside patches. For each tree, a set of patches P_i, local regions in an edge-enhanced feature image (given in subsection 3.3) can be defined as $\{P_i = (F_i, C_i)\}$, where F_i is the appearance of the local feature image, C_i is the label of the patch, $C_i = (l_i, Pose_i)$ where l_i denotes its location inside the object and $Pose_i$ is associated pose class. Whole sets of local cues $\{t(F_i)\}$, in the entire trees associated with particular pose of an object, constitute codebooks or dictionary of particular poses. Specifically, the local cues $t(F_i)$ are binary data given by comparison of two feature values at given paired locations (p_1,q_1) and (p_2,q_2) and defined as:

$$t_{p_1,q_1,p_2,q_2}(F_i) = \begin{cases} 0, & \text{if } F_i(p_1,q_1) < F_i(p_2,q_2) \\ 1, & \text{otherwise} \end{cases} \quad (1)$$

2.1 Building ensemble of trees

In the training phase, construction of trees goes recursively by setting patches as well as paired locations (sampling points) inside each patch. For a given number L of trees, we perform L sessions of training and prepare a set of feature images for training.

The feature image is given by preprocessing (explained in subsection 3.3) the input image to obtain edge-enhancement, while suppressing noise. Since our pose classification is succeeded by model fitting for further precise estimation, total number of pose categories is determined by the resolution and accuracy requirement on the initial pose estimation imposed by the subsequent model fitting process (Tateno, et al., 2010), and the number could be huge (see Section 4). At the beginning of each training session, set of patches are first randomly generated subject to the condition that their locations are inside the object, and sampling points are probabilistically set according to the new scenario given in subsection 3.1. For example, in Figure 2 we have four patches for respective five pose categories, which amount to 20 training images.

A *leaf* node is the one which contains less than a fixed number of patches or its depth is the maximum value a-priori set, and if it contains no patch, we call it *null* or *terminal* node. In the training, we set a maximum depth of node constant among trees, and starting from the *root* node, the node expansion continues until it reaches the maximum depth or terminal node. At each node of a tree, if it is not the *terminal* nor *leaf* node, binary tests (1) are performed for a set of patches inside the node, and they are partitioned into two groups which are respectively fed to two child nodes (Figure 3).

We do not have a strict criterion on the training performance, a criterion on good codebooks being generated. One of reasonable criteria is that many of leaf nodes should have only one patch (i.e., single pose category) so that uniqueness of distributed representations is ensured, and another criterion is the diversity of sampling points so that spatial distribution of query points are not biased to some limited local area of the object. For the second criterion, because of geometrical triangulation principle, it is reasonable to consider that estimated pose category at wide spread positions, many of which supports the same category, is more credible than those from narrowly spread positions.

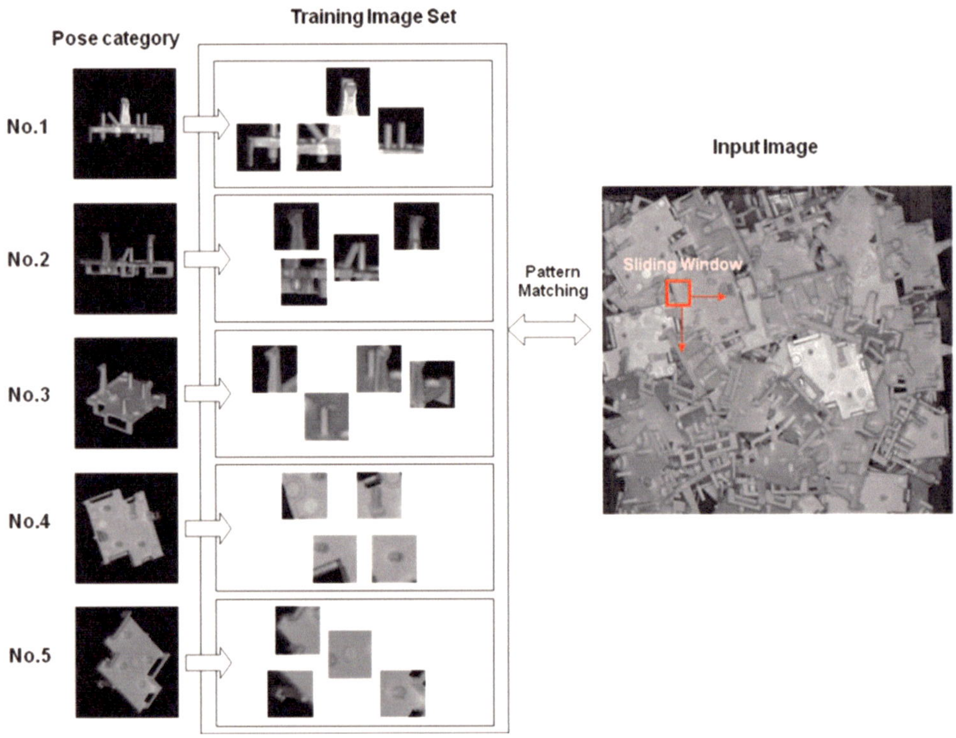

Fig. 2. (left) Schematic patch images for training; (right) Sliding window for matching.

2.2 Detection and pose estimation by ensemble of trees

In sliding window approach, we raster-scan the entire image using a local window of appropriate size, and at each position, parts detection together with pose estimation is done using the ensemble of classifiers .

Decision about the classification is done based on voting the outputs of leaf nodes among trees, followed with thresholding. The voting stage accumulates the supporting, local evidences by collecting outputs of leaf nodes among trees which signify the same pose category. Figure 4 schematically shows concentration of specific pose category as the result of correct voting, and no concentration for other pose classes. For the total number of L trees, voting for class j is performed for each pose category, yielding score $S(j)$ as:

$$S(j)=\sum_{t}^{L}\int_{r}\delta\left(C_{j,t}(r),1\right),$$

where r denotes relative position vector of a patch directing to the center of object, $C_{j,t}(r)$ is 1, if, in the tth tree, class label j is detected by the patch assigned with position vector r, and $C_{j,t}(r)$ is 0, if otherwise. $\delta(a,b)$ is 1 for $a = b$, and 0 for otherwise. In practice, we use the following weighted voting given as:

$$S_r(j)=\sum_{t}^{L}\sum_{k}F(r_k,r)\delta\left(C_{j,t}(r_k),1\right),$$

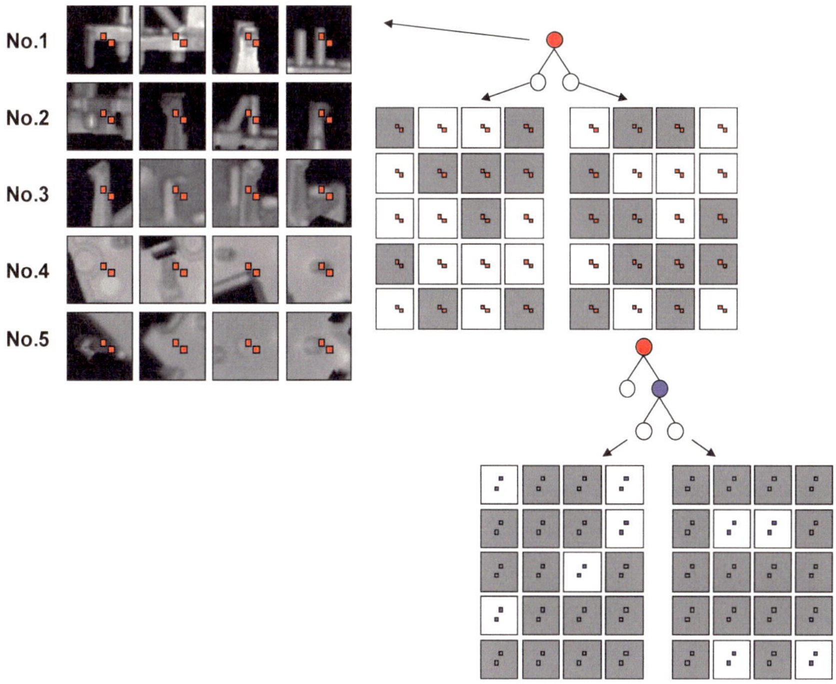

Fig. 3. Patch data partitioning. Based on a comparison of two pixel values, divide a set of patches into two groups with the same content (i.e., left or upper pixel value is larger or not than the other).

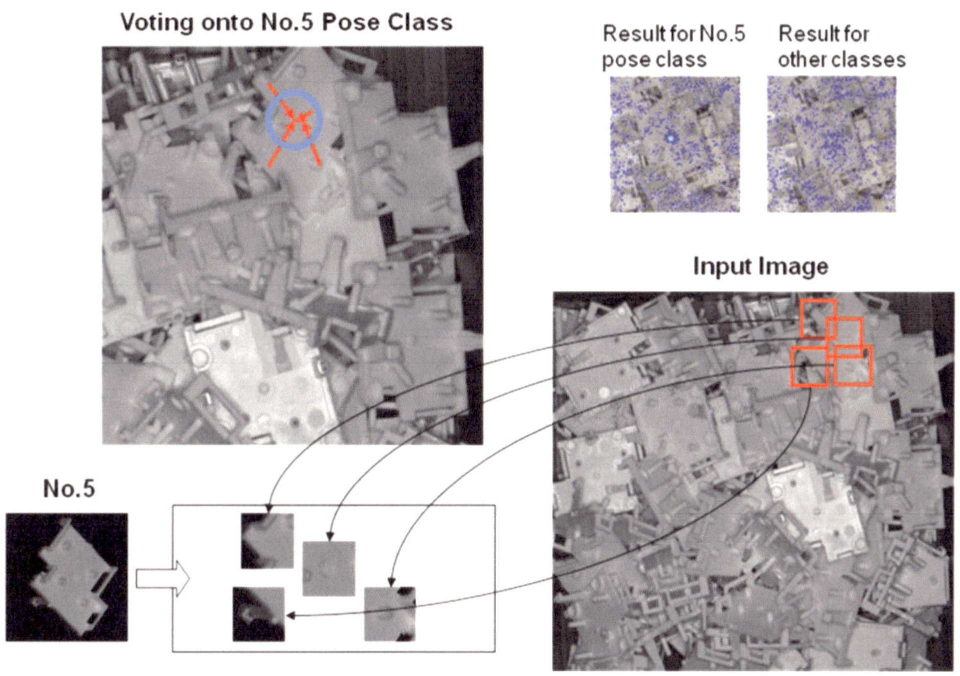

Fig. 4. Observed concentration of voting for correct class.

where $S_r(j)$ is the score for class j at the position r, and r_k is the location of patch k, $F(r_k)$ denotes weighting function with belle shaped envelope. Location of the object detected can be found by taking summation of position vectors r for patches identified as having the same pose category $C_j(r)$ under the conditions that the score $S(j)$ is maximum or ranked K (K is ordinarily set as 1 or 2) or above among other categories and that S is also above a certain threshold S_0. If K is set as 2, we select the most plausible estimate in the subsequent model fitting process (Tateno, et al. 2010).

3. Patch-based approach in ensemble of tree classifiers

First, to deal with three issues given in Section 2 faced in the detection task of piled objects, we incorporate the SRMS scheme (subsection 3.1) in ensemble of classifiers, a class of Hough Forests (Gall & Lempitsky, 2009). This sampling scheme together with CMFV (subsection 3.2) turned out to be very effective for enhancing robustness to occlusion as well as background clutters. The proposed method uses training images composed of only positive data because of SRMS as well as patch data generation inside the object. This is in contrast to Hough Forests which handle both positive and negative data in a supervised learning. In the second, we perform a series of feature extraction (subsection 3.3) for edge enhancement while suppressing noise, namely bilateral filter and gamma correction for smoothing, Laplacian filter for edge extraction, and Gaussian filters for blurring to form a channel of feature images fed to ensemble of trees.

3.1 Spatially restricted and masked sampling (SRMS)

Hough Forests (Gall & Lempitsky, 2009), a class of both random forests and generalized Hough transform, introduced spatial restriction in a form of patch (e.g., local region in a image) so that sampling queries are generated inside respective patches. In addition to this patch-based framework, we introduce here another spatial restriction in a form of mask which is the silhouette of object with particular pose. Mask defined at each node is used to impose probabilistic restriction onto queries so as to be inside the object with a range of poses. In contrast to the proposed approach, Hough Forest does not restrict location of patches in image. Thus, in the training phase to construct trees, each patch is set at random so that its centre position shall be inside the silhouette of objects to be detected, while respective sampling pairs of points are probabilistically set based on the conjoint mask data given as follows.

This combination of locality restrictions in generating sampling queries helps to enhance robustness against occlusion. We define a *conjoint mask* as a conjunction of silhouettes of objects with different poses in the corresponding node. A node in the tree generally contains multiple classes of pose, and at each node, pose categories are partitioned into two sets of data. Those partitioned data are fed to subsequent nodes, where partitioning follow. Here we show several ways of generating the mask data resulting from conjoint of *composite mask* data at each node of a tree. Here, we use the term, *composite mask*, for one of masks in the node. One way of conjoining silhouettes is taking AND operation on them for a given node in the tree. The resulting mask data M is thus given by

$$M = \left(\prod_{k=1}^{N} M_1^k, \prod_{k=1}^{N} M_2^k, \ldots, \prod_{k=1}^{N} M_n^k \right)$$

where N is the number of mask images (i.e., number of pose classes in the current node), $M^k=(M_k^1, M_k^2, \ldots, M_k^N)$,and M_k^j is binary data at the kth location for the jth class of pose categories in the current node, n is the dimension of the mask image (i.e., number of pixels).

Another way of constructing mask data for a given node is to take OR operation on them:

$$M = \left(\sum_{k=1}^{N} M_1^k, \sum_{k=1}^{N} M_2^k, \ldots, \sum_{k=1}^{N} M_n^k \right)$$

Yet another way of constructing mask data in the node is to take ORs of patch data m_k^j within a composite mask image:

$$M = \left(\sum_{k=1}^{N} m_1^k, \sum_{k=1}^{N} m_2^k, \ldots, \sum_{k=1}^{N} m_n^k \right) \tag{2}$$

Here, M in eqn. (2) does not give the conjoint data of mask silhouettes, but it is sufficient for us since sampling pair data are to be generated inside those patches. Next, we show our scheme of probabilistic generation of sampling queries. For conjoint mask data M at each node, we define a probability density function P by simple normalization, given as follows.

$$P = \frac{\left(\sum_{k=1}^{N} m_1^k, \sum_{k=1}^{N} m_2^k, \ldots, \sum_{k=1}^{N} m_n^k \right)}{\sum_{j=1}^{n} \sum_{k=1}^{N} m_j^k} \tag{3}$$

Then we generate pairs of sampling points based on the probability density function (3) which are guaranteed to be inside either of patches. This sampling scheme together with CMFV in the next subsection turns out to be very effective for enhancing robustness to occlusion as well as background clutters.

3.2 Cancellation mechanism for fictitious votes (CMFV)

In automatic assembly line of manufacturing, containers are used for supplying parts. Those containers are usually kinds of trays or boxes made of rectangular planes. In practice, linear portions of a tray cause detection errors if objects to be detected are made up of many or longer linear portions. In such cases, it is not surprising to confuse linear portion of a tray as a part of object to be detected, since our method as well as randomized tree based approaches are based on accumulation of local evidence (e.g., comparison of feature values at two sampling points). This confusion resulting from such degeneration is reminiscent of so-called aperture problem in computer vision.

Proposed cancellation mechanism is intended to alleviate such confusions. As in generalized Hough transform, we perform voting as local evidence accumulation in which each of evidence is obtained through sliding window. In practice, this local evidence accumulation can cause degenerated results which cannot be in principle disambiguated. As a result, we may have excessive concentration of classification results with the same localized category (i.e., posture observed at particular location of the object) at particular locations. In such cases, we consider the result as fictitious. The criterion for this singularity is empirically set as a threshold, assuming appropriate probability density function. For example, if the number of patches used for training is much larger than the total sliding numbers, then we can assume Poisson distribution, and for other cases, binomial distribution. Details about the threshold and probability density will be given in Section 4.

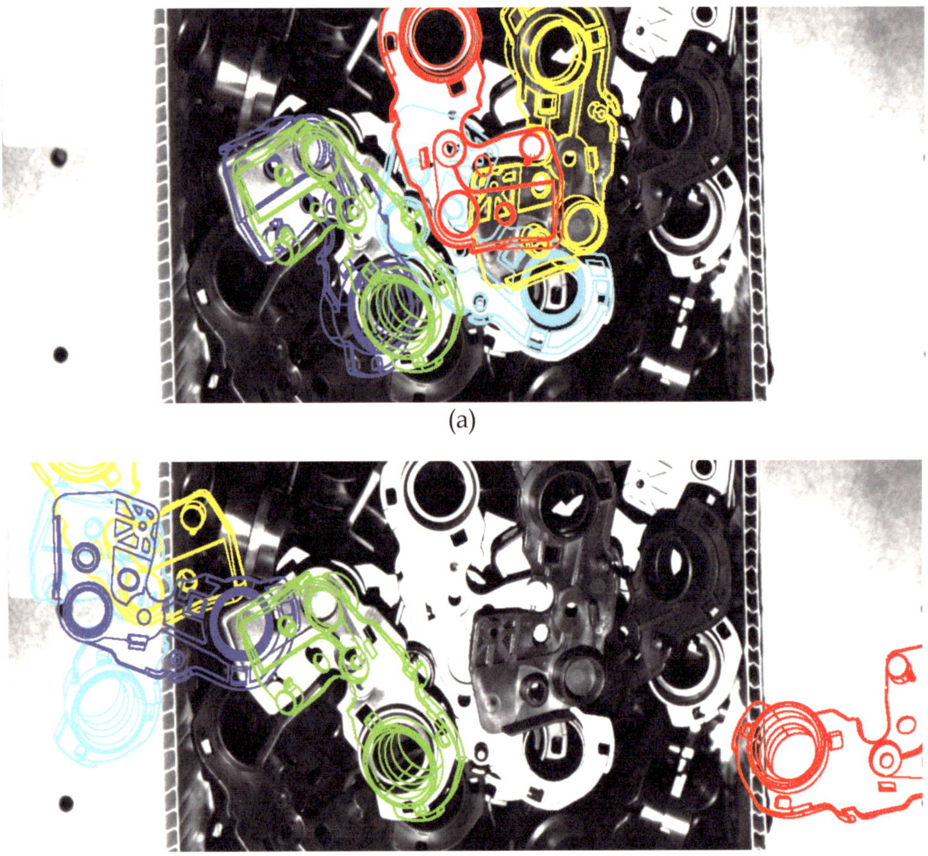

(a)

(b)

Fig. 5. (a) Result with CMFV, (b) Result without CMFV.

3.3 Pre-processing: Feature extraction

We perform a series of pre-processing to extract feature images as input to the ensemble classifiers. A typical feature image is an edge image. This processing includes edge extraction by Laplacian filter, blurring with Gaussian filters, and some other non-linear processing. Examples of extracted feature images are shown in Figure 6. Since edge extraction tends to enhance noise, blurring process is necessary for the suppression of noise, however, it could affect the performance, since contrast of edge image is degraded.

2D features thus obtained with appropriate parameters are very important and they significantly influence the performance of randomized tree-based classifiers. These set of operations turn out to be important for robustness and precision of final results (see Section 4).

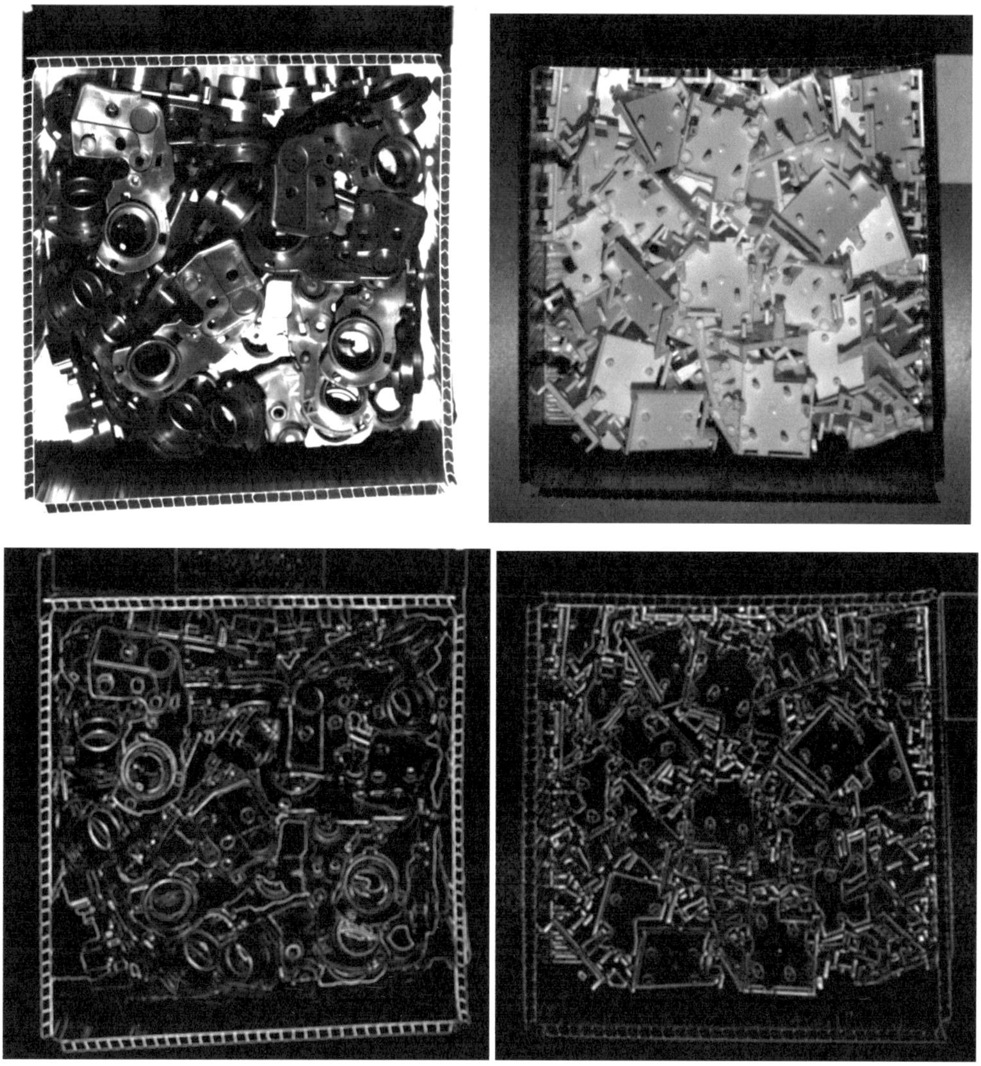

Fig. 6. Feature image obtained by edge extraction and blurring. Upper pictures are input images and lower ones are corresponding feature images as input to the classifier ensemble.

Another possible feature used for the detection task is 2.5 D map (depth map data) obtained by any three dimensional measurement (e.g., stereo vision, triangulation by structured pattern projection, TOF, etc.) method. Various features can be used, in principle, as channels of input to the ensemble classifiers. In this paper we confine to 2D edge-enhanced image as input.

4. Experiments

We used five classes of parts in printers (i.e., inkjet and laser beam printers) for training and testing. Those parts are of plastic mold and many of them are either black or white. Images of parts are taken by Canon EOS Kiss X2. Training image is taken for a single object with

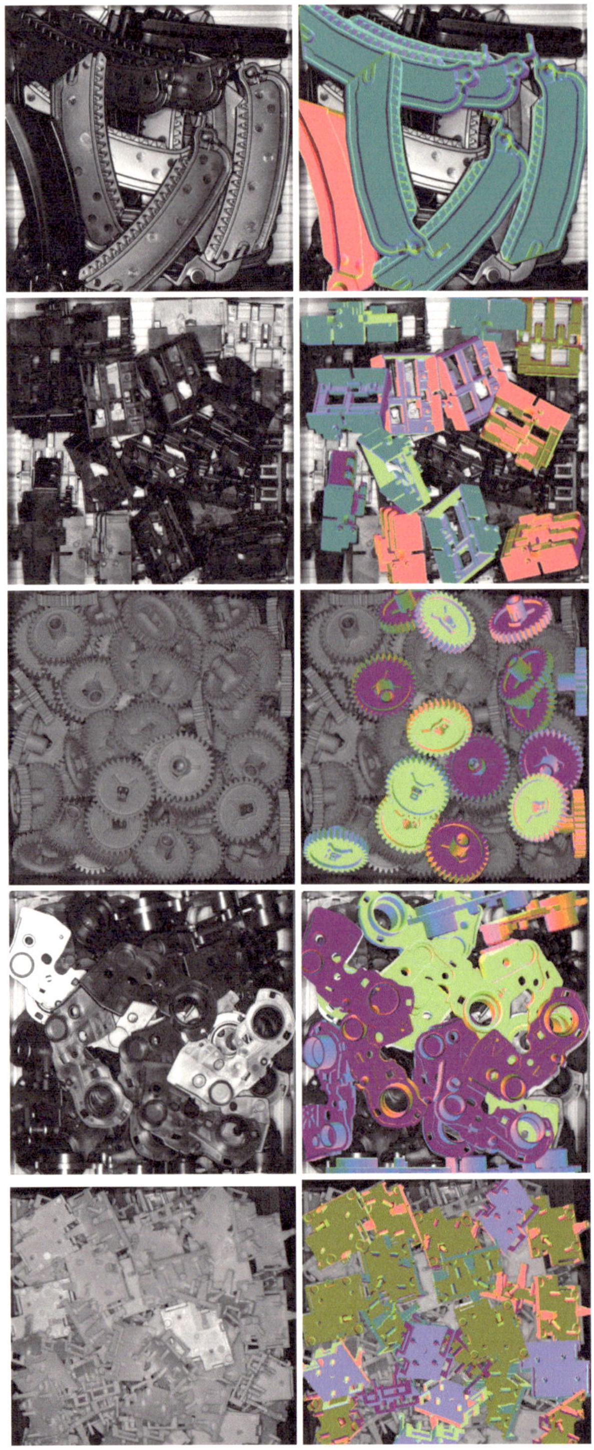

Fig. 7. Detection results for various objects. Correctly detected parts are shown by superimposed CAD image.

particular pose in a flat background (Figure 2 left). The number of pose categories is dependent on the resolution and accuracy requirement as initial estimate of pose which is given as input to the model fitting. Based on experiments, we empirically set the number of basic pose as follows. Pose categories necessary for the initial estimate in the model fitting are found to be defined by every 162 viewpoints evenly distributed on a geodesic dome and it was found necessary to discriminate poses for every 8 degree in-plane rotation, which results in total number of poses, 162 x 45 = 7620. In the training, we set 100 patches for each pose category. The size of input image is 660 x 640 pixels, and the size of patches is 50 x 50 pixels which is set constant for the five parts. Maximum depth of trees is empirically set from 15 to 30, which is dependent on the shape of parts.

Feature extraction was obtained by Laplacian filtering followed with blurring using Gaussian filter as shown in Figure 6. After the feature extraction, ground truth data given by (1) are taken for the respective patch images using queries generated by the probabilistic sampling in subsection 3.1.

4.1 Detection results

We show some detection results for piles of parts in Figure 7 obtained using the proposed SRMS as well as CMFV (Section 3) with the number of trees 32. We indicate here correct detections by superimposing CAD data of corresponding pose category. We do not use depth map data at all in the detection as well as training process.

4.2 Benchmarking

We compared the proposed method with the state-of-the-art, commercially available machine vision software (*HALCON* 9.0 produced by company *MVTec*). Technology related with the reference software can be found in Ulrich et al. (2009), which relies on edge-based fast matching scheme. Here we show some results in Figure 8. As is evident from this figure, it is very difficult for the reference software to detect and estimate 3D pose in the case of parts with white surface properties. Moreover, for black parts in Figure 1 (b) it was entirely unable to detect.

Since our method as well as the reference software *HALCON* in this comparative experiment is 2D-based, it is essentially hard to estimate rotation angle in depth. Our criterion for correct detection is based on the requirement set by model fitting process, which is given by allowable error in position and pose. Maximum allowable error for 'correct' result was given by approximately 10% of size in terms of positional error and approximately 10 deg. in terms of in-plane angle error measured in the image.

Shown in Figure 9 is a kind of 'RPC' curves for varying number of trees in the case of the piled parts shown in the upper picture in Figure 8, the reference software *HALCON* could detect up to only three parts, whereas proposed method could detect increasing number of parts on the order of 20 with growing number of trees.

For the five classes of parts shown in Figure 1 (b), the number of correctly detected parts and average detection time are as follows. It is clear that the proposed method outperforms the reference software *HALCON* in terms of precision and detection time. For fair comparison, we set the same criterion on correct detection.

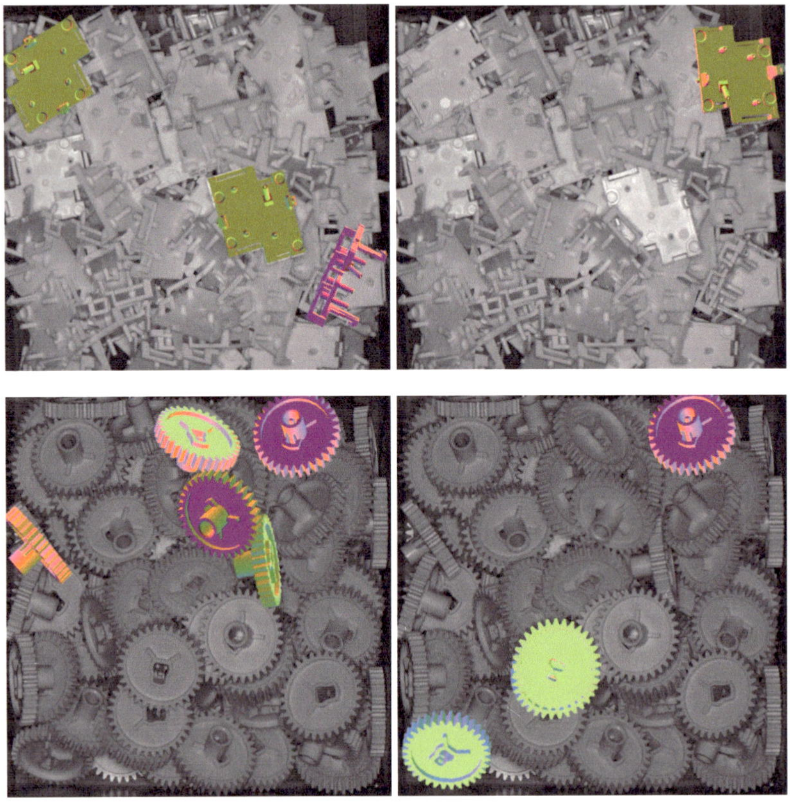

Fig. 8. Results obtained for the reference software *HALCON (9.0)*.

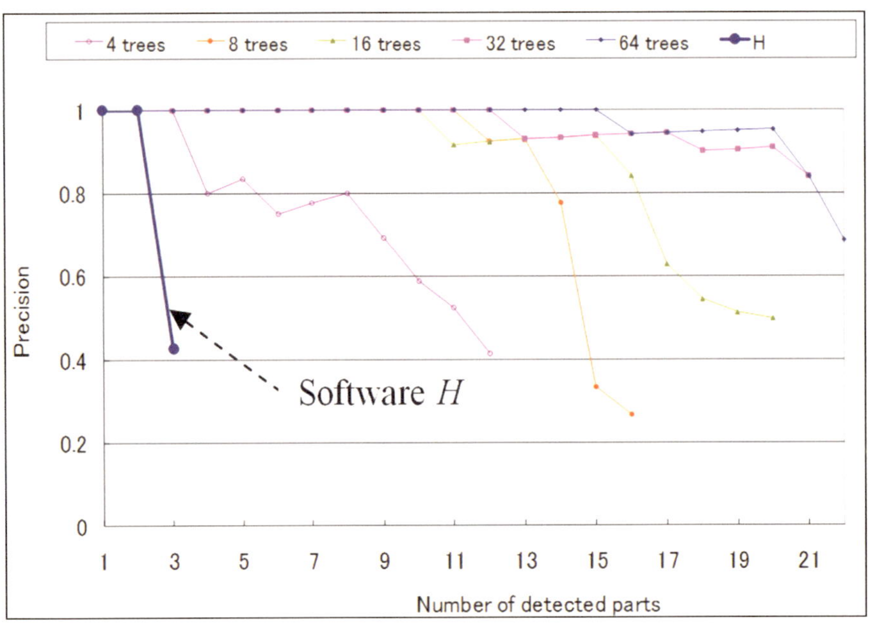

Fig. 9. RPC-like curve obtained for the reference software *HALCON (9.0)*.

	Class 1	Class 2	Class 3	Class 4	Class 5	Detection time (s)
Proposed alg.	20	10	14	10	5	5.8
Ref. Software H	3	0	5	0	0	30.4

In the table above, we show processing times as compared with the reference software (*HALCON*). It clearly shows that the proposed algorithm significantly outperforms the reference software in terms of processing speed.

5. Discussion and conclusions

In this chapter, we proposed a new algorithm based on ensemble of trees for object localization and 3D pose estimation that works for piled parts. We define pose estimation as classification problem which provides initial estimate for the subsequent model fitting processing to obtain further precise estimation.

One important aspect of object detection in the bin-picking task is that it is sufficient to localize and estimate the pose of one 'adequate' object for picking. In fact we used the number of parts detected as measure of 'recall' in the RPC-like curve (Figure 9). The proposed method significantly outperformed the state of the art, commercially available software in terms of precision and processing speed.

6. Acknowledgment

We appreciate T. Saruta, Y. Okuno, and M. Aoba for their efforts in obtaining various results.

7. References

Amit, Y. & Geman, D. (1997) Shape Quantization and Recognition with Randomized Trees, *Neural Computation*, Vol.9 No.7, pp.1545-1588.

Bosch, A., Zisserman, A., & Munoz, X. (2007) Image Classification using Random Forests and Ferns, *Proceedings of ICCV'07*

Drost, B., Ulrich, M., Navab, N., & Ilic, S. (2010) Model Globally, Match Locally and Robust 3D Object Recognition, *Proceedings of CVPR'10*

Gall J. & Lempitsky V. (2009) Class-Specific Hough Forests for Object Detection, *Proceedings of CVPR'09*

Hinterstoisser, S., Benhimane. S., & Navab, N. (2007) N3M: Natural 3D Markers for Real-Time Object Detection and Pose Estimation, *Proceedings of ICCV'07.*

Lepetit, V. & Fua, P. (2006) Keypoint Recognition Using Randomized Trees, *IEEE Transactions on Pattern Analysis and Machine Intelligence*, Vol. 28, No. 9, pp. 1465-1479.

Oshin, O., Gilbert, A., Illingworth, J., & Bowden, R. (2009) Action Recognition using Randomized Ferns, *Proceedings of ICCV2009 Workshop on Video-oriented Object and Event Classification.*

Özuysal, M., Calonder, M., Lepetit, V., & Fua, P. (2010) Fast Keypoint Recognition Using Random Ferns, *IEEE Transactions on Pattern Analysis and Machine Intelligence*, Vol. 32, No. 3, pp. 448-461.

Shotton, J., Fitzgibbon, A., Cook, M., Sharp, T., Finocchio, M., Moore, R., Kipman, A., & Blake, A. (2011) Real-Time Human Pose Recognition in Parts from Single Depth Images, *Proceedings of CVPR'11*

Tateno, K., Kotake, D., & Uchiyama, S. (2010) A Model Fitting Method using Intensity and Range Images for Bin-Picking Applications (in Japanese), *Proceedings of Meeting on Image Recognition & Understanding 2010*

Ulrich, M., Wiedemann, C., & Steger, C. (2009) CAD-Based Recognition of 3D Objects in Monocular Images, *Proceedings of ICRA'09*

Yoshii, H., Okuno, Y., Mitarai, Y., Saruta, T., Mori, K., & Matsugu, M. (2010) Parts Detection Algorithm using Ensemble of Tree Classifiers (in Japanese), *Proceedings of Meeting on Image Recognition & Understanding 2010*

9

Characterization of Complex Industrial Surfaces with Specific Structured Patterns

Yannick Caulier
Fraunhofer Institute for Integrated Circuits IIS, Erlangen, Germany

1. Introduction

Recent researches have demonstrated the importance of structured light patterns for use in the quality control of industrial workpieces. These researches have been focused on the adaptation of the projected light patterns and the direct interpretation of the recorded scenes by means of image content description methods. The novelty of these investigations relies on the fact that the stripe patterns permit at the same time the visual enhancement of the relevant information and a significant reduction of the amount of data to be processed. Such an approach therefore satisfies the major conditions inline inspection systems must fulfill: the robustness in terms of high signal to noise ratio and the low computational costs in order to achieve high inspection throughputs.

The major purposes of this chapter are (i) to give an overview of the actually achieved research results concerning the surface characterization based on the projection and the direct interpretation of structured light patterns, and (ii) to demonstrate that this approach serves the characterization of complex industrial surfaces. The whole quality control process in case of the industrial inspection is addressed. For each main element of the processing chain, a focus on the major achievements is provided: the projection and adaptation of specific stripe patterns (data generation), the segmentation and characterization of these adapted patterns (data processing), and the classification of the corresponding surfaces (data interpretation). This chapter ends by proposing a possible generalization method and gives important further research directions in order to address the inline characterization of complex free-form surfaces.

This chapter is organized into three paragraphs. Paragraph "Data Generation" tackles two possible illumination techniques for the generation of structured patterns. Also the recording of regular patterns in case of complex surface geometries is addressed. The automatic segmentation of disturbed stripe regions is described in paragraph "Data Processing", which also introduces the considered three feature sets for stripe image description. Finally, an application example in case of cylindrical surfaces and its generalization for complex geometries is described in the last paragraph "Data Classification".

In order to consider real-time inline inspection requirements, all the experiments were validated by means of industrial image datasets. Important aspects, such as high robustness against varying recording conditions but also fast data processing for real-time applications are considered.

2. Data generation

The inspection problem is primary tackled under its optical and physical aspects so that the purpose is to define the optimal data generation in case of structured light based surface inspection. This paragraph is therefore dedicated to the optimization possibilities in case of the generation of optimal stripe structures for the inspection of complex industrial surfaces.

At first, the chose of the appropriate illumination technology is addressed. Two different approaches, a "transmission"-based and "collimation"-based are described and compared. It is demonstrated that the latter is more appropriate for the visual enhancement of geometrical surface deformations on semi-reflective surfaces. The generation of adapted, "inverse patterns", is tackled afterwards. It is demonstrated how far pattern adaptation improves the visual interpretation of geometrical complex surfaces.

2.1 Defining the adequate illumination

2.1.1 Generalities

In the optical inspection domain, the observation of surfaces having different reflection coefficients or various geometries for quality control or metrology purposes is done by means of specific lighting approaches. The key point, and common process for all methods, is to *visually enhance* and characterize the relevant information. The chose of the adequate lighting is task dependant and must be defined according to the surface characteristics (reflectivity, geometry).

Within this context, the use of structured light patterns to reveal geometrical and/or textural surface characteristics has a broad range of applications. While deflectometric approaches (fringe structure projection) are dedicated to specular surface inspection, bright- or dark-field methods (projection of collimated light) can be perfectly suited to matt surface quality control (Abouelela, 2005). However, different techniques to generate such light patterns to be projected exist. We might distinguish between two different light projection approaches, a general one called "transmission", and a more specific one named "collimation". The formalism used here is based on the physical generation principle of the stripe patterns. Each illumination is described in detail, so that the geometrical arrangement and the optical properties of the illumination's main elements are tackled.

2.1.2 The transmission and the collimation approaches

The "transmission" based fringe projection technique is the mostly used and developed within the computer vision community. It can consist in the transmission of diffusing light trough a light-transmissive structure or in the transmission of a structured light pattern through a diffusing element. In the last case, structured light patterns can be produced by a LCD (Liquid Crystal Display), a DMD (Digital Micromirror Device), or a DOE (Diffractive Optical Element) device. The principle of the "collimation" is to direct incoming light with, e.g. a 3D fringe selection object (Caulier, 2007), or directional LEDs. Fig. 1 shows the two fringe pattern generation principles with image examples.

Both lighting techniques consist of two different parts: a diffuse illumination and a pattern generation element which filters the light rays using "transmission" or "collimation" techniques. The depicted examples demonstrate that both illuminations lead to similar fringe structures if similar geometrical deformations on specular surfaces are considered. The case of two different deformations is tackled here: dents (concave) and blisters (convex). Even if the images depict different surfaces, the red bold marked surface regions leading to visible fringe perturbations clearly show the similarity of both lighting techniques, in case of the visual enhancement of geometrical structures.

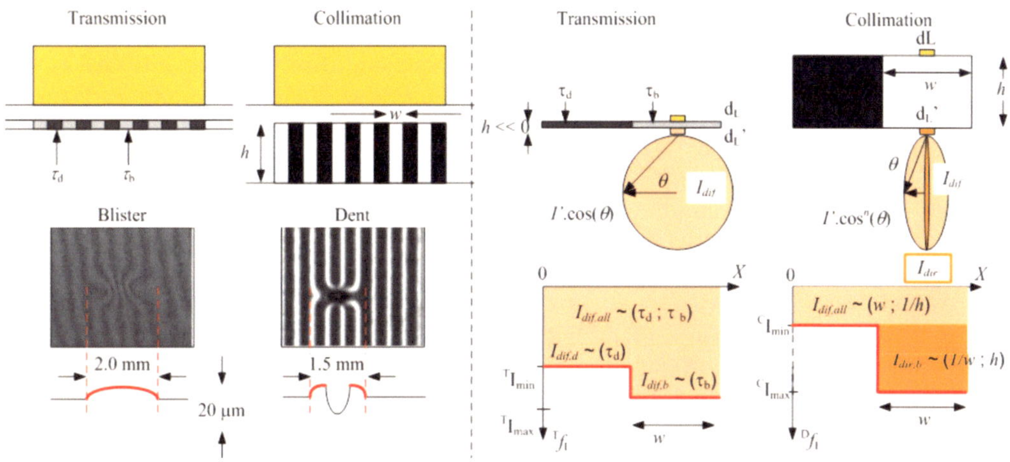

Fig. 1. Principle of the two (i) "transmission" and (ii) "collimation" approaches (left). The theoretical explanation is done for an elementary light source d_L (right). For the transmission technique, the light distribution remains lambertian, whereas for the collimation one, the light intensity profile is more directional. The directionality degree is proportional to the exponent n of the cosine function, where n varies from 0 to infinity. Theoretical intensity profiles ${}^{T}f_l$ and ${}^{D}f_l$ for both lightings are depicted.

However, if both illuminations permit similar visual enhancements, the depicted images in Fig. 1 show that the recorded structures are depicted differently, i.e. that the contrast of the light structures is different. The transmission-based lighting seems to produce "smoother" structures than the collimation-based one. This is a fundamental difference which has a direct influence of the processing methods and also on the considered inspection requirements. In the following, both illuminations are theoretically described.

A first simplification hypothesis consists of considering that the diffuse illumination placed before the pattern generation element is an ideal lambertian light source. Thus, the light profile of an elementary illumination element d_L placed *before* the pattern generation can be modeled by a $cos(\theta)$ function, where θ is the angle between the direction of observation and the normal of d_L. The light profile of an elementary illumination element d_L' *after* the pattern depends of the properties of the structured light filtering element. Light intensity profiles of both models can be expressed with a $cos^n(\theta)$ function, where n is a factor modeling the light directivity. n=1 for the "transmission" approach and n>=1 for the collimation approach, so that higher values of n are synonymous a higher directivity. If the $cos^n(\theta)$ function models the shape of the structured light, the light intensity can be modelled by two one-dimensional functions ${}^{T}f_l(x)$ and ${}^{D}f_l(x)$, where x is the spatial position along the fringe structure.

These profiles depend on fringe pattern geometrical and physical parameters, which are the transmission factors of bright and dark fringes τ_b and τ_d for the "transmission solution", the height h, and the fringe width w for both solutions, where $\tau_b > \tau_d$, $\{\tau_b; \tau_d\}$ in [0:1]) and $h <<$ 1 for the "transmission" one. Both intensity profiles after fringe light source are obtained by summing for elementary surface $d_L'(x)$ at point x the amount of light coming from all neighboring points on the X-axis. The ${}^{T}f_l(x)$ and ${}^{D}f_l(x)$ curves in Fig. 1 vary between two values ${}^{k}I_{min}$, ${}^{k}I_{max}$, where k stands for transmission ^{T} and directed ^{D}. As stated before, the "transmission" lighting consists of a purely diffuse ${}^{T}I_{dif}$, whereas the "collimation" is made of a diffuse ${}^{C}I_{dif}$ and a directional ${}^{C}I_{dir}$ part. ${}^{k}I_{dif,all}$ is a diffuse part induced by *all* the dark and bright structures, while ${}^{k}I_{dif,d}$, ${}^{k}I_{dif,b}$, and ${}^{k}I_{dir,b}$ are diffuse and directed parts induced by the *considered* dark and bright structures. All these values, and so the bright and dark fringe contrast in the images, depend on the illumination specific parameters τ_b, τ_d, w, h.

One of the important aspects of the projected fringe structure is to optimally visually enhance the relevant surface information, whether for direct interpretation (qualitative) or for reconstruction (quantitative) purposes. Optimal fringe interpretation can be obtained when foreground fringe structure is easily distinguishable, i.e. can be segmented from background object structure. Hence, being able to influence the ${}^{k}I_{max}$, ${}^{k}I_{min}$ ratio $R_{b,d}$, but also the ${}^{k}I_{dif}$ and ${}^{k}I_{dir}$ ratio ${}^{k}R_{dif,dir}$ is of major importance. For the theoretical considerations, it can be assumed without loss of generality, that each elementary light source has a diffuse and a directed part. The proportion of each part determines the diffuseness of the directivity of the light source. Thus, the ratios for the "transmission" and "collimation" methods can be expressed as follows (l is a constant and f the light distribution function):

$$
\begin{aligned}
{}^{T}R_{b,d} &= \frac{I_{dif,b} + I_{dif,all}/2}{I_{dir,d} + I_{dif,all}/2} \sim \frac{\tau_b + f(\tau_d;\tau_b)/2}{\tau_d + f(\tau_d;\tau_b)/2} \in [1:l[\\
{}^{T}R^{b}_{dir,dif} &= \frac{0}{I_{dif,all}/2 + I_{dif,b}} = 0 \\
{}^{C}R_{b,d} &= \frac{I_{dir,b} + I_{dif,all}/2}{I_{dif,all}/2} \sim 1 + \frac{f(1/w;h)}{f(w;1/h)} \in [1:\infty[\\
{}^{C}R^{b}_{dir,dif} &= \frac{I_{dir,b}}{I_{dif,all}/2} \sim \frac{f(1/w;h)}{f(w;1/h)} \in [0:\infty[.
\end{aligned}
\tag{1}
$$

Major difference of both lightings is that, whereas for the transmission solution, the diffuse part $I_{dif,all}$ is unavoidable, the second solution offers the possibility to have only the directed light component $I_{dir,b}$, i.e. to strongly reduce the diffuse part by increasing the h/w ratio. This is an interesting property, as it permits to increase the visual enhancement of geometrical structures, especially for semi-reflective surface inspection, and decrease the visual appearance of the surface texture.

For surfaces with a non-negligible diffuse reflecting part, geometrical structures are all the more enhanced for high ratios ${}^{k}R_{b,d}$ and ${}^{k}R_{dir,dif}$. In case of the "transmission" approach, these ratios are theoretically always l-limited, a theoretically infinite ratio can be obtained with the "collimation" approach. This effect is clearly observable in the corresponding images of Fig. 1 representing a dent-like defect. In case of the "collimation" approach, fringe structures, and so the geometrical surface information, are better visually enhanced, than in case of the "transmission" approach. This is the reason why former approaches are used in case of controlled environments, i.e. when optimal lighting conditions are possible, permitting optimal fringe segmentation, in case e.g. of quantitative 3D reconstruction of matt surfaces.

These results show that in case of geometrical information retrieval, and especially for surfaces with a non-negligible diffuse reflecting part, "collimation" approaches are more appropriate. Values of illumination parameters w and h should be determined in accordance to the inspection requirements, as increasing the ratio h/w decreases the amount of additional diffusing perturbing light $I_{dif,all}$, but also decreases the total amount of projected light $I_{dir,b} + I_{dif,all}$.

2.1.3 The considered illumination technique and reference patterns

As the purpose of this chapter is to propose an alternative surface inspection procedure for the inline characterization of complex objects, a robust methodology based of the projection and interpretation of stripe patterns, will be proposed. These investigations are based in images recorded with an industrial inspection system using a "collimation" illumination, see Fig. 2 Therefore, for the rest of the chapter, the "collimation" lighting will be used. The "transmission-one" will only be considered for theoretical considerations.

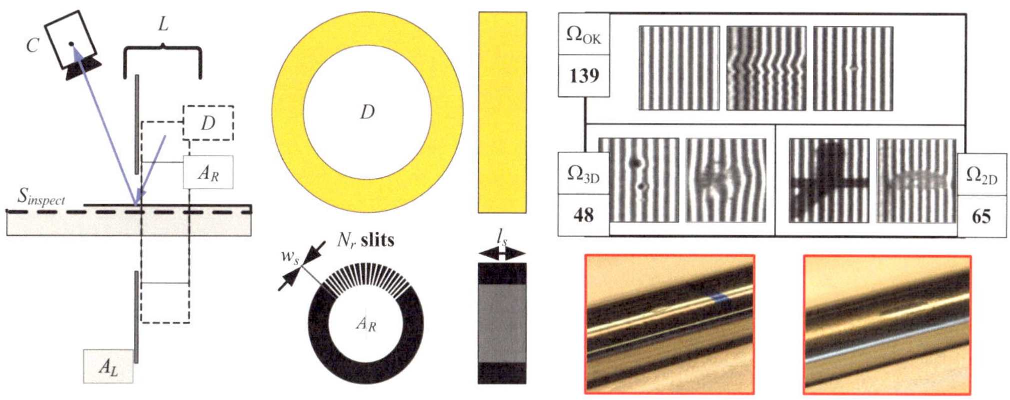

Fig. 2. Considered image datasets (right), recorded with a "collimation" illumination (left), and photos of the cylindrical inspected surfaces (bottom). The illumination L is made of a cylindrical diffusing element D and a slit object A_R to collimation the incoming diffuse light. The line scan camera C records the constant moving surface to be inspected $S_{inspect}$.

Fig. 2 shows some image examples of the considered reference dataset. Each stripe image depicts one type of surface to be characterized, and has been recorded by a special "collimation" illumination producing vertical and periodical stripe structures. The whole set of reference patterns is made of 252 elements manually annotated and classified into three distinct classes Ω_{OK}, Ω_{3D}, and Ω_{2D}. These classes correspond to 139 acceptable surfaces, 48 non-acceptable geometrical defects, and 65 non-acceptable textural defects.

2.2 Generating the appropriate adapted patterns

The second step of the proposed method of this chapter concerns the generation of so called inverse patterns, geometrically adapted to the shape of complex surfaces. Major purpose of this surface inspection methodology, based on the interpretation of regular patterns, is to simplify the processing of the stripe images and therefore to increase the robustness of the proposed approach in case of real-time inline processes. Indeed, as periodical and vertical bright/dark structures have to be processed, the scene interpretation method is equivalent to a matching operation where an a priori known structure, a regular pattern, is compared with an observed one, the pattern being disturbed by a defective surface. The considered approach, i.e. the recording set-up, the inverse pattern generation principle and the recorded images of complex surfaces, is depicted in Fig. 3.

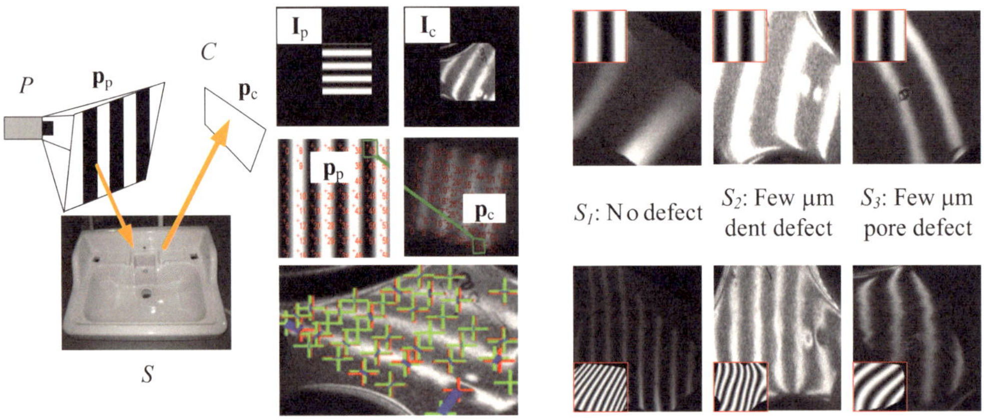

Fig. 3. Considered set-up (left), homography between projector and camera screens (middle) and recorded regular pattern for different complex surfaces (right). The set-up is made of a camera *C* which records the light structure (i) projected by the projector *P* and (ii) reflected by the surface to be inspected *S*. The homography, which permits to link each point $\mathbf{p}_p$ of the projector image $\mathbf{I}_p$ with a point $\mathbf{p}_c$ of the camera image $\mathbf{I}_c$, permits the determination of the adapted patterns, so that vertical and periodical patterns are recorded by the camera.

The provided solution consists of an iterative approach for the determination of the optimal homography linking the projecting screen *P* and the sensor *C* recording complex specular surfaces *S* Inverse image determination permits the projection of an irregular or "inverse" pattern $\mathbf{I}_p$, so that after its projection on the free-form reflective surface, a regular pattern is depicted in the camera image $\mathbf{I}_c$. For $\mathbf{I}_c$ calculation, the following three-step approach was considered in this paper: (i) determination of the position and the size of the structured pattern to be projected, (ii) computation of the screen to camera transformation, and (iii) retrieval of the transformation matrix $\mathbf{H}$, where $\mathbf{I}_c = \mathbf{H} \times \mathbf{I}_p$, permitting the determination of the pattern to be projected. These three steps are addressed in the next paragraphs.

2.2.1 Determining the projector pattern

The size and the position of the structured pattern to be projected have a direct influence on the appearance of the observed structured pattern in the camera image. Thus, pattern position and size in the projector image $\mathbf{I}_p$ have to be adapted in order to reduce noisy effects, as double reflection of the structured pattern or the influence of diffuse lighting. Concerning the investigations presented in this paper, these parameters were determined empirically. The point correspondence is then applied to the depicted projector and camera structured patterns.

2.2.2 Iterative matrix transformation determination

Once the correspondence between projector points $\mathbf{p}_p$ and camera points $\mathbf{p}_c$ is determined, the transformation matrix $\mathbf{H}$ linking the two images can be computed. As point correspondence is done in case of free-form surfaces, $\mathbf{H}$ must be modeled by a polynomial equation of degree r, by means of n corresponding reference points, where $n > n_r$, and n_r is the minimum necessary number of points to retrieve the coefficients of polynomial degree r.

The optimization approach here consists of retrieving optimal degree r and number n of points, in order to minimize the value $e_{c,p}$, which is the Residual Mean Square Error (RMSE) of known camera points $\mathbf{p}_c$ and estimated points ${}^{estim}\mathbf{p}_c$ after applying $\mathbf{H}$ to projector points $\mathbf{p}_p$. The optimization procedure is described in the following equation:

$$\begin{aligned} &\mathrm{H} / \overline{e}_c, p = \mathrm{RMSE}\left(\mathrm{p}_c, {}^{\mathrm{estim}}\mathrm{p}_c\right), \\ &\forall r \in [1:\infty[, \forall n \in [n_r : n_{ref}[\\ &\text{where } n_{ref} = \text{number of correspondance points.} \end{aligned} \tag{2}$$

A two-step procedure is considered here: (i) retrieval of appropriate degree r, its value will depend on the geometrical complexity of the surface ($r = 1$ for the particular case of a planar surface), (ii) determination of optimal number of points n for the transformation. The former serves the determination of the optimal transformation according to the considered free-form surface, the latter permits to consider only robust points, by successively forward selecting the most relevant points. The stopping criterion of forward selection is an a priori defined threshold. An example depicting known points $\mathbf{p}_c$ (green), estimated points ${}^{estim}\mathbf{p}_c$ (red) and error $e_{c,p}$ (blue) is depicted in Fig. 3.

2.2.3 Surface characterization with computed patterns

The evaluation of the method is done visually. For this, three different free-form surfaces with different geometrical complexity are considered. The depicted images in Fig. 3 show how surface visual enhancement is improved by means of the proposed "inverse pattern" method. A geometrical defect is depicted on two of these three surfaces, S_2 and S_3. These results are comparable to the detection of sub-millimeter depths defects using the 3D surface reconstruction with the shape from specular reflection technique, see (Balzer, 2010).

The proposed method permits to project an adapted pattern, so that a regular vertical pattern can be observed in the camera image. As discussed previously, pattern regularity will depend on different parameters, where the geometrical surface complexity is the most determinant.

These results show that the appearance of the regular patterns generated with a "transmission" lighting is similar to the patterns generated by a "collimation" one, see the reference dataset depicted in Fig. 2. This is the reason why, the next paragraphs dedicated to the data processing and classification will tackle the interpretation of the "collimation" patterns. The similarity between both types of patterns will also permit to generalize the results to the automatic inspection of complex surfaces.

3. Data processing

The second part of this chapter involves specific signal and image processing methods. The investigations are focused on multiresolution approaches in the frequency and in the spatial domains. The aim is to retrieve the relevant information from the recorded stripe data, i.e. to segment the perturbed stripe structures synonymous of defective surfaces and to characterize the selected regions by means of appropriate feature-based approaches. These two aspects are tackled in the next paragraphs.

3.1 Free-form segmenting stripe image regions

Once the scene to be characterized has been visually enhanced, the next step consists of the characterization of the stripe regions synonymous of defective surfaces. As the purpose of the characterization is to describe the image regions depicting disturbed stripes, these image parts must be preliminary segmented before being described via feature-based approaches.

The segmentation of stripe structures, so that only the disturbed stripe structure to be characterized is depicted in the pattern, has been addressed is previous researches. It has been demonstrated that for some image content analysis approaches, the segmented images leads to higher classification rates than the images with fixed square sizes (Caulier, 2010). However, hand-segmenting each image region before its classification is, of course, not possible in terms of a fully automatical inspection process, so that unsupervised and adapted segmentation procedures should be defined. Such automatic processes are very often linked with segmentation errors (Unnikrishnan, 2007).

Conventional segmentation processes can be coarsely divided into *contour-based* approaches consisting of determining the transitions between image regions, and *region-based* approaches whose principle is to group image points together with similar characteristics. Although these techniques have been extensively described in the literature, we did not find yet some automatic segmentation approaches of stripe structures. This is in fact a rather complex task, as the depicted defective surfaces by means of stripe patterns are usually not characterized by sharp contours, as these are overlayed by the projected bright/dark stripes. A "simple" solution would be to segment the image patterns to classify by means of sliding overlapping windows of constant sizes. The magnitude of the overlapping regions should be defined according to the specifications of the inspection task.

However, the considered stripe segmentation methodology in this chapter uses an innovative multiscale (wavelet-based) technique permitting the segmentation of free-form relevant image regions. The proposed method, originally developed for the automatic detection of visual saliencies, relies on the assumption that the regions to be detected correspond to denser energy distributions at different scales and frequency subbands. As visual saliencies methods mimic the human visual perception, these are also part of bio-inspired approaches.

3.1.1 The bio-inspired approach

The principle of the bottom-up visual attention model attempts to predict which location in the image will automatically and unconsciously attract the observer's attention towards them. In this biologically-inspired system, an input image is decomposed into a set of multiscale neural ``feature maps'' which extract local spatial discontinuities in the modalities of color, intensity and orientation. Each feature map is endowed with non-linear spatially competitive dynamics, so that the response of a neuron at a given location in a map is modulated by the activity in neighboring neurons. Such contextual modulation, also inspired from recent neurobiological findings, has proven remarkably efficient at extracting salient targets from cluttered backgrounds. All feature maps are then combined into a unique scalar ``saliency map'' which encodes for the salience of a location in the scene irrespectively of the particular feature which detected this location as conspicuous. Fig. 4 depicts the considered model of (Itti, 1998).

3.1.2 Stripe segmentation based on visual saliency map

Concerning the bio-inspired stripe structure segmentation, following reasoning is made. Visual saliency algorithms permit to sample in detail the most relevant features of a scene, i.e. the scene parts containing most important image information. Thus, it is assumed, that in case of the addressed stripe segmentation problematic in this paper, the image parts to be segmented also correspond to the parts with high visual saliency, so that a biological approach can be used as a preliminary segmentation step.

For the purposes of this chapter, we will consider the approach of Itti and Koch (Itti, 1998), encompassing a feature map generation, a center-surround computation, and an across-scale combination until final saliency map generation. The considered method is an image to image transformation, where input data is the stripe structured scene to be segmented and characterized, and output data is the saliency representation of the input scene, which corresponds to high attention degree. Fig. 4 shows the computed saliency maps by means of the considered ITTI approach (Itti, 1998) for three different scene examples belonging to the three considered classes Ω_{OK}, Ω_{3D}, Ω_{2D}.

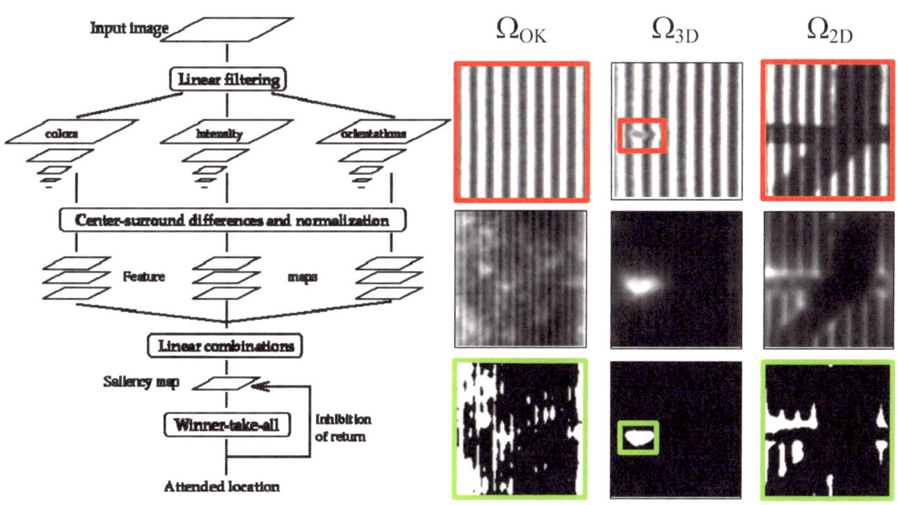

Fig. 4. Considered bio-inspired approach of Itti (Itti, 1998) and examples of saliency maps for three different surface types. The red rectangles correspond to the image regions to be automatically segmented. The green rectangles are the selected regions after binarization of the saliency maps.

The Fig. 4 shows how saliency maps permit to reveal certain stripe structures. If the image contains a salient region, i.e. a locally disturbed pattern, the values of this saliency map region are higher than the surrounding ones. Otherwise, if the pattern remains homogeneous, i.e. corresponds to a non defective surface, no particular image region is revealed by the map. In case of the depicted images, the grey values of the saliency map are higher on the defective regions in case of the Ω_{3D} and the Ω_{2D} images, whereas grey values are more homogeneous for the Ω_{OK} image. A simple binarization of the maps followed shows how far the ITTI approach permits a good object/background differentiation, in particular for locally disturbed patterns. In the following we will see how these saliency representations can be used for the segmentation of stripe patterns.

Saliency maps give a spatial representation the saillancies in an image: the higher the values of a map, the higher the probability that the region (object) differ from the surrounding pixels (background). These maps can therefore be seen as the results of a testing procedure, as the grey value of a pixel map $\mathbf{I}_S(i,j)$ is proportional to the probability $P(O)$ that this pixel belongs to a region to be classified. Thus, the discrimination between object and background classes, Ω_{object} and $\Omega_{\text{background}}$, is equivalent to a classification procedure consisting of the determination of a binarization threshold γ. All pixel whose grey values are higher than γ are classified as object region, whereas all other as background region. This can be stated as follows:

$$\begin{aligned} &\text{if } \mathrm{I}_{s(i,j)} > \gamma \text{ then } \mathrm{I}_s(i,j) \in \Omega_{\text{Object}} \text{ else } \mathrm{I}_s(i.j) \in \Omega_{\text{Background}} \\ &\forall (i,j) \in [n_l, n_c] \\ &\text{where } [n_l, n_c] \text{ is the size of the image } \mathrm{I}_s(i,j) \text{ and } \gamma \in [0,255]. \end{aligned} \tag{3}$$

However, a global binarization procedure also implies falsely classified pixels. The fundamental problem therefore consists of determining the most optimal threshold γ_{optim}, so that most of the image pixels are correctly classified into the classes Ω_{object} and $\Omega_{\text{background}}$. For the following we will consider the classification error rate p_{error} of [Zha01], who permits to evaluate the segmentation process of a map $\mathbf{I}_S$ for a certain threshold γ and the previously described reference image database.

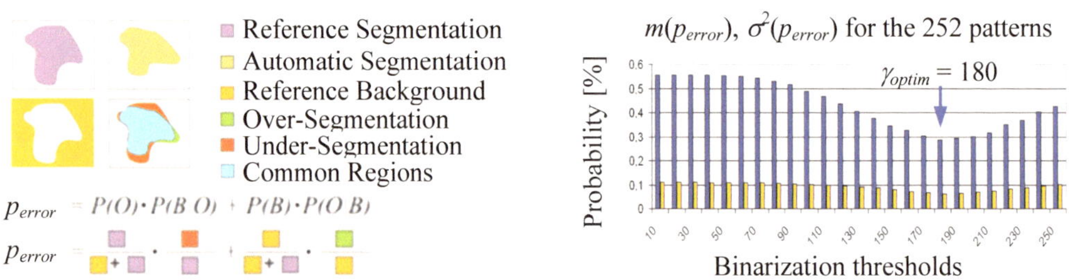

Fig. 5. Computation principle of p_{error} (left) and curves $m(p_{error})$, $\sigma^2(p_{error})$ permitting the determination of γ_{optim} (right).

The consider approach for determining the optimal threshold γ_{optim} consists of computing the error rate p_{error} for all threshold values [0,255], so that optimal threshold will correspond to the lower rates p_{error}. In order to define one threshold for the segmentation of all possible stripe images, the mean $m(p_{error})$ and the variance $\sigma^2(p_{error})$ for the error rates for all 252 images of the reference dataset will be used to evaluate each threshold γ. Fig. 5 depicts show the computation principle of p_{error} and the computed values of $m(p_{error})$ and $\sigma^2(p_{error})$ for the 252 stripe images.

The two curves represent the evolution of $m(p_{error})$ and $\sigma^2(p_{error})$, computed with the considered dataset of 252 stripe images, for each threshold γ. It is noticeable how both curves reach a minimum for the same value of $\gamma_{optim} = 180$. This value, which corresponds to the lowest classification error rate, is the optimal threshold values γ_{optim}.

3.2 Characterization of stripe image regions

Once the regions of interest, i.e. these depicted non-acceptable surface parts, have been segmented, these can be characterized in a second step by means of feature-based approaches. Preliminary investigations showed that optimal stripe characterizations are achieved using adapted and Fourier features. New multiresolution features are introduced in this paragraph. Spatial transformations using Gabor filters are considered. In the following all the considered feature sets are introduced.

3.2.1 The feature based image characterization principle

According to Randen and Husoy (Randen, 1999), it is a-priori not possible to know which textural method is more appropriate for a specific task. A selection of the most appropriate one (methods partially based on work by Wagner and Kueblbeck (Wagner, 1996) and Wagner (Wagner, 1999) was done. Most publications dedicated to *specific* features-based stripe structure characterization are related to fringe pattern identification within the field of interferometric non-destructive inspection. Type and number of described fringe features are task dependent. We may distinguish between spatial-based (Jueptner, 1994; Zhi, 1992), frequency-based (Takeda, 1982; Qian, 2005}, and mixed (wavelet) (Krueger, 2001; Li, 2000) approaches.

The selected methods for the addressed problems in this chapter correlate with the highest recognition rates of the described methods and studies: (i) the *transform* of Weska, (ii) the adapted of Zhi (Zhi, 1992), and (iii) the multiresolution of Mallat (Mallat, 1989). As few publications tackle the problems of image structure characterization, a particular attention was given to the retrieval of specific adapted features. For our purposes, *four geometry* and *two statistic-based features* proposed by (Zhi, 1992) for bright stripe pattern characterization were applied. The contribution to specular surface inspection is achieved through the completion of these features with additional *four specific features*. The novel aspect consists of the use of these features for the characterization of the bright stripes, but also of the dark stripes within the pattern.

These methods being part of general approaches belonging to the main texture families (Tuceryan, 1998) or of specific methods, specially developed for the characterization of image structures. Each of these procedures was optimized by adapting method innate parameters towards the depicted disturbed or non-disturbed stripe pattern (shape, intensity e.g.).

Fig. 6 shows the computation principle for the first two considered features sets, the Fourier and the Stripe ones.

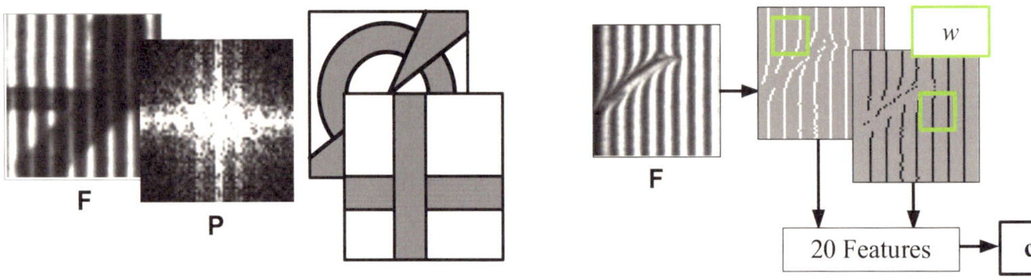

Fig. 6. Computation principle of the Fourier (left) and the Stripe (right) feature sets. The Fourier features correspond to the spectral energy of different regions of the Fourier spectrum. The figure shows the original image **F**, its power spectrum **P** and horizontal, vertical, directional and radial energy regions. The stripe features correspond to the intensity- and geometry-based description of the segmented bright and dark stripe structures. Each feature is first locally computed for different image pixels using a window w. The final feature value for an image **F** is the average of all the locally computed feature values.

3.2.1.1 Fourier textural features

The DFT (Discrete Fourier Transform) made the spectral analysis of discrete images possible, by decomposing the image function into a sum of finite sinusoids. The textural transform approach proposed by Weska (Weszka, 1978) is based on the spectral domain analysis. The features are computed from values in the Fourier spectrum corresponding to different r_F radial and d_θ directional spectral regions. As the characterized stripes have a vertical and periodical structure, further u_F and v_F spectral regions along the u-horizontal and the v-vertical image axes were defined.

The vector computation for the involved Fourier textural analysis method is depicted below:

$$C^F_{r,\theta,v,h} = \left\{C^F_r;C^F_\theta;C^F_v;C^F_h\right\}, N_c = 33\left(=8+10+5+10\right) \quad (4)$$

3.2.1.2 Adapted fringe and stripes features

The considered bottom-up approach is based on preliminary investigations (Caulier, 2008) involving specific geometry-based and intensity-based features according to a two-step procedure, stripe *segmentation* and *characterization*. Each process is characterized by following parameters: the segmentation function f utilized and the image areas **a** that are covered by a local window w sliding over the entire pattern described. Notation for the sub-pixel segmentation peak detectors are b^5 for the ``Blais-and-Rioux" and $c^5$ for the ``Center-of-Mass", see (Caulier, 2008). Hence notations for considered segmentation functions are $f \in \{ b^5; c^5 \}$. Concerning the characterization of the extracted bright and dark regions each feature $c^S_{a(m)}$ represents the average result of an $O_{a(m)}(m)$ operation applied to a bright or dark stripe element. The computation of $O_{a(m)}(m)$ is applied to an image area $a(m)$, whose magnitude is feature dependent.

The new contributions consist of applying the features 9 to 14 defined in (Caulier, 2008) for the bright stripes, also to the dark stripes, so that a total of 20 features are considered here, m

$\in \{ 0,...,19 \}$. Hence, from these 20 features, 8 were specially developed for the considered stripe images, whereas the 12 remaining were described within the context of fringe structure characterization (Zhi, 1992) and adapted for our purposes. It is these 8 features which were used in the industrial application (Caulier, 2008). The description of the stripe feature vector with the different area magnitudes *a(m)* is as follows:

$$C^S_{f,a} : f \in \{f_b ; f_c\} \text{ and } a = \left[a(0),...,a(m),...,a(19)\right],$$
$$a(m) \in \left\{\left[5^2\right];\left[7^2\right];...;\left[17^2\right];\left[M_u \times M_u\right]\right\}, N_c = 20 \tag{5}$$

The maximal value of *a(m)* is determined according to the minimal possible size of the reference stripe patterns which is approximately 20 pixels. As mentioned above, the computation of the stripe feature vector relies on two segmentation functions *f*. Then, each of the stripe feature vector's 20 elements can be computed by means of 8 different area sizes of *a(m)*. Hence, 2×8^{20} stripe feature vectors can be retrieved based on the definition provided in Equation 5. In order to reduce the number of possible feature stripe vectors, and thereby avoiding dimensionality-based problems, a preliminary optimization process to retrieve the most adequate area size of *a(m)* for each feature $c^S_{a(m)}$ is necessary.

When considering the definitions of the 20 operators, we can distinguish between the features whose computation relies on *fixed* and *adapted* image areas *a(m)*, where *m* is the feature index. For fixed areas, the only condition is that the area sizes must be large enough to allow both operators to be applied. For adapted image areas, the most appropriate area size must be defined according to the stripe structures to be characterized.

Preliminary studies (Caulier, 2008) show that an optimal set of image areas $\mathbf{a}^1$, depicted in following Equation, can be defined (maximum possible area size is noted $M^2 = [M_u \times M_v])$:

$$\begin{aligned} &\mathrm{a}^1 : \{\{\mathrm{a}(00)-\mathrm{a}(05);\mathrm{a}(08)-\mathrm{a}(09)\} = \left[M^2\right];...\\ &...\{\mathrm{a}(06)-\mathrm{a}(07);\mathrm{a}(10)-\mathrm{a}(15);\mathrm{a}(18)-\mathrm{a}(19)\} = \left[17^2\right];\\ &...\{\mathrm{a}(16)-\mathrm{a}(17)\} = \left[5^2\right]\} \end{aligned} \tag{6}$$

In order to validate the tests described in (Caulier, 2008), an additional ``non-optimal'' set $\mathbf{a}^2$, complementary of $\mathbf{a}^1$, i.e. $M^2 = [M_u \times M_v]$ value remains, 5^2 and 17^2 values must be exchanged, was also considered.

3.2.2 Proposed Gabor wavelet features

Generalities

Major drawback of the Fourier approach, is that if it permits a good spatial resolution, it is not possible in the same time to have a good resolution in the frequency domain, this phenomena is known as the Heisenberg inequality or uncertainty principle. The purpose is here to investigate how far a multiresolution approach can be used for the classification of stripe image patterns. The considered procedure consists of using the approach of Mallat (Mallat, 1989) and to systematize it to another other wavelet family. The Gabor one will be used. For the Fourier approach, the directional regions of the power spectrum lead to best discrimination results, see paragraph 5.2.1. Hence, using a wavelet decomposition approach, the influence of the combined selection of different frequency regions with the selection of different decomposition levels will be investigated.

For each level, the wavelet coefficients are obtained by convolutions of input image $f(u,v)$ with two one-dimensional filters: h a lowpass filter and g a highpass filter, with $g(n) = (-1)^{1-n} h(1-n)$. A detailed procedure of the image wavelet decomposition can be found in (Mallat, 1989). Fig. 7 describes the wavelet decomposition and reconstruction principle at level r of an image signal $f(u,v)$ with a pyramidal representation according to Mallat (Mallat, 1989).

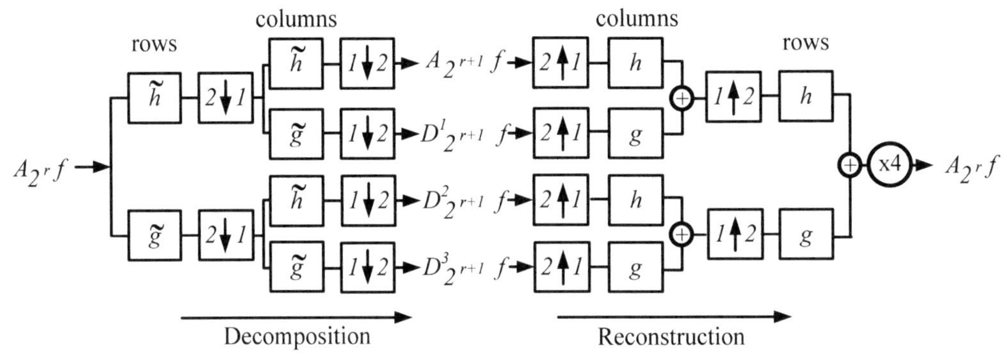

Fig. 7. Decomposition and reconstruction principle of the discrete wavelets transform in the spatial domain according to (Mallat, 1989).

Classification methodology with wavelets

First classification step consists of wavelet decomposing each image. The energy measure of the four subband images $A_{2r}f$, $D^1_{2r}f$, $D^2_{2r}f$ and $D^3_{2r}f$ for each decomposition level r is used as signature for each image. Unlike the Fourier transform, which just utilizes the sine and cosine as basis functions, each new wavelet family brings its own set of new basis functions (the necessary conditions such functions must fulfill are defined in (Mallat, 1989) Section III.A).

As the purpose is to evaluate the classification of stripe image patterns using different types of textural features, the activities are focused on one wavelet family, the Gabor wavelet filters (Kovesi, 2011). This approach was considered to be an appropriate enhancing function, as these filters permit the enhancement of image structures of different shapes, frequencies and orientations. In the following a brief overview of 2D Gabor is provided.

According to the definition of Dunn (Dunn, 1995) which is based on the definition of Daugman (Daugman, 1985), a 2D Gabor filter h is an oriented complex sinusoidal wave h_{sin} modulated by a 2D Gaussian envelope h_{gau}, $h = h_{sin} \times h_{gau}$. Filter main parameters are the wavelength $\lambda = 1/f$, f is the frequency, the standard deviation σ and the orientation α.

Different values of these parameters permit the elaboration of different filters of different shapes, sizes and directions.

For the purpose of this chapter, the Gabor filter definition given by Kovesi will be considered. The author defines three output images, $\mathbf{I_r}$, $\mathbf{I_i}$, $\mathbf{I_a}$, results of the convolution of the input image $\mathbf{I_{in}}$ with the (i) real part h_r of the 2D filter h, the (ii) imaginary part h_i of the 2D filter h and the (iii) amplitude of both real and imaginary images. Filter description according to Kovesi (Kovesi, 2011) for the x-direction is provided by the following equation (for the sake of simplicity, the x-direction was considered, same equations hold for the y-direction by replacing the x with y):

$$f_r = cos\left(\frac{2.\pi}{\lambda}\chi\cdot\right)\cdot e^{-\left(\frac{x^2}{\sigma_x^2}\right)}\cdot f_\alpha \quad f_i = sin\left(\frac{2.\pi}{\lambda}\chi\cdot\right)e^{-\left(\frac{x^2}{\sigma_x^2}\right)}\cdot f_\alpha$$
$$\text{where } \sigma_x = \lambda\cdot k_x \qquad (7)$$
$$\mathrm{I_r} = \mathrm{I_{in}} * f_r \quad \mathrm{I_i} = \mathrm{I_{in}} * f_i \quad \mathrm{I_a^2} = \mathrm{I_r^2} + \mathrm{I_i^2}$$

f_α is a rotating function defined for an angle α in degrees (an angle of 0 gives a filter that responds to vertical features. The scale factors k_x and the filter σ_x relative to the wavelength of the filter. This is done so that the shapes of the filters are invariant to the scale. k_x controls the shape of the filter in the x-direction.

According to equation 7, λ permits to regulate the modulation of the *cos* and *sin* waves with the Gaussian envelope. For λ=1 the filtering is equivalent to an image blurring with a Gaussian kernel, and therefore reveals low image frequencies. Higher values of λ and of k permit image filtering with Gaussian kernel modulated with *cos* or *sin* envelopes, which is equivalent to convolve the image with second or first derivative filter kernels. Thus, in case of high image frequencies enhancement and if the edge information is considered as important images signatures, a variation of the variance σ of the Gaussian kernel between $\sigma \in [2:4]$ seems to be an adequate choice.

Therefore, for the purposes of this chapter, the Gabor filterbank is made of Gabor filter values, l = 2 and k=2, so that σ= 4, two orientations, α = 0° and α = 90°. The *sin* function was considered. These filters were applied for the computation of image features. Four decomposition levels r = { 2, 3, 4, 5 }, corresponding to object sizes in the image of { 4, 8, 16, 32 } were considered. Hence, the resolution given by the considered stripe structure which is $d_{P,px}$=8 pixels is taken into consideration, see the images in Fig. 2. It is assumed that the resolution level r=3 could bring interesting classification results.

Classification methodology with wavelets simple decomposition

This first classification procedure uses all computed approximation and detail images during Mallat's decomposition algorithm using the Gabor wavelet. First element of Wavelet's simple decomposition feature vector $\mathbf{c}_{WS}$ equals the energy value of original image $\mathbf{F}$. Then, the remaining elements of $\mathbf{c}_{WS}$ are filled with the energy values of the approximated and all detail images obtained by decomposing the image $\mathbf{F}$. Following expression of feature vector $\mathbf{c}_{WS} \in R^{N_c}$ is obtained by decomposing image $\mathbf{F}$ until level r with Gabor wavelet:

$$c_{WS} = \{E(0,f), E(1,A), E(j,D^1), E(j,D^2), E(j,D^3),$$
$$c_{WS} \in R^{N_c}, \quad N_c = 2 + 3r \tag{8}$$
$$\text{with } j = \{1,...,r\}, \quad r \in Z$$

Where $E(0,f)$ is the energy of original image pattern **F**.

Classification with wavelets generalized decomposition

A shortcoming of the conventional dyadic wavelet transform is that it does not benefit from possibly useful features that can be obtained by further decomposing the high frequency subbands. An improvement therefore consists of decomposing also the high frequencies (Coifman, 1992; Tikkanen; 1997, Nasir, 2002; Cohen1997) i.e. in computing the approximation and detail images not only for the approximation image $A_{2^\wedge(r+1)}f$ at each level r but also for the detail images $D^1_{2^\wedge(r+1)}f$, $D^2_{2^\wedge(r+1)}f$, and $D^3_{2^\wedge(r+1)}f$ ($r > 1$, $r \in Z$). This general wavelet decomposition approach will be named the *generalized multiresolution decomposition* also named wavelet packet analysis (Tikkanen, 1997).

A feature vector **c** is made of the energy values of all approximation and detail images computed during the generalized multiresolution decomposition. First element of **c** equals the energy value of original image **F**. Following expression feature vectors of Wavelet's generalized decomposition is obtained:

$$c_{WG} = \left(E(0,f), \{E(j,k,A)\}, \{E(j,k,D^1)\}, \{E(j,k,D^2)\}, \{E(j,k,D^3)\} \right)$$
$$c_{WG} \in R^{N_c}, N_c = 1 + \sum_{j=0}^{r} 4^j \tag{9}$$
$$\text{with } j = \{1,...,r\}, \quad k = \{1,...,4^j\} \quad r \in Z$$

Where $E(0,f)$ is the energy of the original image pattern **F** and $E(j,k,A)$, $E(j,k,D^1)$, $E(j,k,D^2)$, $E(j,k,D^3)$ are the energies at level j of the subband image numbers k.

Classification with subband wavelet generalized decomposition

This further investigation consists of the evaluation of each subband separately, in order to figure out how far the selection of a particular subband can lead to an improvement of the reached classification rates. For this purpose the images are classified each by means of the four subband images obtained with a generalized wavelet decomposition approach. In order to really estimate which of the four coefficients contain the most discriminating information, we consider the *generalized* wavelet decomposition for the decomposition level r=3, as this level correspond to the size of the considered period of the reference images, which is $2^3 = 8$.

Doing this, four feature vectors are defined. $\mathbf{c}_{SWG,(D2,3,A)}$ contains the energy of all approximation images $A_{2^\wedge(r+1)}f$ up to level r=3 using Gabor wavelet and the *generalized* Wavelet decomposition. Feature vectors $\mathbf{c}_{SWG,(3,A)}$, $\mathbf{c}_{SWG,(3,D1)}$, $\mathbf{c}_{SWG,(3,D2)}$ and $\mathbf{c}_{SWG,(3,D3)}$ are filled with the energy of all detail images $D^1_{2^\wedge(r+1)}f$, $D^2_{2^\wedge(r+1)}f$, and $D^3_{2^\wedge(r+1)}f$ up to level r=3 using

Gabor wavelet and the *generalized* Wavelet decomposition. Following expressions of feature vector using the Wavelet's subband generalized decomposition are obtained:

$$\begin{aligned}
c_{SWG,(3,A)} &= \left(\left\{E\left(j,k,A\right)\right\}\right) \in R^{N_c} \\
c_{SWG,\left(3,D^1\right)} &= \left(\left\{E\left(j,k,D^1\right)\right\}\right) \in R^{N_c} \\
c_{SWG,\left(3,D^2\right)} &= \left(\left\{E\left(j,k,D^2\right)\right\}\right) \in R^{N_c} \\
c_{SWG,\left(3,D^3\right)} &= \left(\left\{E\left(j,k,D^3\right)\right\}\right) \in R^{N_c} \\
N_c &= \sum_{j=1}^{r} 4^{j-1} \\
\text{with} \quad j = \left\{1,\ldots,r\right\}, &\quad k = \left\{1,\ldots,4^{j-1}\right\} \quad r \in Z
\end{aligned} \tag{10}$$

Where $E(j,k,A)$, $E(j,k,D^1)$, $E(j,k,D^2)$, $E(j,k,D^3)$ are the energies at level j of the subband image numbers k.

4. Data interpretation

The third and last part of this chapter is dedicated to the interpretation, i.e. classification, of the generated metadata by means of data mining techniques. The general purpose is to retrieve, for a specific inspection task, the optimal processing chain, i.e. leading to high detection and low false alarm rates. The proposed procedure uses a classification factor as an evaluation criterion for the direct evaluation and comparison of different data processing approaches.

It is demonstrated that this procedure can be applied for the evaluation of any image processing tasks. In a first step, the method is validated for a certain surface shape within the context of a specific inspection purpose. In a second step, the case of complex free-form structure characterization is used for the generalization of the method. The considered reference stripe image dataset was introduced at the beginning of this chapter in paragraph 3.1.3.

4.1 Optimal processing chain determination

As described at the beginning of this chapter, the purpose is to demonstrate how the projection of adapted structured patterns can be used for the inline inspection of complex reflective surfaces. Within the context of industrial non destructive testing, an inspection system has been developed for the characterization of cylindrical specular surfaces (Caulier, 2009). The principle relies on the combination of a collimation-based illumination and line scanning cameras to record and process the reflected regular patterns. Fig. 2 shows some images examples depicting defective cylindrical surface parts.

The system's requirements are automatically to classify the surfaces of the recorded cylindrical objects into non-defective, defective geometrical or defective textural. This three-class problem consists of determining the most optimal processing chain based on the feature-based description of the recorded stripe image. The aims are here (i) to determine most appropriate stripe segmentation approach, (ii) to evaluate the relevance of the proposed feature sets, (iii) to retrieve the most optimal one, and (iv) to generalize the approach and the results obtained with the inspection of cylindrical surfaces to the interpretation of complex surfaces.

The problem is therefore tackled in two steps. At first, the lowest complexity problem of most appropriated processing chain retrieval within the context of regular stripe characterization is addressed. For this, in order to retrieve the most appropriate feature set, the previously described Fourier and Stripe features are evaluated by means of three different *image sets*, three different *classifiers*, and one *classification methodology*. Thus, for each of the four feature sets, *nine* different pattern analysis procedures were considered.

Then, the generalization is done by using the most optimal features defined for the characterization of "simple" surface for the interpretation of complex objects. As the aim is also to retrieve the most appropriate feature sets, here also different processing chains will be evaluated. However, as the purpose is the generalization to complex surfaces, the investigations will be focused on (i) the compare of *nine* different image sets taken reference databases, and on (ii) the involving of special feature subset selection (FSS) methods based on the previously defined optimal classifier.

4.2 Optimal processing chain for cylindrical surfaces

As defined previously, concerning the *image sets*, three segmentation approaches were considered: (i) a free-form approach as described in the paragraph 4.1, *image sets* Φ_{ff}, and (ii) fixed sizes of 64x64 pixels and 128x128 pixels, with *image sets* Φ_{64^2}, Φ_{128^2}. The patterns were recorded with the industrial system (Cau2007c), see the examples in Fig. 2.

In order to address a general stripe pattern characterization approach, the proposed empirical method involved three different classification principles, all using a specific technique or a particular configuration. Following classifiers were involved: The non-parametric Nearest-Neighbor approach, **1-NN** for k=1, **3-NN** for k=3, and, the parametric Naive Bayes **NB**, applied to the previously cited classifiers. As to the *classification methodology*, and a stratified 10-fold cross validation (Kohavi, 1995), were chosen.

All the results are shown in Tables 1 and 2. These tables list the stripe patterns' (i) classification using Fourier's, the Stripe's and the Gabor wavelet's features.

Different generalities can be done concerning the segmentation process, the classification approach and of course the feature sets, on the basis of the results listed in both tables.

At first, free-form segmentation approaches seem to be more appropriated for Wavelet and Stripe features. The common aspect of both approaches is to use the average results of spatial filters applied to the reference images for feature vector construction. Free-form segmentation permits to retrieve only the disturbed parts of the whole stripe structures to be characterized, and to discard the undisturbed ones in case of defective surface characterization, and contrariwise to retain only the undisturbed part in case of non defective surface characterization, see Fig. 4. In other word, free-form segmentation permits a first spatial discrimination between disturbed and undisturbed stripe regions. The assumption is therefore that, such segmentation techniques are more appropriate to classification methods using spatial filters for feature vector computation.

	Fourier vectors	Image sets			Stripe vectors	Image sets		
		Φ_{ff}	Φ_{64^2}	Φ_{128^2}		Φ_{ff}	Φ_{64^2}	Φ_{128^2}
NB	c_F	63,4	75,4	61,5	$c_{SISC,g1,BR5}$	84,9	84,5	78,3
	$c_{F,(r)}$	58,3	67,2	57,8	$c_{SISC,g1,CM5}$	83,7	82,8	60,0
	$c_{F,(\theta)}$	66,4	80,4	55,8	$c_{SISC,g2,BR5}$	79,6	75,8	61,4
	$c_{F,(v)}$	49,2	79,3	58,8	$c_{SISC,g2,CM5}$	77,6	73,8	57,1
	$c_{F,(u)}$	54,8	41,0	36,2				
1-NN	c_F	79,2	84,9	84,4	$c_{SISC,g1,BR5}$	88,1	85,3	8,2
	$c_{F,(r)}$	72,7	69,8	65,7	$c_{SISC,g1,CM5}$	**88,9**	81,6	82,8
	$c_{F,(\theta)}$	78,2	**87,9**	83,6	$c_{SISC,g2,BR5}$	86,1	85,9	83,5
	$c_{F,(v)}$	76,9	79,2	75,9	$c_{SISC,g2,CM5}$	84,6	78,4	78,8
	$c_{F,(u)}$	65,0	75,6	61,5				
3-NN	c_F	76,2	58,3	81,3	$c_{SISC,g1,BR5}$	86,5	84,3	80,5
	$c_{F,(r)}$	68,6	70,6	65,8	$c_{SISC,g1,CM5}$	84,5	77,7	80,9
	$c_{F,(\theta)}$	75,8	81,2	82,5	$c_{SISC,g2,BR5}$	83,5	81,7	81,1
	$c_{F,(v)}$	74,3	77,8	75,7	$c_{SISC,g2,CM5}$	82,2	78,8	74,8
	$c_{F,(u)}$	63,9	75,6	64,7				

Table 1. Rates of correctly classified patterns of the three image sets Φ_{ff} Φ_{64^2}, Φ_{128^2} for the Fourier and the adapted Stripe features.

		Image sets				Image sets				Image sets		
	Vector	Φ_{ff}	Φ_{64^2}	Φ_{128^2}		Φ_{ff}	Φ_{64^2}	Φ_{128^2}		Φ_{ff}	Φ_{64^2}	Φ_{128^2}
NB	$c_{W,S,(D2,2)}$	53,7	53,7	51,1	$c_{W,G,(D2,2)}$	58,1	56,4	52,1	$c_{SW,G,(D2,3,A)}$	64,9	59,2	48,5
	$c_{W,S,(D2,3)}$	78,5	71,2	63,1	$c_{W,G,(D2,3)}$	64,2	58,2	51,6	$c_{SW,G,(D2,3,D1)}$	78,3	75,4	66,0
	$c_{W,S,(D2,4)}$	83,9	80,3	77,4	$c_{W,G,(D2,4)}$	65,1	61,8	75,4	$c_{SW,G,(D2,4,D2)}$	60,0	58,6	55,1
	$c_{W,S,(D2,5)}$	**86,5**	83,2	61,8	$c_{W,G,(D2,5)}$	70,1	64,5	57,3	$c_{SW,G,(D2,5,D3)}$	85,7	77,3	71,2
1-NN	$c_{W,S,(D2,2)}$	74,1	71,6	73,9	$c_{W,G,(D2,2)}$	72,0	71,4	71,6	$c_{SW,G,(D2,3,A)}$	76,4	75,2	70,2
	$c_{W,S,(D2,3)}$	82,6	79,2	75,2	$c_{W,G,(D2,3)}$	**84,2**	82,8	75,2	$c_{SW,G,(D2,3,D1)}$	78,7	75,5	66,3
	$c_{W,S,(D2,4)}$	76,5	74,5	74,0	$c_{W,G,(D2,4)}$	83,0	82,1	78,4	$c_{SW,G,(D2,4,D2)}$	80,9	81,6	81,4
	$c_{W,S,(D2,5)}$	76,0	76,2	75,2	$c_{W,G,(D2,5)}$	81,0	81,6	79,4	$c_{SW,G,(D2,5,D3)}$	82,3	79,4	72,6
3-NN	$c_{W,S,(D2,2)}$	70,4	72,4	69,5	$c_{W,G,(D2,2)}$	78,6	74,5	68,7	$c_{SW,G,(D2,3,A)}$	78,4	76,1	68,4
	$c_{W,S,(D2,3)}$	75,0	75,0	75,5	$c_{W,G,(D2,3)}$	80,1	76,4	72,0	$c_{SW,G,(D2,3,D1)}$	79,3	75,5	68,5
	$c_{W,S,(D2,4)}$	80,1	76,0	77,6	$c_{W,G,(D2,4)}$	83,7	81,2	78,8	$c_{SW,G,(D2,4,D2)}$	81,9	81,7	81,4
	$c_{W,S,(D2,5)}$	75,4	75,4	74,8	$c_{W,G,(D2,5)}$	81,2	79,2	76,9	$c_{SW,G,(D2,5,D3)}$	**82,4**	82,0	74,6

Table 2. Rates of correctly classified patterns of the three image sets Φ_{ff} Φ_{64^2}, Φ_{128^2} for the Gabor Wavelet features.

Concerning the use of adapted images, we remark that better classification rates are reached for fixed image sizes, and especially for the Φ_{64^2} image set. In fact, the Fourier spectrum is applied to images whose sizes equal a power of two, so that in case of variable image sizes a padding with zeros is necessary. Hence, as far as Fourier features are concerned, the image region containing the stripe to be characterized but also the surrounding image region contains important discriminative information.

Then concerning the classifiers, Tables 1 and 2 show that in general best classification rates were obtained using the 1-NN classifier for the three consider feature types. For the Wavelet features, the most appropriate classifiers depend on the involved feature vector computation techniques. For the Fourier features, we remark that optimal results were reached for the reduced feature set $c_{F,(\theta)}$ involving only the ten directionality power spectrum features. For this configuration, the classification rates are similar to those obtained with the whole 33 features and even better.

To conclude, it is observable that wavelet features do not seem to be the more appropriate in case of the considered classification task, i.e. in comparison with the Fourier and the Stripe techniques. Classification rates with Wavelet features are the lowest and no particular classifier seems to be more appropriate, as for each of the three considered wavelet feature computation techniques, a different classifier leads to highest classification rates.

4.3 Generalization to complex surfaces

4.3.1 Generalization of the reference image dataset

The generalization concerns the projection of adapted patterns in order to observe regular patterns, which can be characterized with the previously described methodology. Major advantage of the adapted pattern projection method, compared to similar 3D reconstruction ones, see (Ihrke, 2008; Balzer, 2010), is to visually enhance geometrical defects on complex free-form surfaces, for the purpose of qualitative direct surface characterization.

However, in case of inline inspection systems perturbing effects leading to non optimal surface characterization by means of adapted projected pattern interpretation may occur. Major perturbing factors are due to imperfect recording conditions or surface characteristics. In case of workpiece positioning above the defined moving tolerances e.g., additional noise such as defocusing or light glares, will perturb the reference point determination. Also in case of small surfaces, whose size does not permit to project enough stripe patterns, the point determination might not be optimal. In addition, certain surfaces, such as e.g. several layered light transmitting ones, may also lead to sub-optimal visual appearance of the stripe in the recording sensor.

This signifies that in case of a generalization of the proposed surface characterization method based on the projection of stripe patterns, the reference dataset must be enlarged in order to cover all possible regular patterns to be characterized.

The task is not to enumerate all possible regular stripe pattern disturbances. This would hardly be possible. Hence, it is preferable to focus our investigations on a restricted and predefined number of *not perfectly vertical* and *not perfectly periodical* stripe patterns.

Further pattern structures have therefore to be defined. The easiest and simplest way consists of using the considered set introduced in paragraph 4.2 and to ``transform'' or "adapt" them, so that these can be used for the characterization of free-form surfaces.

Thus, the stripe-illumination-based complex surface inspection task will be addressed by means of different image sets: The reference initial set Φ_0^0 previously introduced, and eight further derived sets. The four sets Φ_1^1- Φ_1^4 correspond to the warping of all patterns of Φ_1^1 with increasing projective transformations. The four sets Φ_2^1- Φ_2^4 correspond to the warping of all patterns of Φ_0^0 with increasing cylindrical transformations. Two projective -1- and cylindrical -2- transformations have been considered. All sets are made of 252 patterns.

Fig. 8 shows 3 of the 9 considered stripe image data sets. Φ_0^0 is the reference set where the stripe structures are periodical and vertical, Φ_1^4 is the set corresponding to the warped patterns of set Φ_0^0 with a maximum perspective distortion -1- and Φ_2^4 is the set corresponding to the warped patterns of set Φ_0^0 with a maximum cylindrical distortion -2-.

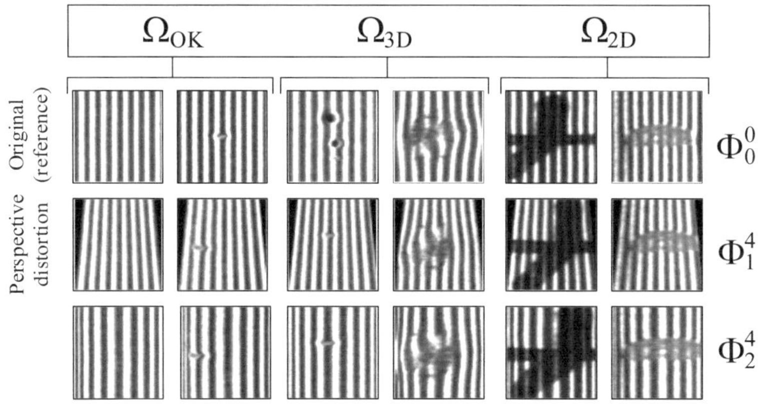

Fig. 8. Reference patterns for the classification of free-form rough and specular surfaces. The figure shows 6 image examples taken from the three different pattern sets $\Phi_0{}^0$, $\Phi_1{}^4$ and $\Phi_2{}^4$.

These image patterns correspond to three different complex surfaces illuminated with an adapted pattern: -0- for surfaces ideally depicted, and -1- and -2- for surfaces inducing perspective and cylindrical distortions.

$\Phi_1{}^4$ and $\Phi_2{}^4$ corresponds to patterns with a maximal perturbation of type -1- and a maximal perturbation of type -2-. These patterns have been simulated by transforming patterns $\Phi_0{}^0$ [The first number is an indices the second is an exponent] with perspective and cylindrical distortions. All the patterns have a size of 64 x 64 pixel.

4.3.2 Optimal processing chain for increased surface complexity

This paragraph addresses the procedure for the determination of optimal feature subsets using feature evaluation, grouping, fusing, and selection in case of the general inspection problem stated in this paper, i.e. the inspection of complex objects using structured illumination. It has been demonstrated in the previous paragraph 4 that two feature families, the 33 Fourier and the 20 adapted stripe features, lead to best classification rates. In order to define to appropriate features sets in case of the generalization to complex surfaces, feature subset selection (FSS) methods are evaluated by considering the generalized reference databases.

The question is to what extend an appropriate fusion and selection of the Fourier and the adapted stripe feature sets can lead to a better quality control of the complex surfaces? For this purpose, two group of three feature vectors will be considered. First group consists of vectors $\mathbf{c}^F{}_{r,\theta,v,u}$, made of the 33 Fourier features, $\mathbf{c}^F{}_\theta$, made of the 10 directional Fourier features, and $\mathbf{c}^S$ consisting of the 20 adapted features. Second group encompasses vectors $\mathbf{c}^F{}_{r,\theta,v,u}{}^S$, made of the 33 Fourier features and the 20 adapted stripe features, $\mathbf{c}^F{}_\theta{}^S$ made of the 10 directional Fourier features and the 20 adapted stripe features, and ${}^{1\text{-NN}}\mathbf{c}^F{}_\theta{}^S$ made of the selected features of vector $\mathbf{c}^F{}_\theta{}^S$ using a 1-NN-wrapper-based FSS method. Classification results using the three two feature vector groups and the two distortion types are depicted in Fig. 9.

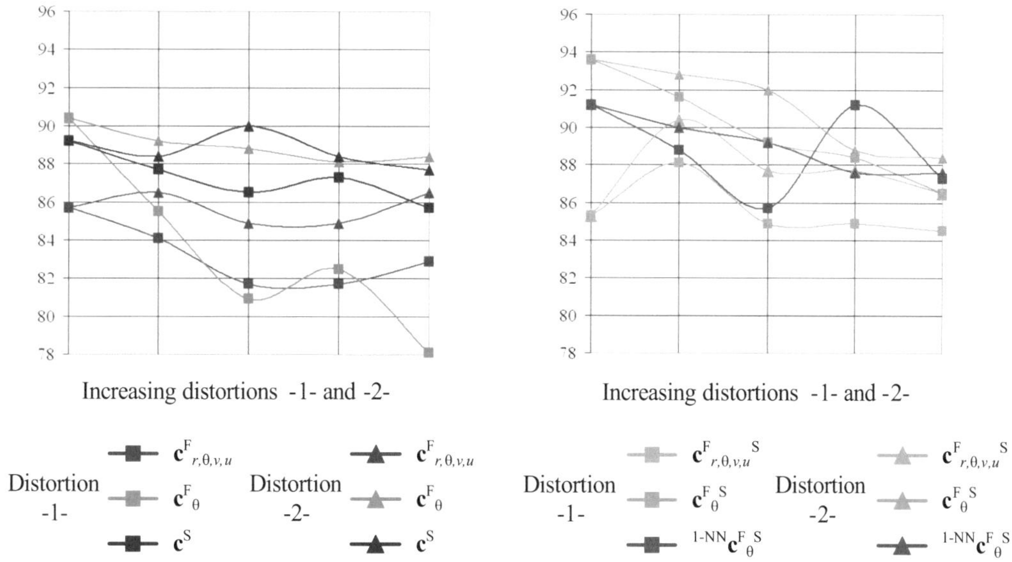

Fig. 9. Classification rates for image distortions of type -1- and of type -2- by means of the two groups of feature vectors $\{ \mathbf{c}^{F}_{r,\theta,v,u}; \mathbf{c}^{F}_{\theta}; \mathbf{c}^{S} \}$ and $\{ \mathbf{c}^{F}_{r,\theta,v,u}{}^{S}; \mathbf{c}^{F}_{\theta}{}^{S}; {}^{1\text{-NN}}\mathbf{c}^{F}_{\theta}{}^{S} \}$.

The detection rates were computed for different image sets and correspond to increasing distortions of type -1- and of type -2-. Left to right values: detection rates for image set Φ_0^0 to image sets Φ_1^4 and Φ_1^2. On the whole, the reported classification rates in Fig. 9 (right) are higher than those depicted in Fig. 9 (left). This shows the importance of determining the adequate features with the appropriate feature selection processes. In case of the considered inspection task, optimal features sets are 10 directional Fourier features and the 20 adapted stripes feature, whereas the optimal feature selection method is the wrapper-based 1-NN approach. However, if the feature fusion permits to reach higher classification rates of approximately 2 % (difference between the maximal detection rates of both considered graphics), the FSS method does not improve the classification rates, as similar or even lower classification results are observed when the FSS method is applied.

The last investigation is dedicated to a more detailed depiction of the considered FSS method, in order to determine the relevant features.

4.3.3 Optimal features in case of the generalization approach

In order to determine the most relevant features the influence of increasing distortions of type -1- and of type -2- on the number and types of selected features is investigates. Table 3 shows the results in case of a wrapper 1-NN approach and a 10-fold cross-validation.

	Type -1- distortion						Type -2- distortion				
Feature set	Φ_0^0	Φ_1^4	Φ_2^4	Φ_3^4	Φ_4^4		Φ_0^0	Φ_1^4	Φ_2^4	Φ_3^4	Φ_4^4
$N_{c,sub}$	90	90	95	107	108		90	99	107	116	107
$c^S(00)$	0	0	0	0	0		0	0	1	0	0
$c^S(01)$	1	1	0	0	3		1	1	0	1	0
$c^S(02)$	0	0	0	0	0		0	0	2	0	0
$c^S(03)$	0	0	1	2	3		0	0	3	2	3
$c^S(04)$	**3**	**3**	**2**	**8^{*}**	**2**		3	3	4	5	3
$c^S(05)$	4	4	1	3	3		4	2	1	3	5
$c^S(06)$	2	2	4	1	0		**2**	**4**	**1**	**5**	**8^{*}**
$c^S(07)$	**9^{**}**	**9^{**}**	**3**	**8^{*}**	**3**		**9^{**}**	**8^{*}**	**7**	**6**	**5**
$c^S(08)$	6	6	6	3	3		6	6	6	6	6
$c^S(09)$	7	7	4	6	4		7	5	4	4	2
$c^S(10)$	7	7	6	7	5		7	4	6	6	1
$c^S(11)$	0	0	3	5	4		0	0	1	1	0
$c^S(12)$	6	6	3	3	5		6	6	2	6	1
$c^S(13)$	**2**	**2**	**6**	**8^{*}**	**10^{***}**		**2**	**5**	**5**	**8^{*}**	**4**
$c^S(14)$	0	0	1	0	7		0	4	4	4	3
$c^S(15)$	0	0	3	2	7		0	5	0	2	6
$c^S(16)$	**9^{**}**	**9^{**}**	**6**	**6**	**9^{**}**		**9^{**}**	**4**	**6**	**3**	**4**
$c^S(17)$	**2**	**2**	**5**	**10^{***}**	**6**		**2**	**1**	**4**	**5**	**8^{*}**
$c^S(18)$	1	1	5	4	6		1	3	0	4	6
$c^S(19)$	**6**	**6**	**8^{*}**	**10^{***}**	**9^{**}**		6	7	7	7	4
$c^F_\theta(0)$	0	0	2	1	0		0	0	0	0	1
$c^F_\theta(1)$	0	0	2	1	1		0	1	3	3	2
$c^F_\theta(2)$	5	5	3	4	1		5	7	4	5	5
$c^F_\theta(3)$	4	4	5	2	3		4	2	5	2	3
$c^F_\theta(4)$	4	4	4	1	3		**4**	**5**	**9**	**6**	**6**
$c^F_\theta(5)$	**10^{***}**	**10^{***}**	**10^{***}**	**10^{***}**	**10^{***}**		**10^{***}**	**10^{***}**	**10^{***}**	**10^{***}**	**10^{***}**
$c^F_\theta(6)$	0	0	0	1	0		0	2	4	0	0
$c^F_\theta(7)$	1	1	2	1	0		1	2	2	5	4
$c^F_\theta(8)$	1	1	0	0	1		1	1	5	3	1
$c^F_\theta(9)$	0	0	0	0	0		0	1	1	4	6

Table 3. Selected features when a wrapper 1-NN approach is used for increasing distortion of type -1- and -2-. The maximum number of times a feature can be selected is 10. The variables $N_{c,sub}$ on the left give the total number of selected features after the 10 runs. The 10 time, 9 time and 8 time selected features are marked with ***, ** and *. Results for all relevant features are marked in bold.

An important parameter is the variable $N_{c,sub}$, which is the total number of selected features after the 10 runs of the 10-fold cross-validation. As 10 is the maximum number of times a feature can be selected, $N_{c,sub}$ / 10 is the average measure of feature relevance. For both tables, increasing the distortion of the bright/dark structures, leads to an increase of the necessary relevant features.

A general remark for both tables concerns the types and the number of selected features, which are approximately the same. It appears that approximately seven features, i.e. only a fourth of the initial 30 ones, are relevant. Most of the selected features are adapted ones, whereas mainly the directional 90° Fourier features have a strong relevance.

It is also noticeable, that feature relevance is related to the bright/dark structure distortion degree. As an example, in case of both tables, the importance of feature $c^S(13)$ is proportional to the distortion degree, whereas the contrary is observed for feature $c^S(07)$.

5. Conclusion

The chapter addressed the inspection by means of structure lighting of complex surfaces within the context of industrial inline quality control processes. The whole processing chain was considered by first tackling the structured light generation, then the processing of the acquired images, and finally the classification of the segmented and characterized structured light patterns.

At first, the generation of appropriate structured light patterns has been tackled. From the two described illumination techniques, the "transmission" and the "collimation" one, it has been demonstrated that the latter is more appropriate for the characterization of geometrical structures as such a lighting technology permits to reduce the diffuse part of the reflected light. As the aim of the chapter is to define a general approach for complex surface interpretation by means of structured lighting, a general and adapted method has been presented. The principle consists of recording and processing regular patterns by projecting patterns which are adapted to the complex geometries of the surfaces under inspection. The generation of adapted patterns uses coded light to determine the homography linking the screen and camera points for pattern adaptation. Different images examples validate the proposed procedure.

The second part was dedicated to the adapted segmentation of the generated structured patterns. The originality of the proposed method relies on the use of a bio-inspired approach to compute saliency maps. It has been showed how such maps permit to reveal disturbed regions of stripe patterns synonymous of defective surface parts. The proposed free-form segmentation procedure consists of a saliency map generation preceding a binarization. It has been showed how a supervised approach permits to compute the most optimal threshold for all the considered reference images. The segmented stripe patterns were characterized afterwards by means of different features, whose innate parameters were adapted to the special task of stripe pattern interpretation. Two features sets lead to highest classification rates, the Fourier and the adapted Stripe features.

The third and last part addressed the definition of the most optimal stripe structures processing chain by using the classification rate as evaluation criteria. The purpose is to tackle the inspection of complex surfaces. At first, a reference industrial inspection system using a "collimation" lighting for generating regular patterns, has been considered. By means of the classification rate, it has been demonstrated that the most optimal chain consists of (i) defining 64 x 64 pixels fixed size patterns, (ii) processing them with Fourier and adapted Stripe features, and (iii) classifying the computed vector with a 1-NN classifier.

In a second part, the general inspection task consisting of interpreting complex surface has been tackled. The most optimal processing chain for complex surface interpretation was defined (i) by generalizing the reference dataset to more complex surfaces and (ii) by retrieving the adapted feature sets using a wrapper-based selection feature procedure. The results showed that only a certain number of features are relevant, and that reduced but appropriate features permit to reach classification rates for complex surfaces similar to rates obtained with more "simple" geometries.

To conclude, this chapter proposed a feature-based surface characterization methodology based on the direct interpretation of regular patterns, adapted to the geometrical complexity of the surface. This approach permits to reach higher positioning tolerances in case of real-time surface inspection. As no depth information is computed, the proposed inspection procedure is more dedicated to binary decisions, i.e. whether the surface is defective or not. However, such detailed characterization is not necessary in case of the real-time industrial surface inspection. A preliminary determination of the projected pattern characteristics in accordance to the smallest critical defects to be detected is sufficient in this context.

6. References

Abouelela, A., Hazem M. and Eldeeb, H. and Wahdan, A. and Nassar, S. M.. (2005). Automated vision system for localizing structural defects in textile fabrics. *Pattern Recognition Letters*, Vol. 26, No. 10, pp. 1435-1443.

Balzer, J.; Werling, S. (2010). Principles of Shape from specular reflections. *International Journal of Measurement*. Vol. 43, No. 10. pp. 1305-1317.

Caulier, Y.; Spinnler, K.; Bourennane, S.; Wittenberg T.. (2007) New structured Illumination Technique for the Inspection of High Reflective Surfaces. *EURASIP Journal on Image and Video Processing*.

Caulier, Y.; Spinnler, K.; Bourennane, S.; Wittenberg T.. (2008). Specific Features for the Analysis of Fringe Images. *Optical Engineering Journal*, Vol. 47, No. 5. pp. 057201-01-057201-11.

Caulier, Y. Andreas Goldschmidt, Spinnler, K.; Arnold, M. Automatic detection of surface and structural defects on reflecting workpieces. *Photonik international*. Vol. 2.

Caulier, Y.; Bourennane, S.. Visually Inspecting Reflective Surfaces: A Feature-Based Approach. (2010). *Transactions on Pattern Analysis and Machine Intelligence (TPAMI)*, 2010.

Cohen, A.; Daubechies, I.; Feauveau, J.-C. (1997). Biorthogonal bases of compactly supported wavelets. Communications on Pure and Applied Mathematics. Vol. 45, No. 5, pp. 485–560.

Coifman, R. and Meyer, Y. and Wickerhauser, V. (1992). Wavelet analysis and Signal Processing, In: *Wavelets and their applications*. Ruskai, M. B., pp. 53-178, Jones and Bartlett Publischers, 978-0867202250.

Daugman, J.. (1985). Uncertainty relation for resolution in space, spatial frequency and orientation optimized by two-dimensional visual cortical filter. *Optical Society of America*. Vol. 2, pp. 1160-1169.

Dunn, D. (1995). Optimal Gabor Filters for Texture Segmentation. *IEEE Trans. in Image Processing*. Vol. 7., No. 4, pp. 947-964.

Ihrke, I. and Kutulakos, K. N. and Lensch, H. P. A. and Magnor, M. and Heidrich, W.. (2008). State of the Art in Transparent and Specular Object Reconstruction. *EUROGRAPHICS 2008 STAR - State of the art report*, 2008.

Itti, L.; Koch, C.; Niebur, E.. (1998) A model of saliency-based visual attention for rapid scene analysis. *IEEE Transactions on Pattern Analysis and Machine Intelligence*, Vol. 20, No. 11, pp. 1254-1259.

Jüptner, W. and Kreis, T. and Mieth, Ulrike and Osten, Wolfgang. (1994). Application of neural networks and knowledge-based systems for automatic identification of fault-indicating fringe patterns. *Proceedings of SPIE, Photomechanics*. Warsaw, Poland, May, 1994.

Kohavi, R.. A Study of Cross-Validation and Bootstrap for Accuracy Estimation and Model Selection. *Proceedings of Proceedings of the 14th international joint conference on Artificial intelligence*, 1-55860-363-8, San Francisco, 1995.

Kovesi, P. D. (2011). MATLAB and Octave Functions for Computer Vision and Image Processing. In: *Centre for Exploration Targeting, School of Earth and Environment*, Available from: http://www.csse.uwa.edu.au/~pk/Research/MatlabFns.

Krüger, S.; Wernicke, G.; Osten W.; Kayser, D.; Demoli, N.; Gruber, H.. (2001) Fault detection and feature analysis in interferometric fringe patterns by the application of wavelet filters in convolution processors. *Journal of Electronic Imaging*, Vol. 10, No. 1, pp. 228-233.

Li, X.. (2000). Wavelet transform for detection of partial fringe patterns induced by defects in nondestructive testing of holographic interferometry and electronic speckle pattern interferometry. *Optical Engineering Journal*. Vol. 39, No. 10, pp. 2821.

Mallat S.. (1989). A Theory for Multiresolution Signal Decomposition: The wavelet Representation. *IEEE Transaction of Pattern Anal. and Machine Intell.*, Vol. 11, No. 7, pp. 674-693.

Rajpoot, N. M.. Texture Classification Using Discriminant Wavelet Packets Subbands. *Proceedings of the 2002 45th Midwest Symposium on Circuits and Systems*, 0-7803-7523-8, 4-7. August 2002.

Qian, K.; Seah, H. S.; Asundi, A. (2005). Fault detection by interferometric fringe pattern analysis using windowed Fourier transform. *Measurement Science and Technology*, Vol. 15, pp 1582-1587.

Randen, T.; Husoy, J. H.. (1999). Filtering for Texture Classification: A Comparative Study. *IEEE Transactions on Pattern Analysis and Machine Intelligence*, Vol. 21, No. 4, pp. 291-310.

Takeda, M.; Ina, H.; Kobayashi, S.. (1982). Fourier-transform method of fringe-pattern analysis for computer-based topography and interferometry. Journal. Optical. Society of America. Vol. 72, No. 1, pp. 156-160.

Tikkanen, P. E.; Sellin, L. C.. Wavelet and Wavelet Packet Decomposition of RR and RT max interval time series, *Proceedings of the 19th Annual International Conference of the IEEE Engineering in Medicine and Biology Society*, 0-7803-4262-3, Chicago, IL., USA, 30 October-2 November 1997.

Tuceryan, M.; Jain, A.. (1998). Texture Analysis, In: *The Handbook of Pattern Recognition and Computer Vision (2nd Edition)*, Chen, C. H.; Pau, L. F.; Wang, P. S. P., pp. 207-248.

Unnikrishnan, R.; Pantofaru, C.; Hebert, M.. (2007). Toward Objective Evaluation of Image Segmentation Algorithms. *IEEE Transaction of Pattern Anal. and Machine Intell.*, Vol. 29, No. 6, pp. 929-944.

Wagner, T.; Kueblbeck, C.. (1996). Automatic Configuration of Systems for Surface Inspection. *Machine Vision Application in Industrial Inspection*, Vol. 3029, No. 1, pp. 128-138.

Weszka, J.. (1978). A survey of threshold selection techniques. *Journal of Computer Graphics and Image Processing*. Vol. 7, No. 2, pp. 259-265.

Wagner, T.. (1999). Automatische Konfiguration von Bildverarbeitungssysteme. University of Erlangen-Nürnberg, Germany.

[Zha01] A review of recent evaluation methods for image segmentation. Technical report, Department of Electronic Engineering, Tsinghua University, Beijing, China, 2001.

Zhi, H.; Johansson, R. B. (1992). Interpretation and Classification of Fringe Patterns, *Proceedings of 11th International Conference on Image, Speech and Signal Analysis*, 0-8186-2920-7, The Hague, Netherlands, 30 August-3 September 1992.

10

Discontinuity Detection from Inflection of Otsu's Threshold in Derivative of Scale-Space

Rahul Walia[1], David Suter[2] and Raymond A. Jarvis[1]
[1]*Monash University*
[2]*Adelaide University*
Australia

1. Introduction

Discontinuity detection is studied across disciplines of thermodynamics, chemistry, geology, manufacturing, equipment maintenance, signal processing, computer architecture (bit recognition), finance (jump processes to model markets) and image processing. In image processing an edge is often modeled as a discontinuity (Lindeberg, 1998). Hence discontinuity detection, can provide edge information for image analysis with applications in robotic vision, medical imaging, tomography and surveillance etc. Scale-Space theory (Koenderink, 1984; Lindeberg, 1994; Witkin, 1983), is a framework for multi-scale analysis of function/image. While there are non-Gaussian Scale-Space representations (Duits et al., 2003), this article is confined to widely accepted Gaussian Scale-Space (Babaud et al., 1986; Lindeberg, 1994). Existing Scale-Space literature is focused mainly on developing Scale-Space theory with a view to:

a. Study impact on underlying signals/images (Babaud et al., 1986; Koenderink, 1984; Lindeberg, 1998; Romeny, 1994; Witkin, 1983)
b. Determine appropriate scale(s) relevant to the image/signal(Lindeberg, 1994).
c. Extract information and knowledge to develop applications like feature detection, feature classification, image segmentation, image matching, motion estimation, shape computation and object recognition etc.
d. Correlate the Scale-Space framework with biological vision (Hubel & Wiesel, 1987; Koenderink & Doorn, 1992; Koenderink & van Doorn, 1987; Young, 1987).

Current Scale-Space literature, does not adequately explore the statistical component of Scale-Space. There are contextual applications of various statistical parameters (Rodriguez, 2006; Sakai & Imiya, 2009; Zagal et al., 2000), in contemporary Scale-Space research, but they are limited in scope to specific applications or/and statistics of image features like blob volume, clusters, thresholds etc. Researchers would be well-assisted if some theoretical basis were available for statistical assumptions in Scale-Space. In this article, we present a theoretical foundation for some statistical assumptions with regard to the derivative of a discontinuity in Scale-Space. A discontinuity has an infinitesimal existence in Scale-Space, which leads to the assumption of continuity of underlying image/function in any conventional Scale-Space analysis (Koenderink, 1984; Lindeberg, 1994). This article reveals that even though Scale-Space eliminates discontinuity at infinitesimal scale, the Probability

Density Function (PDF) of the derivative of a discontinuity retains its unbalanced bimodality in Scale-Space. This chapter makes following theoretical contributions:

1. Derivation of Probability Density Function (PDF) and Cumulative Distribution Function (CDF) for the derivative of a discontinuity in Scale-Space (Theorem 2).
2. Proof of bimodality (Theorem 3) and unbalance (Theorem 4) of the PDF of the derivative of a discontinuity in Scale-Space.
3. Proof that the Otsu's Threshold (OT) (Otsu, 1979) owing to its sensitivity to unbalanced and bimodal PDFs has different patterns in Scale-Space based on the presence / absence of a discontinuity:
    a. Transient Increase: Discontinuity present.
    b. Monotone Decrease: Discontinuity absent.

The above mentioned theoretical results, are then applied for a simultaneous solution of following problems in image processing (Figure 1):

1. Scale appropriate to the discontinuity.
2. Threshold appropriate to the discontinuity.
3. Boundaries of entities in images.

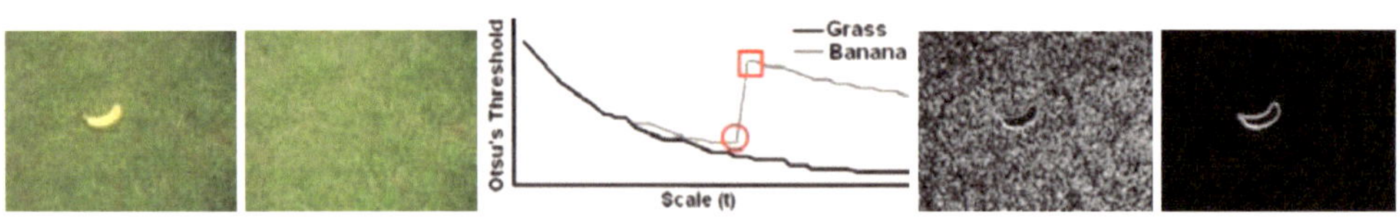

(a) (b) (c) (d) (e)

Fig. 1. OT Patterns in the absence and presence of a discontinuity (boundary). (a) **Banana**: Image with discontinuity (boundary).(b) **Grass**: Image without discontinuity (boundary). (c) OT Plots against Scale for "Grass" and "Banana". □ and ○ : Upper and Lower points of inflection. (d), (e) Segmentation at lower and upper points of inflection respectively.

2. Statistical distributions of a Gaussian function

The term Gradient Magnitude in Scale-Space (GMSS) will be used hereon to represent

a. *In 1-D Non-Discrete Functions* The derivative of the Scale-Space representation of the functions
b. *In 2-D Discrete Images* The magnitude of the gradient (computed by Sobel operator) of Scale-Space representation of the images.

In this section, the statistical distributions of the GMSS of a discontinuity will be derived. The reason for doing so is to show that the PDF of the GMSS of a discontinuity is bimodal and unbalanced i.e. the probability of one mode far exceeds the probability of the other mode. A discontinuity is mathematically represented as a step function. Consequently the derivative of the discontinuity is a Dirac's Delta (Khuri, 2004) as shown in Figure 2. Convolution of the Dirac delta (δ)[1] with a Gaussian function will result in GMSS of a discontinuity.

[1] Since a Dirac's Delta is zero everywhere except at one point, therefore its PDF will be of the type shown in Figure 2. The PDF of a Dirac's Delta exists only at two points i.e. at ($x = 0$ and $x = \infty$) and is bimodal and unbalanced.

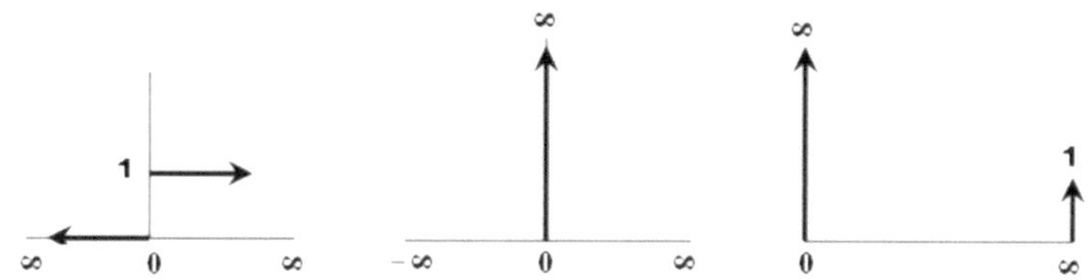

Fig. 2. Unit step function Dirac Delta PDF of Dirac Delta

Theorem 1. The GMSS $L'(x;t)$ of a Step function is given by a Gaussian Kernel G(x;t):

$$L'(x;t) = G(x;t) \qquad where$$

$$G(x;t) = \frac{e^{-\frac{x^2}{2t}}}{\sqrt{2\pi t}} : \text{Gaussian Kernel at Scale (t)} \tag{1}$$

$$(-N < x < N : N \to \infty) : \text{Domain of the Function}$$

Proof. For a Step function $H(x)$, the Scale-Space (SS) representation $L(x;t)$ is given by

$$H(x) = \left\{ \begin{matrix} 1 \; if \quad N > x > 0 \\ 0 \; if \; -N < x < 0 \end{matrix} \right\} \tag{2}$$

$$L(x;t) = H(x) \otimes G(x;t) \tag{3}$$

Convolution commutates with differentiation and the derivative of a Step function is a Dirac Delta ($\hat{\delta}(x)$) (Khuri, 2004) (Figure 2). Consequently:

$$L'(x;t) = \frac{d}{dx}[H(x) \otimes G(x;t)] = \frac{d}{dx}[H(x)] \otimes G(x;t) = \hat{\delta}(x) \otimes G(x;t)$$

$$where \qquad \hat{\delta}(x) = H'(x) = \left\{ \begin{matrix} \infty \; if \; x = 0 \\ 0 \;\; if \; x \neq 0 \end{matrix} \right\} \tag{4}$$

A Gaussian Kernel reduces to Dirac Delta at zero Scale i.e ($\hat{\delta}(x) = G(x;0)$), therefore (4) is equivalent to the convolution of two Gaussian Functions with scales ($t_1 = 0, t_2 = t$). Since $[G(x; t_1 + t_2) = G(x;t_1) \otimes G(x;t_2)]$ (Lindeberg, 1994), therefore (4) simplifies to:

$$L'(x;t) = G(x;t) \quad where \quad x \in 2N \qquad \square$$

Theorem 1 simplifies the GMSS of a discontinuity to a Gaussian Function, which in turn allows formulation of the statistical characteristics of the GMSS of a discontinuity.

2.1 PDF of a Gaussian function

Theorem 2. A continuous random variable **g** which takes the values $g \in (\frac{1}{\sqrt{2\pi t}}, \frac{1}{\sqrt{2\pi t}} e^{\frac{-N^2}{2t}} : N \to \infty)$, given by a Gaussian function $G(x;t)$ has following statistical distributions:

$$PDF: \qquad f_g(g) = \frac{t}{Ng\sqrt{-t log_e(2\pi t g^2)}} \tag{5}$$

$$CDF: \qquad F_g(g) = \frac{\sqrt{-t log_e(2\pi t g^2)}}{N} \tag{6}$$

Proof. Gaussian function ($G(x;t)$) is symmetric and provides a one to one, monotonic and inverse mapping between x and g, in each half of the Cartesian plane. Therefore a *uniformly distributed* random variable ($\mathbf{X}$) which takes the values $x \in (0, N : N \to \infty)$ in the positive spatial domain of the Gaussian function can derive the PDF and CDF for $\mathbf{g}$. The uniformly distributed PDF of ($\mathbf{X}$) is given by:

$$f_x(x) = \frac{1}{N} \text{ where } x \in \{0, N : N \to \infty\} \tag{7}$$

The equivalent PDF and the domain for the gaussian variable ($\mathbf{g}$) is given by:

$$f_g(g) \text{ where } g \in \left\{ \frac{1}{\sqrt{2\pi t}}, \frac{1}{\sqrt{2\pi t}} e^{\frac{-N^2}{2t}} : N \to \infty \right\} \tag{8}$$

The probabilities for both the random variables are equal in the mapped ranges :

$$\begin{aligned} \int_{\frac{1}{\sqrt{2\pi t}}}^{g} f_g(g)\mathrm{d}g = \int_0^x f_x(x)\mathrm{d}x \quad where \\ x = G^{-1}(g;t) = \sqrt{-t\log_e(2\pi t g^2)} \end{aligned} \tag{9}$$

Introducing a change of variable from x to g in the right hand side of (9) and solving:

$$\int_{\frac{1}{\sqrt{2\pi t}}}^{g} f_g(g)\mathrm{d}g = \int_{\frac{1}{\sqrt{2\pi t}}}^{g} f_x(x) \left\| \frac{d[G^{-1}(g;t)]}{dg} \right\| \mathrm{d}g \tag{10}$$

$$f_g(g) = \frac{t}{Ng\sqrt{-tlog_e(2\pi t g^2)}} \tag{11}$$

The CDF [2] can be computed by integrating the PDF (11):

$$P[g \leq \mathbf{g} \leq \sqrt{2\pi t}] = F_g(g) = \frac{\sqrt{-tlog_e(2\pi t g^2)}}{N}$$ □

Alternate proof of PDF($F_g(g)$), can be provided by replacing the value of (x) from (9) in the CDF ($F_x(x) = x/N$) of uniformly distributed variable $\mathbf{X}$ and then differentiating it w.r.t (g), which would provide expression (11).

2.2 Bimodality of the PDF of a Gaussian function

Theorem 3. The PDF $f_g(g)$ of a Gaussian function is bimodal.

Proof. The bimodality of the PDF can be proved by the existence of exactly one minima in the PDF ((Eisenberger, 1964; Kemperman, 1991; Schilling et al., 2002)). For a point (g_0) belonging to the domain of the PDF to be a minima, it's first derivative should be zero ($f_g^{'}(g_0) = 0$) and

[2] The correctness of the expression can easily be verified by replacing the term ($\sqrt{-tlog_e(2\pi t g^2)}$) in CDF with x from (9). This gives ($P[0 \geq \mathbf{X} \geq x] - F_x(x) = x/N$) which is the expression of a CDF of a uniformly distributed variable ($\mathbf{X}$)

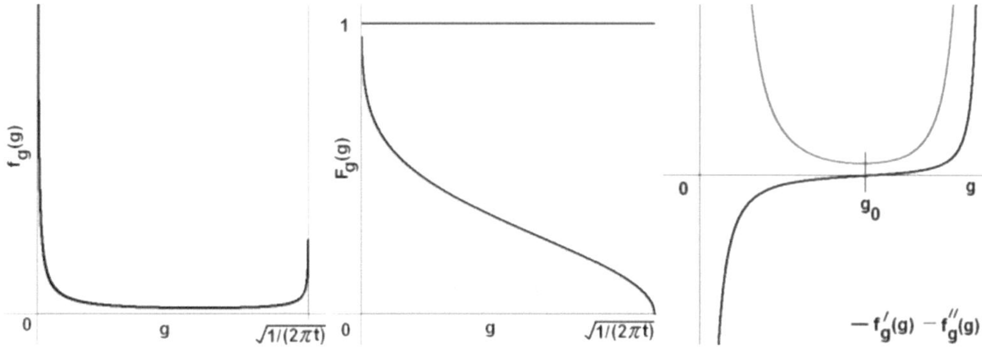

(a) PDF (b) CDF (c) Derivatives of the PDF

Fig. 3. PDF, CDF and derivatives (w.r.t. g) of the PDF of a Gaussian Function. The PDF has an inverse J-Shape. The derivatives show that only one minimum exists in the PDF making it bimodal.

it's second derivative should be positive ($f_g''(g_0) > 0$). The first and the second derivatives (with respect to g) of the PDF are given by:

$$f_g'(g) = \frac{t^2}{Ng^2(G^{-1}(g;t))^3} - \frac{t}{Ng^2G^{-1}(g;t)} \tag{12}$$

$$f_g''(g) = \frac{3t^3}{Ng^3(G^{-1}(g;t))^5} - \frac{3t^2}{Ng^3(G^{-1}(g;t))^3} + \frac{2t}{Ng^3(G^{-1}(g;t))} \tag{13}$$

The minima g_0 in the PDF can be located from the root(s) of (12)

$$\begin{aligned} \frac{t^2}{g_0^2(G^{-1}(g_0;t))^3} &= \frac{t}{g_0^2G^{-1}(g_0;t)} \\ \Rightarrow g_0 &= \frac{1}{\sqrt{2e\pi t}} \quad \forall \; t > 0 \end{aligned} \tag{14}$$

Existence of only one root for the first derivative of the PDF, implies that only one extrema exists in the PDF. Substituting the value of g_0 from (14) into (13) and solving

$$f_g''(g_0) = 4\sqrt{2}t^2(e\pi)^{\frac{3}{2}} > 0 \tag{15}$$

From (14) and (15), g_0 is the (only) minima in the PDF, therefore it establishes the bimodality of the PDF (Figure 3). □

2.3 Unbalance in modes of the PDF of a Gaussian function

Theorem 4. The bimodal PDF $f_g(g)$ of the GMSS of a Step Discontinuity is unbalanced, i.e the probability of one mode is much greater than the other.

$$F_g(g_0) << 1 - F_g(g_0) \tag{16}$$

Proof. The probabilities of the two modes separated at g_0 can be computed from CDF (6)

$$\begin{aligned} P(g \geq g_0) &= F_g(g_0) \\ &= \frac{g_0}{N}\sqrt{-t\log_e\left(2\pi t g_0^2\right)} = \left(N\sqrt{2e\pi}\right)^{-1} \end{aligned} \tag{17}$$

$$P(g < g_0) = 1 - F_g(g_0) \tag{18}$$

Dividing (17) by (18) gives the ratio of the probabilities of the two modes:

$$\frac{F_g(g_0)}{1 - F_g(g_0)} = \begin{matrix} Limit \\ N \to \infty \end{matrix} \frac{1}{N\sqrt{2e\pi} - 1} << 1 \tag{19}$$

□

2.4 Scale life of the GMSS of a discontinuity

Theorem 5. The Scale-Life (SL) of a discontinuity (with a magnitude A), i.e the interval of scales ($t \in (0, SL)$) within which the discontinuity can be statistically identified by the unbalanced bimodality of the PDF of the GMSS of a discontinuity is given by:

$$SL = \frac{A^2}{2\pi\epsilon^2} \quad where \quad \epsilon = \text{Upper Bound of error} \tag{20}$$

Proof. From Theorem 1 for a discontinuity with a magnitude A, the GMSS will be given by:

$$L'(x;t) = \frac{Ae^{-\frac{x^2}{2t}}}{\sqrt{2\pi t}} \tag{21}$$

In a manner similar to proof of Theorem 2, it can be shown that the PDF ($f_g(g)$) of the GMSS (21), will be defined in the interval $g \in (0, A/\sqrt{2\pi t})$ with the second mode existing at $(A/\sqrt{2\pi t})$. For this mode to be identifiable as a separate mode it should be greater than or equal to (ϵ) i.e.

$$\frac{A}{\sqrt{2\pi t}} \geq \epsilon \Rightarrow t \leq \frac{A^2}{2\pi\epsilon^2} \tag{22}$$

□

The concept of an infinitesimal existence of a discontinuity in Scale-Space/Heat Equation is acknowledged by research community (Gonzalez-Velasco, 1995; Lindeberg, 1994; Widder, 1975), but seldom defined. Theorem 5 provides one (amongst plausibly many) rigorous definition of the life of a discontinuity, derived from (and therefore limited to) the statistics of the GMSS of a discontinuity.

Implication of (ϵ): Any discrete application of the theoretical results would invoke the upper bound of error(ϵ), and therefore needs to be understood in the context of discretization in general, and selection of histogram bin size in specific (in images/signals). The selection of bin size inadvertently defines (ϵ) and is dependent on:

1. Physical limitations of the sensor/hardware: E.g.a camera might be able to distinguish 8, 64 or 256 intensity levels depending on 3,6 or 8 bit representation. The upper bound of error (as measured with respect to absolute ambient intensity) for a 8 bit representation will be much lower than that of 3 bit representation.

2. Accuracy desired by the user: Even though the sensor is capable of higher precision (or lower error), an algorithm/user might require a lower precision, wherein the upper bound of error is artificially set at a higher value.

Hence the ability of algorithms to capture the unbalance and bimodality of the PDF within the Scale-Life will depend on the precision of the hardware as well as the bin size of the histogram.

2.5 Comments

The theoretical results of this section can be perceived to be at slight variance with the assumptions and models of a discontinuity in conventional scale-space, and the reasons for this variance will be discussed in this subsection. The widely accepted norm of *ignoring a discontinuity in Scale-Space and analyzing the underlying signal/image as if it were continuous*, can be attributed to the following factors:

1. *Requirement of Scale-Space framework,* to comply with the principles of homogeneity and isotropy, necessitates the framework to remain uncommitted to a gaussian scale. Consequently modeling a discontinuity as done in heat equation, would result in violation of the fundamental requirements of Scale-Space.
2. *Inadequate Information:* Most of the problems of Computer Vision, are related to identifying the presence/absence of a discontinuity followed by a contextual analysis of the discontinuity. In the absence of this basic information about the presence of a discontinuity, much less its properties like the magnitude and location of the discontinuity, it is difficult to model the transient presence of a discontinuity in Scale-Space.
3. *Absence of appropriate model:* Even if the location and magnitude of the discontinuity were available, a model to represent the discontinuity is difficult to prepare, because it leads to a lot of unanswered questions like how long does the discontinuity last? and, how to model the transfer from a discontinuous state to a continuous state?
4. *Mathematical simplification:* Theoretically a discontinuity disappears at an infinitesimal scale, therefore by ignoring this infinitesimal scale, a continuous model of a discontinuity can be mathematically justified.

The text (Theorems 1, 2 and 5) so far, is not meant to contradict or discredit existing conventions of Scale-Space , but to present a mathematically valid alternate representation of the Scale-Space. As an illustration of alternate (to Scale-Space) representations of a discontinuity, consider the heat equation (Gonzalez-Velasco, 1995; Widder, 1975). In heat equation a discontinuity may be explicitly modeled in following mathematically valid conventions:

1. As a Neumann Boundary Condition which specifies the rate of temperature (equivalent to intensity) change at the boundary.
2. As a heat source in space delimited by boundaries.

These alternate models adopted by a broader theoretical framework of heat equations, illustrate the need to represent a discontinuity in forms other than the one adopted in conventional Scale-Space theory. The PDF of the GMSS of a discontinuity (Theorems 1, 2 and 5)is an alternate representation of a discontinuity, which is unaltered in the Scale-Life (Theorem 5) including the zeroth scale. The unbalanced bimodality of the PDF can be applied homogeneously across the scales for the Scale-Life of a discontinuity, without the ambiguity of modeling or ignoring a transition from a discontinuous state to a continuous state.

This section provides a mathematical expression for the the PDF of a Gaussian Function, with a universal applicability for disciplines employing Gaussian Functions e.g. inverse problems of heat equation, chemical diffusion and Scale-Space Theory. The unbalanced bimodality of the PDF of a Gaussian Function facilitates interpretation of the second mode of the PDF as a statistical outlier. Consequently any problem of a discontinuity detection can be reformulated as a statistical problem of outlier detection. The sections hereon can be viewed as one application of the general results of this section, wherein a statistical parameter (OT) sensitive to outlier data, is used to detect a discontinuity in images.

3. OT: Unbalanced histograms

In this section a general review of OT will be presented and an expression of OT for unbalanced bimodal PDF will be developed for 1-Dimensional function, with a view to accommodate the GMSS of a discontinuity. OT is statistically generated from a normalized

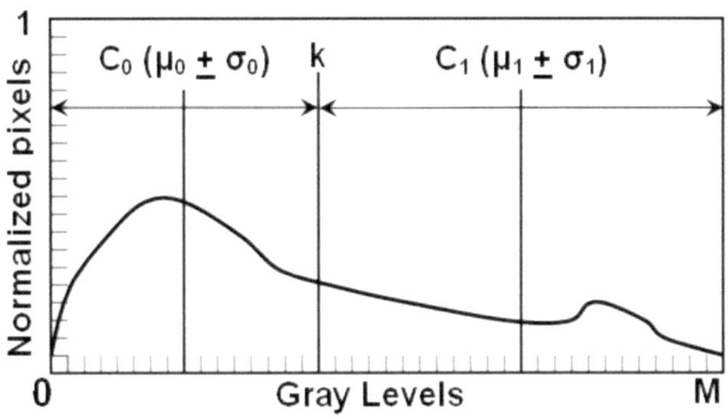

Fig. 4. Schema: Otsu's Threshold (which maximizes the Between Class Variance) in a histogram.

histogram with M bins corresponding to M gray levels in an image (Figure 4). Each bin represents the percentage of the pixels in the image with corresponding gray level. This normalized histogram is bifurcated into two classes C_0 and C_1 at a hypothetical threshold (k). The hypothetical threshold (k), Means (μ_0, μ_1) and Standard Deviations (σ_0, σ_1) of two classes are shown in Fig 4. The maximum of Between Class Variance (BCV) determines the appropriate threshold (OT). BCV ν_B is defined by 23:

$$\nu_B = \omega_0(\mu_0 - \mu_T)^2 + \omega_1(\mu_1 - \mu_T)^2 \quad \textit{where} \tag{23}$$

$$\omega_0 = \sum_{i=0}^{k} p_i, \quad \omega_1 = \sum_{i=k+1}^{M} p_i \tag{24}$$

0th *order Cumulative Moment for* C_0 *and* C_1

$$\mu_T = \sum_{i=0}^{M} ip_i, \gamma_k = \sum_{i=0}^{k} ip_i \tag{25}$$

1st *order Cumulative Moment up to M and k*

$$\mu_0 = \frac{\gamma_k}{\mu_T}, \mu_1 = \frac{\mu_T - \gamma_k}{\mu_T} \qquad (26)$$

Mean Gray Levels for C_0 and C_1 respectively

$$p_i = \frac{n_i}{N_T} : \text{Normalized probability at gray level } i \qquad (27)$$

n_i, N_T : *No of pixels at gray level i, Total pixels*

Theorem 6. OT (which maximizes BCV) is obtained at gray level (k*) defined by:

$$k^* = \frac{\mu_0 + \mu_1}{2} \qquad (28)$$

Proof. Differentiate v_B (23) with respect to gray levels (k) and equate to zero. For details see Lin (2003). □

The proof (Lin, 2003) is for histograms, but the results can easily be generalized to continuous PDFs. One solution of Theorem 6 is when the OT exists at the function/image mean.

Corollary 1. The maximum of BCV is obtained at the image mean (μ_T) if and only if the probabilities of the two classes are equal:

$$k^* = \mu_T \quad \Longleftrightarrow \omega_0 = \omega_1 = 0.5 \qquad (29)$$

Proof (If). Substituting $\omega_0 = \omega_1 = 0.5$ in μ_T (25):

$$\mu_T = \mu_0\omega_0 + \mu_1\omega_1$$
$$\mu_T = 0.5\mu_0 + 0.5\mu_1 \quad \square$$
$$\mu_T = k^* \quad \text{from (28)}$$

Proof (Only if). Equating μ_T (25) to k^* (28).

$$\frac{\mu_0 + \mu_1}{2} = \mu_0\omega_0 + \mu_1\omega_1$$
$$\text{substituting } \omega_0 = 1 - \omega_1 \text{ from (24)} \quad \square$$
$$\omega_0 = \omega_1 = 0.5$$

Corollary 1 allows analysis of the OT in terms of the function/image mean, without constructing a PDF/histogram. Some of the plausible distributions mentioned by Lin (2003) where Corollary 1 is applicable are *unimodal, perfectly balanced bimodal and unbalanced bimodal*. Corollary 1 can be tailored to a PDF containing two linearly separable classes with unbalanced probabilities (Figure 5). This is done with a view to develop an expression for OT, applicable to the PDF of a Gaussian function.

Let a random variable $\mathbf{Q} = \{q \in (0, q_{end})\}$ with PDF $f_q(q)$ be composed of two populations (NC, IC) which are linearly separable at $(q = \psi)$ having distributions $f_{nc}(q)$ and $f_{ic}(q)$ respectively, such that

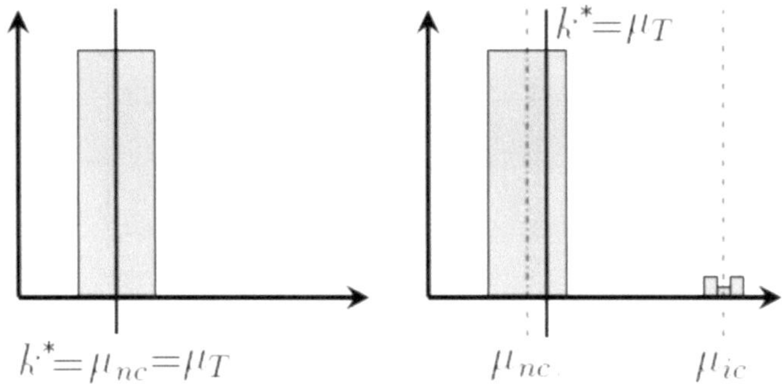

Fig. 5. OT Schema: (Left) Symmetric PDF and (Right) Unbalanced dual-class PDF (larger class is Symmetric about its mean)

$$f_q(q) = \left\{ \begin{array}{ll} m_{nc} f_{nc}(q) & if \quad q < \psi \\ m_{ic} f_{ic}(q) & else \end{array} \right\} \quad where \tag{30}$$

a.	$\psi \in (0, q_{end})$ Point of linear separation of two classes
b.	$m_{ic} = \frac{q_{end} - \psi}{q_{end}}, m_{nc} = \frac{\psi}{q_{end}}$ Probabilities of the two classes
c.	μ_{ic}, μ_{nc} Averages of the two classes
d.	$m_{ic} << m_{nc}$ PDF is Unbalanced
e.	$\mu_{ic} m_{ic} + \mu_{nc} m_{nc} = \mu_T$ Average value of the variable q
f.	$\mu_{ic} > \mu_{nc}$ Order of classes
g.	$f_{nc}(\mu_{nc} - q) = f_{nc}(\mu_{nc} + q)$ $\forall (q < \mu_{nc}) : f_{nc}(q) \neq 0$ $f_{nc}(q)$ is symmetric about μ_{nc}

Equation (30) presents a PDF which is a super-set of the PDF of a Gaussian function with following salient features:

1. It is unbalanced.
2. It is not strictly Bimodal, but accommodates bimodal PDFs.
3. Has an additional requirement of symmetry of the first mode about its mean, which is satisfied by a Gaussian function's PDF under limiting conditions (proof follows in Corollary 2).

Theorem 7. For a PDF of the kind (30), the OT is given by:

$$k^* = \mu_{nc} + m_{ic}(\mu_{ic} - \mu_{nc})$$

Proof. First consider that the PDF consists of *only NC*, i.e:

$$(f_q(q) = f_{nc}) \iff (\psi = q_{end}, \mu_T = \mu_{nc} \text{ and } m_{ic} = 0)$$

The probability of two halves of the PDF separated at a mean value ($k^* = \mu_T$) is ($\omega_0 = \omega_1 = 0.5$) (Figure 5 Left). Therefore from Corollary 1:

$$k^* = \mu_{nc} = \mu_T \tag{31}$$

If class (ic) with a very small probability $(m_{ic} << m_{nc})$ is added to this distribution (Figure 5 Right), then the OT and the PDF mean (μ_T) will change slightly, because for a very small change in μ_T, applicability of Corollary 1 will persist (Lin, 2003):

$$\begin{aligned} k^* &= \mu_T = m_{nc}\mu_{nc} + m_{ic}\mu_{ic} \\ &= \mu_{nc} + m_{ic}(\mu_{ic} - \mu_{nc}) \end{aligned} \tag{32}$$

□

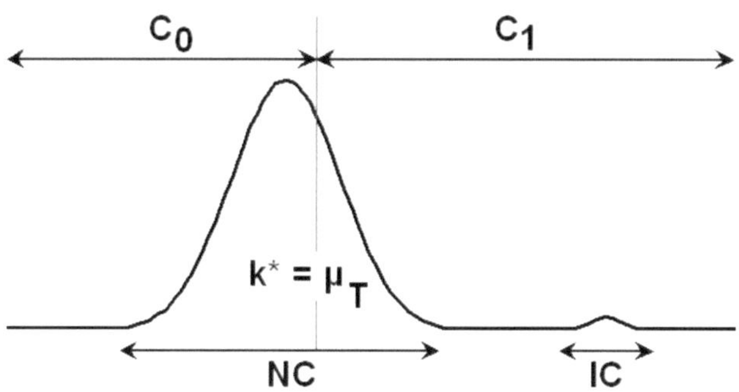

Fig. 6. Schematic illustration of two sets of classes :(C_0 and C_1) and (NC, IC) in an unbalanced PDF.

Schematic illustration of two sets of classes as discussed so far: (C_0 and C_1) and (NC, IC) in an unbalanced PDF is shown in Fig 6. The First set (IC and NC), appears as two separate distributions in the PDF. The second set of classes (C_0 and C_1) exists due to the hypothetical bifurcation of the PDF at OT. Equation (30) imposes less rigorous conditions for application of Theorem 7 for unbalanced PDFs as compared to (Lin, 2003). The PDF need not be strictly bimodal as long as it is unbalanced and the larger class (NC) is symmetric about its mean. In the next section, Theorem 7, will be adapted specifically for the PDF of a Gaussian function. The unique inverse J Shape (Figure 3) of the PDF of the Gaussian function implies that the PDF is concentrated around a value of zero. Thus by providing a rigorous definition of zero ϵ and its associated spatial domain ($\delta(t)$), following simplifications of Theorem 7 can be achieved:

1. Definition of point ψ in (30) and consequently linear separation of the PDF of the Gaussian Function into IC and NC.
2. Expressions for average values and the probabilities of the two classes IC and NC.
3. Proof of PDF's i.e. unbalance $m_{ic} << m_{nc}$.
4. Elimination of need to prove symmetry of bigger class (NC). Proof is provided in the next section.

4. OT for a Gaussian function

Formal definition of zero (ϵ) and the corresponding spatial domain $\delta(t)$, is obtained from Cauchy's Epsilon-Delta ($\epsilon, \delta(t)$) definition (Felscher, 2000) as applied to limit of $G(x;t)$ when $x \to \infty$.

Theorem 8. If ($\epsilon, \delta(t) \in \Re^+$) represent the real and positive upper bounds of error, for the Gaussian function ($G(x;t)$) and the associated spatial variable (x) respectively, where ϵ can be made infinitesimally small and $\delta(t)$ depends continuously on ϵ and scale (t), then the limit of the Gaussian function when $x \to \infty$ is given by:

$$\underset{x \to \infty}{Lt} G(x;t) = 0 \tag{33}$$

Alternatively for a given ϵ and a scale (t) there exists a $\delta(t)$, such that for all x belonging to the interval $(\|\delta(t)\|, \|\infty\|]$, the Gaussian function takes a value less than ϵ:

$$\exists \delta(t) : \forall x \in (\|\infty\| > \|x\| > \|\delta(t)\|) \Rightarrow \|G(x;t)\| < \epsilon \tag{34}$$

Proof. It is trivial to show from the definition of G(x;t) (1), that for a given ϵ following value of $\delta(t)$ provides the interval $(\|\delta(t)\|, \|\infty\|]$ for which $(G(x;t) < \epsilon)$:

$$\delta(t) = \sqrt{-t \log_e (2\pi t \epsilon^2)} \tag{35}$$

□

Magnitude (A) instead of unity, will change (35) to:

$$\delta(t) = \sqrt{-t \log_e (\frac{2\pi t \epsilon^2}{A^2})} \tag{36}$$

Graph of $\delta(t)$ from (36) at various ratios of (ϵ / A) for a Step Discontinuity is shown in Figure 7.

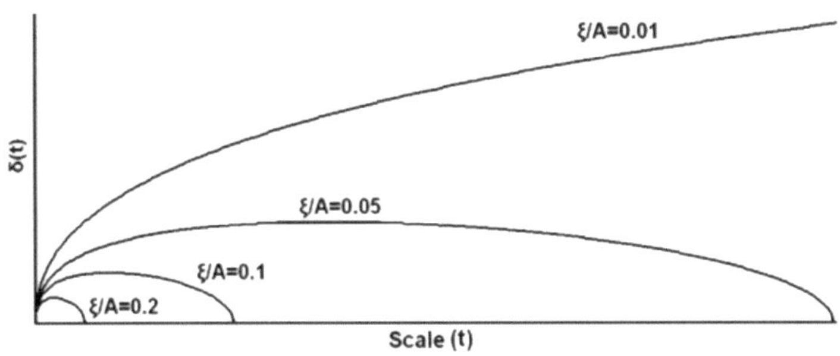

Fig. 7. Graph of $\delta(t)$ from (36) at various ratios of (ϵ / A) for the GMSS of a Discontinuity.

The graphs show that a maximum exists in each plot. At this point IC and NC can be defined in the context of the Gaussian Function with the help of ϵ.

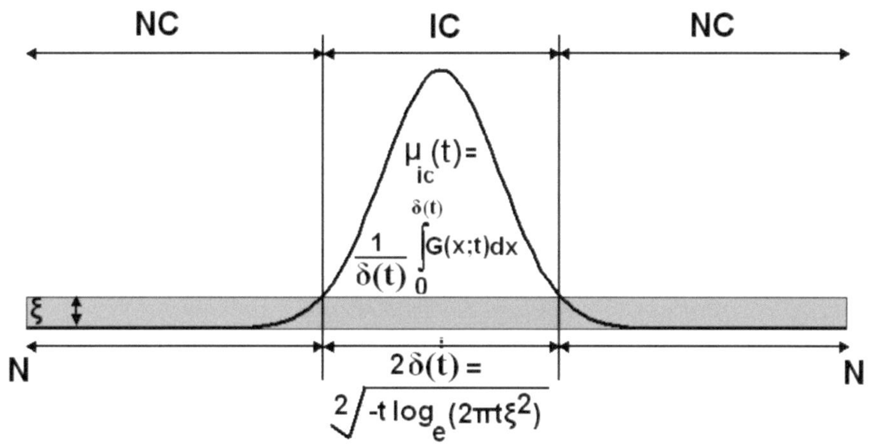

Fig. 8. GMSS of a Unit step (Gaussian) function along with various parameters of IC and NC.

4.1 Definition and statistics: IC and NC

Definition 1. Interface Class (IC) at scale (t) is defined as the spatial domain of the Gaussian Function where the value of the Gaussian Function is greater than the upper bound of error (ϵ).

$$\begin{aligned} IC(t) &= \{x \in N : G(x;t) > \epsilon\} \Rightarrow \\ IC(t) &= \{x \in N : x < \delta(t)\} \end{aligned} \tag{37}$$

Definition 2. Non-interface Class (NC) at scale (t) is defined as the spatial domain of the Gaussian Function where the value of the Gaussian Function is lesser than or equal to the upper bound of error (ϵ).

$$\begin{aligned} NC(t) &= \{x \in N : G(x;t) \leq \epsilon\} \Rightarrow \\ NC(t) &= \{x \in N : x \geq \delta(t)\} \end{aligned} \tag{38}$$

Figure 8 depicts the two classes IC and NC in the context of a Gaussian Function. Based on the above definitions the statistics of IC and NC can be determined.

Class	Statistic
NC	Mean: $\mu_{nc}(t) = 0$ Probability: $m_{nc}(t) = \dfrac{N - \delta(t)}{N}$
IC	Mean: $\mu_{ic}(t) = \dfrac{1}{\delta(t)} \int_0^{\delta(t)} G(x;t)dx$ Probability: $m_{ic}(t) = \dfrac{\delta(t)}{N}$

(39)

4.2 Applicability of theorem 7

Corollary 2. Bifurcating the Gaussian function $(G(x;t))$ or it's PDF $(f_g(g))$ at $(\psi = \epsilon)$, results in two classes IC and NC such that:

1. Two classes have an unbalanced probability.

$$P[g < \epsilon] >> P[g \geq \epsilon] \tag{40}$$

2. The larger class NC can be assumed to be symmetric about its mean μ_{nc}

Proof (Unbalance). Since $G(x;t)$ provides a one to one, monotone mapping from x to g and x is uniformly distributed over the interval $(0, N : N \to \infty)$, therefore the ratio of probability of NC to IC is given by:

$$\frac{P[g < \epsilon]}{P[g \geq \epsilon]} = \frac{P[x > \delta(t)]}{P[x \leq \delta(t)]} = \underset{N \to \infty}{Lt} \frac{N - \delta(t)}{\delta(t)} >> 1 \tag{41}$$

□

Proof (Assumption of Symmetry of bigger class). The domain $g < \epsilon$ of the NC in the PDF is infinitesimally small, lesser than the upper bound of error and consequently immeasurable. The PDF of (NC) can be computed by applying the limits $(g \to 0^+)$ to (5).

$$\underset{(g \to 0^+)}{Lim} f_g(g) = \underset{(g \to 0^+)}{Lim} \frac{t}{Ng\sqrt{-t\log_e(2\pi t g^2)}} = \frac{\sqrt{t}}{N} \underset{(g \to 0^+)}{Lim} \frac{1/g}{\sqrt{\log_e(2\pi t g^2)^{-1}}} \tag{42}$$

Equation (42) is of the form $(\frac{\infty}{\infty})$, therefore a simplification of (42) is possible by the application of *L'Hopital's Rule*, i.e. differentiating both the numerator and the denominator w.r.t (g).

$$\frac{\sqrt{t}}{N} \underset{(g \to 0^+)}{Lim} \frac{\frac{\mathrm{d}}{\mathrm{dg}}(\frac{1}{g})}{\frac{\mathrm{d}}{\mathrm{dg}}(\sqrt{\log_e(2\pi t g^2)^{-1}})} = \frac{\sqrt{t}}{N} \underset{(g \to 0^+)}{Lim} \frac{\sqrt{\log_e(2\pi t g^2)^{-1}}}{g} = \infty \tag{43}$$

Since the PDF $(f_g(g))$ has a value of infinity in an infinitesimal interval $(g < \epsilon)$, therefore the PDF in the interval $(g < \epsilon)$ can be approximated by a Dirac Delta. The Dirac Delta is the limiting case of the Symmetric Gaussian Function (with zero standard deviation), therefore assumption of symmetry of the PDF of the NC is justified. □

Corollary 2 implies that Theorem 7 is applicable for a Gaussian function where the IC and NC are separated at a Gaussian value $(G(x;t) = \epsilon)$ or at equivalent space coordinate $(x = \delta(t))$.

Theorem 9. The OT, for the GMSS of a step function is given by:

$$k^*(t) = \frac{1}{N} Erf[\delta(t)] \quad where \quad Erf[\delta(t)] = \frac{1}{\sqrt{\pi}} \int_0^{\delta(t)} e^{-p^2} \mathrm{d}p \tag{44}$$

Proof. Replacing $(\mu_{nc}(t) = 0)$ from (39) in Theorem 7:

$$k^*(t) = m_{ic}(t)\mu_{ic}(t) \tag{45}$$

Substituting $\mu_{ic}(t)$ and $(m_{ic}(t))$ from (39) and $(G(x;t))$ from (1) into (45) and solving:

$$k^*(t) = \frac{\delta(t)}{N\delta(t)} \int_o^{\delta(t)} \frac{e^{-\frac{x^2}{2t}}}{\sqrt{2\pi t}} \mathrm{d}x = \frac{1}{N} Erf[\delta(t)]$$

□

Plots of OT and $\delta(t)$ from Theorem 9 at various ratios of (ϵ/A) are shown in Figure 9. Both the plots $(k^*(t), \delta(t))$ contain a maximum, which can also be verified by differentiating (44) and (36) w.r.t scale (t).

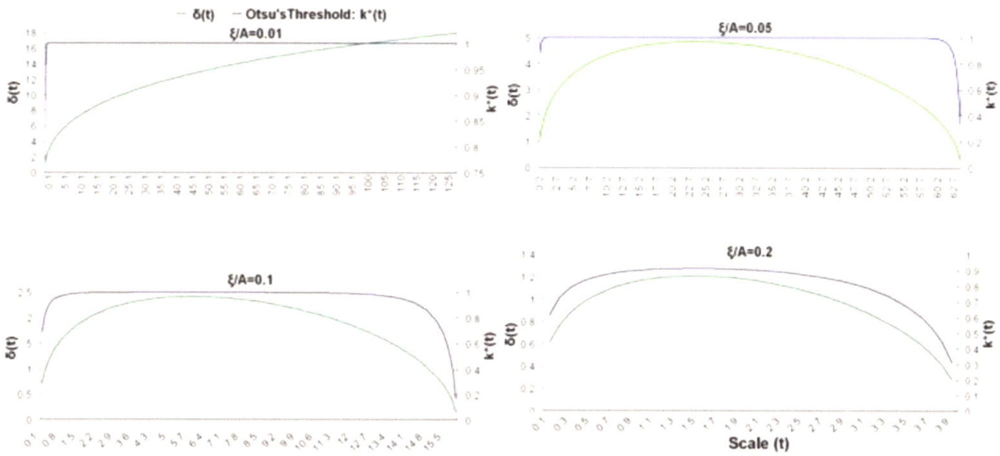

Fig. 9. Comparative Graphs of OT $[k * (t)]$ and $[\delta(t)]$ against Scale at various ratios of ϵ/A.

5. OT for continuous functions

Theorem 10. The OT for the GMSS of a continuous signal is monotonically decreasing i.e.

$$k^*(t) > k^*(t + \Delta t) \quad \forall \quad \Delta t > 0 \tag{46}$$

Proof. Due to the Central Limit Theorem, the PDF of the GMSS of a continuous function can be approximated by a normal (and consequently symmetric) distribution. Therefore from Corollary 1:

$$k^*(t) = \mu_T(t) \quad and \quad k^*(t + \Delta t) = \mu_T(t + \Delta t) \tag{47}$$

For continuous functions, the Maximum Principle is valid Gonzalez-Velasco (1995); Lindeberg (1994); Widder (1975). Due to the Maximum principle the GMSS of the continuous function (and its mean) will be monotonically decreasing with increase of scale (Babaud et al. (1986); Lindeberg (1994; 1998)), i.e. :

$$\mu_T(t) > \mu_T(t + \Delta t) \quad \forall \quad \Delta t > 0 \tag{48}$$

Combining (47) and (48)

$$k^*(t) > k^*(t + \Delta t) \quad \forall \quad \Delta t > 0 \qquad \square$$

Comparison of Theorem 9 with Theorem 10 reveals the contrasting patterns for OT when traced against scale:

1. Transient Increase: When a discontinuity exists.
2. Monotone Decrease: When a function is continuous.

6. Heuristic and algorithm for simultaneous scale, threshold and discontinuity detection

In this section, the theoretical results from non-discrete 1-Dimensional functions of the previous sections will be applied to the analysis of discrete 2-Dimensional images. Specifically the interface between the entities will be detected. The ideology underpinning this section is that the edges in images can be broadly classified as

a. *Boundary-Edges:* These edges correspond to interface between two entities of a 3D physical world, when projected onto the 2D image surface as a consequence of the process of image capture (e.g. an edge located at the interface of foreground-background). Since Boundary-Edges exist at the interface of heterogeneous surfaces or/and processes, the discrete gradient (Sobel) of Boundary-Edges is computed from dissimilar neighborhoods leading to a high intensity gradient and also spatial Scarcity, rendering Boundary-Edges similar to IC of the 1D functions.
b. *Non-Boundary-Edges:* In contrast, the Non-Boundary-Edges owe their discrete gradient computation to homogeneous neighborhood, resulting in low gradient and high spatial probability which is similar to NC of the 1D functions.

This apparent similarity, has empirical support from contemporary literature (Bhanu & Faugeras, 1982; Lin, 2003; Medina Carnicer & Madrid Cuevas , 2008; Rosin, 2001) wherein existence of unbalanced histograms for the derivative of images have been reported. The similarity of the histograms and the associated statistics (probability and mean) of IC and NC with Boundary-Edges and Non-Boundary edges respectively, allows for the development of a heuristic to extrapolate Theorem 9 and Theorem 10 for detection of Boundary-Edges. The problem of Boundary-Edges identification can be subdivided into:

1. Finding the optimum Scale
2. Finding the optimum Threshold at the Scale
3. Locating the interface, as the discontinuity travels in scale-space (Lindeberg, 1994).

To locate the Boundary-Edges, the following heuristic has been evolved which identifies both the scale and threshold appropriate to the interface using OT.

6.1 Heuristic

In the presence of an inflection in the plot of OT (calculated for the GMSS of an image) against incremental scale, it can be assumed that a discontinuity due to a foreground-background interface exists in the image. This discontinuity can be identified by thresholding the GMSS of the image at the scale and OT corresponding to the upper point of inflection in the plot of OT.

Justification: There are three aspects of the Heuristic i.e. presence of discontinuity, appropriate scale and appropriate threshold, which need to be justified individually:

1. *Presence of a discontinuity:* The presence of an inflection only in the presence of discontinuity has been shown via Theorem 9 and Theorem 10 for functions with and without a discontinuity respectively.

2. *Scale at upper point of inflection:* There is no universally agreed definition of appropriate scale; hence the justification of appropriate scale is qualitative rather than mathematical. The upper point of inflection is the scale appropriate for the discontinuity owing to following reasons:
   (a) Lower point of inflection wrongly classifies the pixels as belonging to IC (Figure 10). Hence for all scales lower than the lower point of inflection, false classification as IC is a strong possibility.
   (b) Experimentally and theoretically (most of the graphs of Figure 9), it has been observed that often from the lower to the upper point of inflection there is only small difference of scales. Hence attempting to locate the scale between the upper and lower points of inflection is mostly futile.
   (c) Scales greater than the scale at the upper point of inflection can give comparable results for the discontinuity identification as the scale at upper point of inflection, but at some scale greater than the upper point of inflection the IC will cease to exist. In the absence of a priori information of this scale where IC ceases to exist, using a scale greater than the one identified by the upper point of inflection, runs the risk of attempting to locate IC at a scale at which the IC does not exist. Therefore the upper point of inflection is the best scale for IC detection.
3. *OT at upper point of inflection as the threshold:*
   (a) OT at the upper point of inflection has been chosen as the threshold as it corresponds to the scale appropriate to the discontinuity.
   (b) Minimum False positives: The upper point of inflection represents the highest threshold intensity (utilizing OT in Scale-Space). Since the average value of the IC is greater than the rest of the image, therefore the highest threshold results in lowest false classification of the pixels as IC.

The application of the heuristics is demonstrated in Figure 10, wherein the Boundary-Edges have been located by thresholding at scale and OT corresponding to upper point of inflection[3]. Figure 10(a) depicts a synthetic image comprising of background only. A foreground is added to the texture of Figure 10(a) as shown in Figure 10(c) resulting in Figure 10(b). OT plotted against incremental scale for the GMSS of Figures 10(a) and 10(b) results in OT graphs shown in Figure 10(d), wherein the plot corresponding to the background only image has a monotonic decay in contrast to the image with a foreground which shows an inflection. Thresholding the image with the foreground (Figure 10(b)) at scale and OT corresponding to the upper point of inflection identifies the foreground-background interface.

6.2 Algorithm and results

Based on the Heuristic of the OT for discontinuity detection, a simple algorithm comprising of following steps can locate Boundary-Edges in images Walia & Jarvis (2009):

[3] The detection of the appropriate scale is not for a general discontinuity but conditional to the presence of a specific discontinuity. The discontinuity should be due to interface and therefore have an unbalanced histogram similar to the PDF in(30). The scale *cannot* be identified using the method presented here for discontinuities which are not due to interfaces.

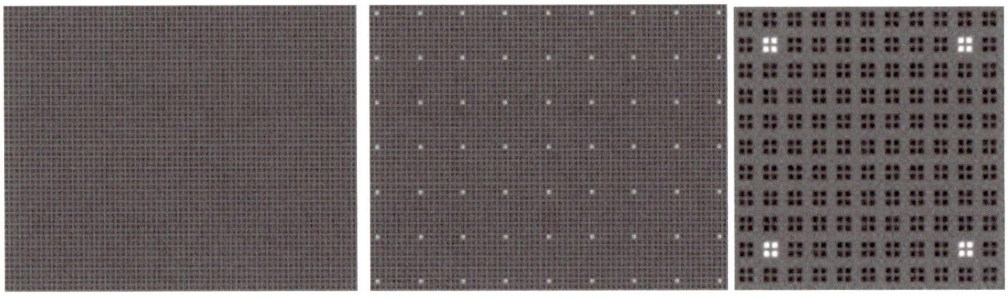

(a) Syn 1: Background Only (b) Syn 2: Foreground present (c) Close Up of Syn 2

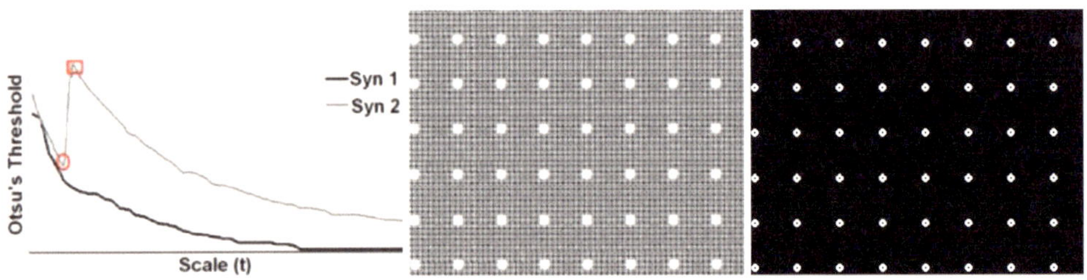

(d) Plots of OT for (a) and (b) (e) Segmentation: Lower point of inflection (f) Segmentation: Upper point of inflection

Fig. 10. Illustration of Heuristic presented on Synthetic Images. (d) OT plots indicate a monotonic decay and an inflection in the absence and presence of a foreground (and consequent Boundary-Edge) respectively. (f) Upper point of inflection detects Boundary-Edges.

Algorithm 1: Simultaneous detection of scale, discontinuity and threshold in images

Compute the Sobel derivative of the input image;
while *Not end of Scale Range* **do**
  Convolve the Sobel derivative of the image, with a Gaussian Kernel of current scale;
  Compute histogram;
  Compute and record OT at current scale;
if *Increment of OT in plot against scale exists* **then**
  Identify the (scale, OT) pair at which the OT attains a maximum;
  Convolve the Sobel derivative of the input image with the scale identified;
  Threshold at the OT identified ;
else
  Output: No Discontinuity;

This chapter has discussed theoretical concepts behind the evolution of statistics of a discontinuity. It is difficult to provide a comprehensive comparison between the Algorithm 1 and other contemporary research because usage of *statistics* of Scale-Space representation of derivative of image/functions to identify discontinuity is a novel proposition. (Lindeberg,

1998) has done extensive research in the 1990's on automatic scale selection with applications in edge detection which can be used to compare the theory presented in this chapter. The differences with the approach of (Lindeberg, 1998) are :

1. Assumption of continuity of an edge in Scale-Space by Lindeberg. As a consequence occasionally (Lindeberg, 1994) relies on maximum principle, which does not hold when a discontinuity is present (Gonzalez-Velasco, 1995).
2. Lindeberg exploits individual properties (like edge strength, non-maxima suppression, blob volume, scale-normalized gradient magnitude, directional derivatives etc) of local spatial features (like blob, edge and ridge) in images, to rank and identify various features. The theory and algorithm of this article utilize, the collective statistics of image/signal to identify discontinuity and are therefore immune to errors (Walia & Jarvis, 2009) arising out of local spatial considerations. Both the approaches have contextual relevance.

OT is a well known method for segmenting images and therefore provides a good benchmark to compare the performance of the algorithm presented here. A comparison of the segmentations based on Algorithm 1 with OT is shown in Figure 12. Algorithm 1 had scale increments of 0.1, and histogram comprised of 255 bins. Figure 12 shows images having well defined IC. When OT is applied at zeroth scale, the probability $m_{ic}(0)$ of IC is very small, therefore the OT instead of segmenting IC from NC, segments the NC at approximately the mean of NC ($\mu_{nc}(0)$). In comparison tracing the OT in the GMSS of the images, results in identifying both the scale and threshold appropriate for identifying IC.

The images set used in Figure 12 originate from eclectic sources, without a ground truth so a simple measure was chosen to compare Algorithm 1 with Otsu's Algorithm (Otsu, 1979). The thresholded results of the two algorithms were stored as binary images (Figure 11(b) and Figure 11(c)). The difference in the number of positives (Figure 11(c)) between the Otsu's algorithm and Algorithm 1 expressed as a percentage of the total pixels in the image determined the improvement in the boundary classification. The average improvement in classification was computed for the dataset of images shown in Figure 12. Algorithm 1 had on an average 31.3% better classification of the foreground-background boundary owing to a reduction in false positives as compared to Otsu's Algorithm.

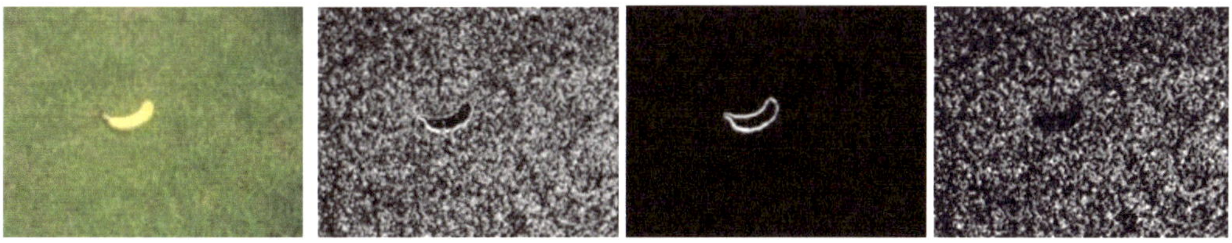

(a) Input (b) Otsu's Algorithm (c) Algorithm 1 (d) Eliminated False positives

Fig. 11. Algorithm comparison: Number of eliminated false positives (d=b-c) expressed as a percentage of the total pixels provides the improvement in boundary detection.

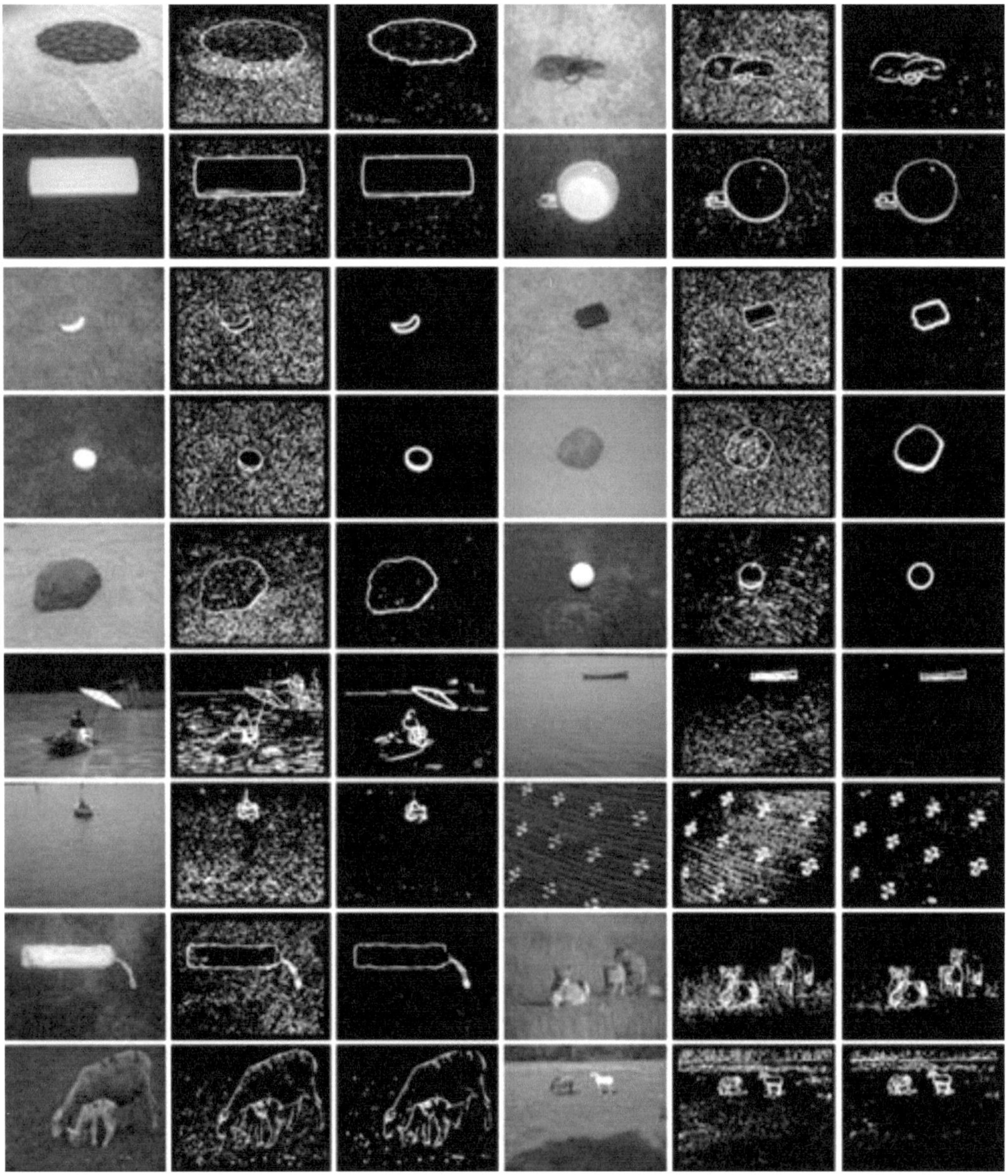

Fig. 12. Segmentation comparison . *Column 1 and 4:* Images with obvious boundaries.(some from Berkeley Dataset Martin et al. (2001)). *Column 2 and 5:*Thresholded by OT. *Column 3 and 6:* Thresholded at Scale and OT corresponding to inflection of OT.

7. Conclusion

Theoretical framework along with one application for characterization and identification of a discontinuity in Scale-Space framework for the derivative of image/function has been presented. This chapter shows that the PDF of the derivative of a discontinuity is unbalanced and bimodal in Scale-Space. By taking the derivative of functions/images, the discontinuities are formulated as outliers with higher average value and low probability. Since OT is a statistical parameter sensitive to the outliers (smaller mode in bimodal distribution) in a given data set, therefore it can detect and locate discontinuities. It is likely that many statistical parameters sensitive to outliers would exhibit similar response to discontinuities in scale-space.

8. References

Babaud, J., Witkin, A. P., Baudin, M. & Duda, R. O. (1986). Uniqueness of the gaussian kernel for scale-space filtering, *IEEE Trans. Pattern Anal. Mach. Intell.* 8(1): 26–33.

Bhanu, B. & Faugeras, O. D. (1982). Segmentation of images having unimodal distributions, *Pattern Analysis and Machine Intelligence, IEEE Transactions on* PAMI-4(4): 408 –419.

Duits, R., Felsberg, M., Florack, L. & Platel, B. (2003). α scale spaces on a bounded domain, *Lecture Notes in Computer Science* 2695: 494–510.

Eisenberger, I. (1964). Genesis of bimodal distributions, *Technometrics* 6(4): pp. 357–363.

Felscher, W. (2000). Bolzano, cauchy, epsilon, delta, *The American Mathematical Monthly* 107(9): 844–862.

Gonzalez-Velasco, E. (1995). *Fourier Analysis and Boundary Value Problems.*, Academic Press Limited.

Hubel, D. H. & Wiesel, T. N. (1987). Brain mechanisms of vision, *Nature* .

Kemperman, J. H. B. (1991). Mixtures with a limited number of modal intervals, *The Annals of Statistics* 19(4): pp. 2120–2144.

Khuri, A. I. (2004). Applications of dirac's delta function in statistics, *International Journal of Mathematical Education in Science and Technology* 35: 185–195.

Koenderink, J. (1984). The structure of images, *Biological Cybernetics* 50: 363–370. 10.1007/BF00336961.

Koenderink, J. J. & Doorn, A. J. V. (1992). Receptive field assembly pattern specificity, *Journal of Visual Communication and Image Representation* 3(1): 1 – 12.

Koenderink, J. J. & van Doorn, A. J. (1987). Representation of local geometry in the visual system, *Biol. Cybern.* 55(6): 367–375.

Lin, K. C. (2003). Fast image thresholding by finding the zero(s) of the first derivative of between-class variance, *Mach. Vision Appl.* 13(5-6): 254–262.

Lindeberg, T. (1994). *Scale Space Theory in Computer Vision.*, The Kluwer International Series in Engineering and Computer Science., Kluwer Academic Publishers, Netherlands.

Lindeberg, T. (1998). Feature detection with automatic scale selection, *International Journal of Computer Vision* 30: 79–116. 10.1023/A:1008045108935.

Martin, D., Fowlkes, C., Tal, D. & Malik, J. (2001). A database of human segmented natural images and its application to evaluating segmentation algorithms and measuring ecological statistics, *Proc. 8th Int'l Conf. Computer Vision*, Vol. 2, pp. 416–423.

Medina Carnicer, R. & Madrid Cuevas, F.J. (2008). Unimodal thresholding for edge detection, *Pattern Recognition* 41(7): 2337–2346.

Otsu, N. (1979). A threshold selection method from grey-level histograms, *SMC* 9(1): 62–66.

Rodriguez, R. (2006). A strategy for blood vessels segmentation based on the threshold which combines statistical and scale space filter: Application to the study of angiogenesis, *Computer Methods and Programs in Biomedicine* 82(1): 1–9.

Romeny, B. M. H. (1994). *Geometry-Driven Diffusion in Computer Vision.*, Kluwer Academic Publishers., Kluwer Academic Publishers, Netherlands.

Rosin, P. L. (2001). Unimodal thresholding, *Pattern Recognition* 34(11): 2083–2096.

Sakai, T. & Imiya, A. (2009). Unsupervised cluster discovery using statistics in scale space, *Engineering Applications of Artificial Intelligence* 22(1): 92 – 100.

Schilling, M. F., Watkins, A. E. & Watkins, W. (2002). Is human height bimodal?, *The American Statistician* 56(3): 223–229.

Walia, R. & Jarvis, R. (2009). Structure, scale-space and decay of otsu's threshold in images for foreground/background discrimination, *VISSAPP (2)*, pp. 120–128.

Widder, D. (1975). *The Heat Equation.*, Academic Press inc.

Witkin, A. P. (1983). Scale-space filtering, *IJCAI'83: Proceedings of the Eighth international joint conference on Artificial intelligence*, Morgan Kaufmann Publishers Inc., San Francisco, CA, USA, pp. 1019–1022.

Young, R. A. (1987). The gaussian derivative model for spatial vision: I. retinal mechanisms., *Spatial vision* 2(4): 273–293.

Zagal, J. C., Björkman, E., Lindeberg, T. & Roland, P. E. (2000). Significance determination for the scale-space primal sketch by comparison of statistics of scale-space blob volumes computed from pet signals vs. residual noise, *NeuroImage* 11(5, Supplement 1): S493.

11

Reflectance Modeling in Machine Vision: Applications in Image Analysis and Synthesis

Robin Gruna[1] and Stephan Irgenfried[2]

[1]*Fraunhofer Institut of Optronics, System Technologies and Image Exploitation IOSB, Karlsruhe*

[2]*Institute for Process Control and Robotics (IPR), Karlsruhe Institute of Technology KIT Germany*

1. Introduction

Taking images of objects under different illumination and viewing directions and analyzing them has long been an active research area in both machine vision and computer graphics. While computer graphics aims to synthesize realistic images from abstract scene descriptions, machine vision is concerned with the problem of deducing properties of a scene based on its interaction with light. For this reason, many algorithms from both disciplines rely on an accurate measurement and modeling of how light reflects off surfaces in a physically correct way.

In this chapter we show how machine vision for automated visual inspection can greatly benefit from reflectance measuring and modeling in the context of image analysis and synthesis.

From the viewpoint of image synthesis, reflectance measurement and modeling of real-world materials can be used to simulate machine vision systems by synthesizing images with computer graphics. The design process of machine vision systems requires a lot of domain specific experience and is also often based on the "trial and error" principle. Many aspects have to be taken into account and often the construction of a prototype system is inevitable. Finding the right camera position(s) and achieving a satisfying illumination of the inspected objects and surfaces is a difficult process as well as the training of the classification and decision algorithms. Simulation of machine vision systems using computer graphics can support and shorten this process and even lead to better results than using a manual setup.

Traditionally, computer graphics systems are designed to create images that are presented to human eyes. The goal is, to make the viewer judge the images as, e.g., believable, logical or beautiful. But for machine vision systems, the physical credibility is the most important factor. To achieve this goal, several areas of computer graphics have to be investigated under the perspective, what is required to create synthetic images that can be used as ground truth data for image processing, classification and decision algorithms?

Crucial parameters for machine vision systems are the reflection properties of the object surfaces in the scene, which we will focus on in this work. Modeling, how light behaves when it hits an object surface has been an important research area in computer graphics since the beginning. Measuring, e.g., the Bidirectional Reflectance Function (BRDF) of real world materials and fitting the mathematical models to the data opened up the path to very realistic looking images and also introduced more physical correctness to synthetic images.

We present a machine vision simulator that can create synthetic images showing the inspected objects under varying perspectives and illumination conditions and compare them with images captured using a real camera. The synthetic images and real world images are compared on an image pair basis, but also the results of different image processing algorithms applied to the images are investigated. To achieve a high degree of realism, we use camera calibration to calculate the intrinsic and extrinsic parameters of the real camera and use those parameters in the camera model of the simulator. Our simulator uses measured BRDF-values applied to CAD models to achieve a high quality in simulating diffuse and specular behavior for isotropic and even anisotropic materials.

From the viewpoint of image analysis, reflectance measurement is performed by capturing illumination series. Illumination series are obtained by taking images of objects under varying illumination directions providing the input for various machine vision algorithms like photometric stereo and algorithms for image fusion. Since illumination series contain much more information about the reflectance properties of the illuminated surface than a single intensity image, they facilitate image analysis for defect detection. This is demonstrated by means of a machine vision application for surface inspection.

However, capturing a large number of images with densely sampled illumination directions results in high-dimensional feature vectors, which induce high computational costs and reduce classification accuracy. In order to reduce the dimensionality of the measured reflectance data, reflectance modeling can be used for feature extraction. By fitting parametric models to the data obtained from recording illumination series, the fitted model parameters represent meaningful features of the reflectance data. We show how model-based extraction of reflectance features can be used for material-based segmentation of printed circuit boards (PCBs).

2. Reflectance acquisition and modeling

2.1 Measuring reflectance

Over the last decades, many different setups and devices have been developed to either measure the reflection properties of a material directly or to capture an image series and derive the reflective behavior of the inspected material or a group of materials from them. The approaches differ a lot in their mechanical, optical and electrical setup, depending on the type of material and the material parameters to be acquired. They range from large outdoor equipment and room-filling robot-based equipment to tiny handheld devices. Measurement times can range from seconds to many hours. In the following, we'll give an short introduction on the different configurations known from literature and then present our own measurement device.

Two main distinctions between the measurement setups can be made: First, how many pairs of incident and outgoing angles of the light on a point on the sample surface are captured at a time? And second, is only one point on the sample surface or multiple points measured? Sampling of multiple points on the object surface is necessary to acquire an SVBRDF (Spatially Varying BRDF) and BTF (Bidirectional Texture Function).

2.1.1 Devices for measuring the reflectance properties of a material

One representative example in the field of the table based gonioreflectometers is the work of Murray-Coleman & Smith (1990), who designed a gonireflectometer with a single light source and one detector movable along arcs in the hemisphere above a rotatable sample area. Similar did White et al. (1998) with improved accuracy. To capture the reflection for more than one angle pair at a time, curved mirrors are widely used, like in the works of Ward (1992), Dana (2001), Ghosh et al. (2007), Mukaigawa et al. (2007). To add more mechanical degrees of freedom in the positioning of the material probe, robot arms have been used, e.g., Dana et al. (1999), Sattler et al. (2003), Hünerhoff et al. (2006). Debevec et al. (2000) constructed a light stage with multiple light sources and cameras at fixed positions to acquire the reflectance field of a human face. Wrapping the material probe around a cylinder, Marschner et al. (2000) simplified the mechanical BRDF measurement setup to measure isotropic BRDFs. Ngan et al. (2005) extended this setup making the cylinder itself rotatable to also measure anisotropic BRDF. Gardner et al. (2003) developed a linear, tabletop scanner-like device to caputure the spatially varying BRDF of a material. An interesting work, that shows how to capture SVBRF and BTF without mechanically moving camera and light source by using a kaleidoscope is the one by Han & Perlin (2003). Using an array of 151 digital cameras arranged in form of a dome above the material sample, Mueller et al. (2005) presented their measurement device, similar to Malzbender et al. (2001). With the goal of developing a small handheld device to measure BRDFs, Ben-Ezra et al. (2008) created an LED based device using the LEDs as lightsource and detector. More camera based handheld devices to measure SVBRDFs were presented by Dong et al. (2010) and Ren et al. (2011). Measuring reflectance is also an interesting topic for other research areas. A large outdoor field goniometer to capture the BRDF of earth and plants for remote sensing applications was presented by Sandmeier (2000). Jackett & Frith (2009) designed a device to measure the reflection properties of road surfaces to improve the safety and sustainability of road lighting.

2.1.2 Devices for capturing illumination series

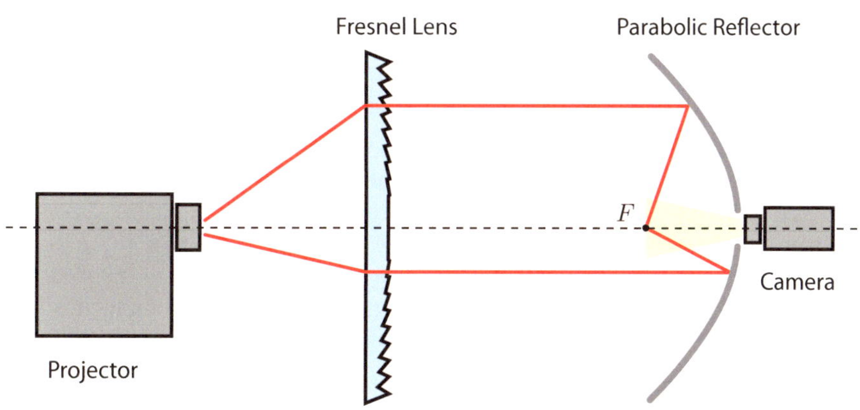

Fig. 1. Device for variably illuminating an object from different directions. A digital LCD projector, a Fresnel lens, a parabolic reflector with a center hole and a digital camera are aligned along their optical axes. By placing the optical center of the projector at the focal point of the Fresnel lens, all emitted light rays intersect at the focal point F of the reflector.

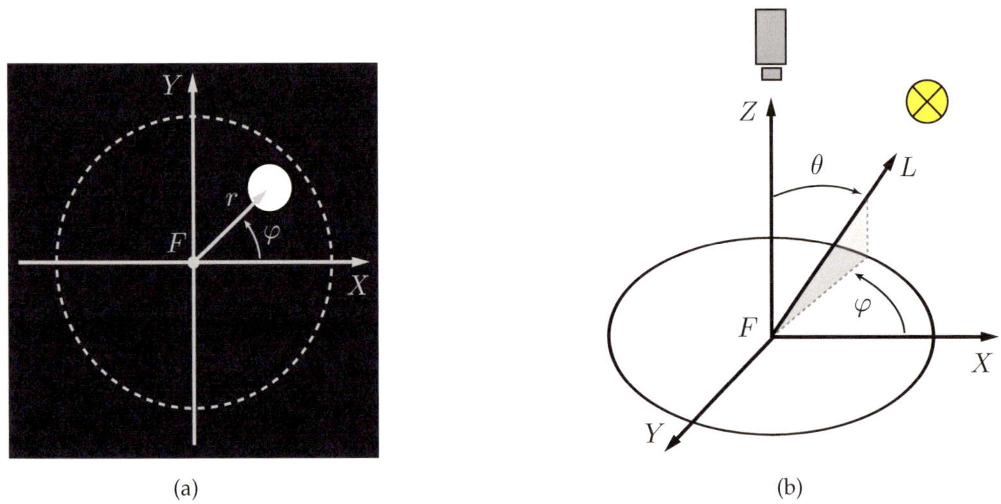

Fig. 2. (a) Illumination pattern used in our experiments. The circular illumination spot is parametrized using polar coordinates r and φ. The pole of the coordinate system is aligned with the focal point F of the parabolic reflector. (b) Projecting the illumination pattern from Figure (a) with the illumination device shown in Figure 1 approximates a distant illumination whose direction is described by the azimuthal angle φ and polar angle θ with origin F. The relationship between the illumination pattern and the illumination direction is given through Equation (1).

Closely related to the approaches described above to measure reflectance functions of material samples are devices to record images of a scene as a whole under varied illumination directions. In the field of machine vision, this set of images is commonly referred to as *illumination series* (Puente León (1997); Lindner & Puente León (2007)). In the following, we present an acquisition device to capture illumination series of small objects. The proposed illumination device is shown in Figure 1. The optical components are: a digital LCD projector, a Fresnel lens, a parabolic reflector with a center hole and a digital camera. All components are aligned along their optical axes.

The projector serves as programmable light source, which allows to control the relative radiance along the emitted light rays independently. Assuming a pinhole model for the projector, each projector pixel can be thought of as the source of a single ray of light that emanates from the optical center of the projector. By placing the projector at the focal point of the Fresnel lens, the light rays are collimated by the lens and converge at the focal point F of the parabolic reflector. As a consequence, the illumination direction of light rays incident to the focal point F can be controlled by projecting spatially-varying illumination patterns.

To image objects under different illumination directions, the object is placed at the focal point F of the parabolic reflector and the reflected radiance is captured by the camera attached to the center hole of the reflector. Although the proposed illumination device allows to project arbitrary complex illumination patterns, in our first experiments we consider a simple binary illumination pattern depicted in Figure 2. The illumination pattern consists of a single fixed-size circular spot of white pixels with all other pixels being black. Due to the spatial extent of the illumination spot, the illumination pattern leads to a bundle of light rays rather

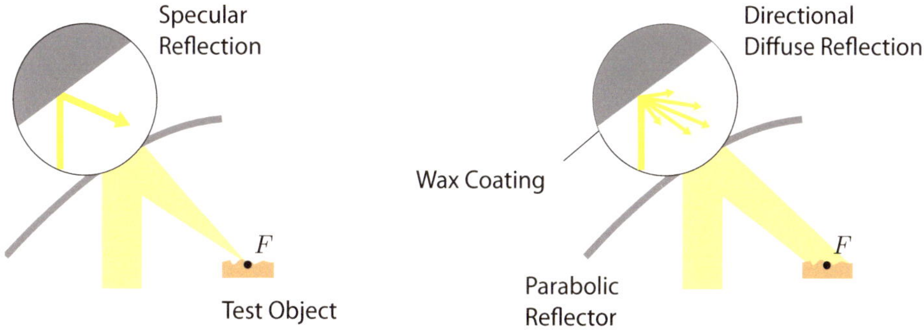

Fig. 3. Projected ray bundle reflected at the surface of the parabolic reflector. A thin wax layer is used to roughen the specular surface of the reflector in order to obtain a diffuse directed reflection and to broaden the illuminated area size at F.

than a single ray. However, each individual light ray of the bundle follows the laws of reflection and refraction and hence a cone-shaped illumination is incident to F (see Figure 3).

By establishing a Cartesian coordinate system with its origin at the focal point F of the parabolic reflector and its Z-axis aligned with the optical axis of the device and pointing into the direction of the camera, we are able to parametrize the illumination pattern and so the resulting illumination direction incident to F as depicted in Figure 2. The position of the illumination spot is considered in the X-Y-plane and can alternatively be represented by its radial coordinate r and angular coordinate φ in polar coordinates. The resulting illumination is then described by the illumination vector $L(\varphi, \theta)$, which points in the direction of the cone-shaped illumination. The angular coordinates of $L(\varphi, \theta)$ are derived as follows: the azimuthal coordinate φ equates the angular coordinate of the illumination pattern and the polar coordinate θ is determined by

$$\theta := \arctan \frac{4fr}{4f^2 - r^2}, \tag{1}$$

where f denotes the focal length of the parabolic reflector.

Note that given highly accurate optical components and a perfectly calibrated setup, all emitted light rays by the projector would intersect at a small point at the focus F of the parabolic reflector. Clearly, this is not practical for illuminating objects with spatial extent. Moving the test object slightly behind or in front the focal point would result in a spatial extension of the directed illumination, however, as consequence, the position of the incident illumination will vary dependent on its direction. To account for this problem, we use a parabolic reflector that is not perfectly specular, having instead a matte surface appearance. Hence, the specular reflection at the reflector surface becomes partly diffuse and the incident ray bundles from the projector are reflected in a small solid angle toward the focal point F. This results in a broadening of the area for which the direction of the incident illumination can be varied. To enhance this effect, we use wax based transparent dulling spray to coat the surface of the reflector in order to further increase the diffuse component of the reflection (see Figure 3). However, care must be taken to keep the proportion between diffuse and specular reflection right, since the wax coating can cause undesirable interreflections and, as consequence, the directional property of the illumination decreases. By applying the dulling spray in a thin layer, we yield a circular area of radius $\approx 15\,\text{mm}$ around F of nearly homogeneous irradiance for which we can control the direction of the illumination. In a similar approach in Peers et al. (2006), heated acrylic is used to roughen the texture of a hemispherical mirror in order to obtain a more diffuse reflection. Another way to solve the problem of inhomogeneous illumination is to use prospective shading correction techniques as employed by Jehle et al. (2010).

2.2 Modeling reflectance

A lot of research work has been published in the field of modeling reflectance. It was soon recognized in the early days of computer graphics that the way light reflects off a surface is a key point in calculating light paths through a scene. Most of the models were invented to reproduce the appearance of a material, that existing models could not model well enough by that time. The models have been designed taking into account or neglecting some physical constraints, e.g., energy conservation, helmhotz reciprocity, polarization or sub-surface scattering. Often, they are a trade-off between being intuitive, accurate and computationally demanding.

2.2.1 Analytical models

Torrance & Sparrow (1967) laid a cornerstone for analytical models with their work about surfaces consisting of small microfacets. Modeling the geometry of those microfacets and their distribution was subject of many follow-up works, like Blinn (1977) who combined this theory and the ideas of Phong (1975). Cook & Torrance (1982) suggested a reflection model based on their previous work and added a Fresnel term to better model effects that appear at grazing angles and also models the dependency of reflection on the wavelength. Kajiya (1985) developed an anisotropic reflection model for continuous surfaces directly based on the Kirchhoff approximation for reflection properties of rough surfaces. In their work, Poulin et al. (1990) modeled anisotropy by using small cylinders. Hanrahan & Krueger (1993) presented work on modeling complex reflection due to subsurface scattering in layered surfaces. To reduce the number of coefficients in the previous physically based models, Schlick (1994) proposed his rational fraction approximation method. Oren & Nayar (1994) suggested to model rough surfaces through small Lambertian faces. Ashikhmin et al. (2000) presented a BRDF generator, that can calculate a BRDF based on a general microfacet distribution. Recently, Kurt et al. (2010) suggested a new anisotropic BRDF model based on a normalized microfacet distribution.

Other analytical reflection models have also been built upon the theory of physical optics, like in the work of He et al. (1991), which was extended to also model anisotropy by Stam (1999).

2.2.2 Empirical models

The empirical model most known in the world of computer graphics is the model proposed by Phong (1975) and improved by Lewis (1994) to increase physical plausibility. Neumann et al. (1999) combined it with parts of analytical models and derived their own BRDF model for metallic reflection. Ashikhmin & Shirley (2000) extended the model to also model anisotropic behaviour. A widely known work in modeling anisotropic reflection is Ward (1992). Other approaches purely focus on finding a mathematical representation for the shape of a material's BRDF. So did Schröder & Sweldens (1995) with wavelets, Sillion et al. (1991) and Westin et al. (1992) with sherical harmonics, Koenderink et al. (1996) with Zernike polynomials, Lafortune et al. (1997) with a set of non-linear primitive functions, Kautz & McCool (1999) with separable approximations based on singular value decomposition and normalized decomposition, Lensch et al. (2001) by expressing a spatially varying BRDF through a linear combination of basis functions achieved by analyzing the SVBRDF of multiple measured materials which in turn inspired Matusik et al. (2003) to derive their data-driven reflectance model based on the measurements of more than 130 real world materials. Another work in the area of data-driven reflectance models is Dong et al. (2010), who present an SVBRDF bootstrapping approach for the data aquired using their own handheld measurement device to speed up the aquisition and processing of material properties significantly.

2.2.3 Model fitting

One question, that has been a challenge to many authors in the field of modeling reflection is: How can an analytical or empirical model be fit to the measured BRDF data of a material? In the following, we'll give a short description on the fitting process of the Lafortune reflection model and refer the reader to the original publications for further details. Lensch et al. (2001) defined the isotropic version of the Lafortune model with multiple lobes to be fitted in the following form:

$$f_r(\hat{u},\hat{v}) = \rho_d + \sum_i [C_{x,i}(u_x v_x + u_y v_y) + C_{z,i} u_z v_z]^{n_i} \quad (2)$$

with $\hat{u},\hat{v}$ being light and viewing directions, ρ_d denoting the diffuse component. C_x, C_z define the direction of the lobe i, n denotes the width of the lobe. A widely used method in the literature to perform a non-linear optimization to fit the model parameters to approximate the measured BRDF data (Lafortune et al. (1997)) or sampled radiance (Lensch et al. (2001)) is to use the Levenberg-Marquardt algorithm (Marquardt (1963))and to define the error between the model prediction and the measured value for a given light and view angular constellation as the objective function to be minimized. The objective function proposed by Ngan et al. (2005) is the mean squared error between the model value for a given parameter vector and the measured value.

3. Applications of reflectance measurements and models in machine vision

3.1 Synthetic ground truth data for machine vision applications

Design, prototpying, testing and tuning of machine vision systems often requires many manual steps, time and domain specific knowledge. During the development phases of such a system, the use of computer graphics and computer generated synthetic images can simplify, speed up and improve this process. Different aspects of such a simulation system were also subject of related work, e.g. Reiner et al. (2008) focused on modeling different real world luminainares, Chessa et al. (2011) proposed a system for creation of synthetic ground truth data for stereo vision systems using virtual reality technology. In contrast to this related work, our work mainly deals with modeling the reflectance properties of materials as good as possible by using the methods described in Section 2.2. Simulating object surface appearance correctly under different viewing and illumination conditions is one of the fundemental requirements to such a simulation system to be able to create ground truth data for machine vision applications. In the following we will demonstrate the benefits of using synthetic images in the process of developing a machine vision system.

3.1.1 Scene modeling

Using computer graphics, not only the objects to be inspected can be modeled, but also the whole scene including lightsources and camera(s). For the positioning of the camera(s) in the scene, it is important to know, if all necessary parts of the objects to be inspected are inside the field of view of the camera and are also not blocked by other objects or the object itself (Khawaja et al. (1996)), e.g., for inspection of threads. Adding such a constraint to the optimiziation process of the system, many mechanically possible setups can be omitted in an early design phase. Figure 4 shows the scene preview of our simulation software. In real world applications, the positioning of camera and illumination is often constrained by mechanical limitations as part of a production line or by other parts of the machine vision system, e.g., housings to block extraneous light, which can also be included in the simulation.

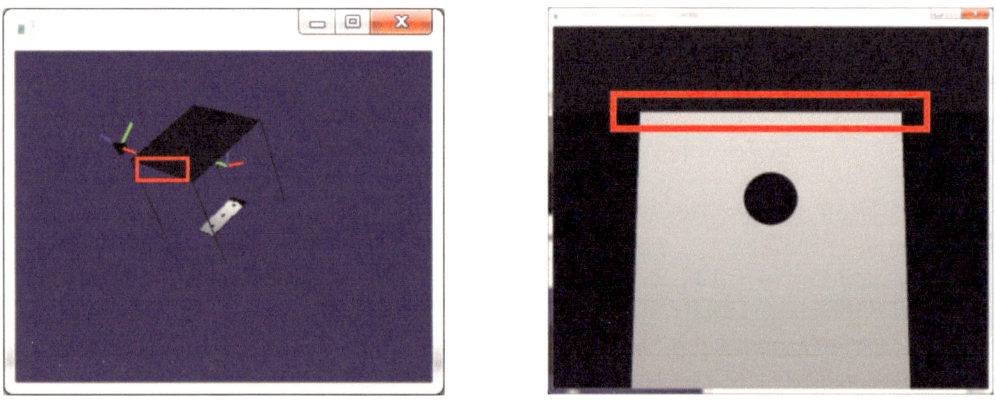

(a) (b)

Fig. 4. (a) Preview of the whole scene (b) Camera view. The red area marks the edge of the housing that blocks parts of the object from being visible.

3.1.2 Automated creation of synthetic image series

A major advantage of using simulation for the development of a machine vision system is the possibility to automatically simulate known variations of the scene and the objects as shown in Figure 5. One can imagine, that the combination of possible camera-, illumination- and object variations leads to an enourmous number of system conditions to be evaluated to find the optimal setup. Building a prototype system to check all of them would be a long lasting procedure and, due to mechanical restrictions in positioning of the components, it would be very hard to create reproducable results. Here we see a great potential to use the power and flexibility of computer graphics to simulate these variations and on the one hand, provide valuable information to the machine vision expert, on the other hand automatically optimize the system configuration based on objective functions.

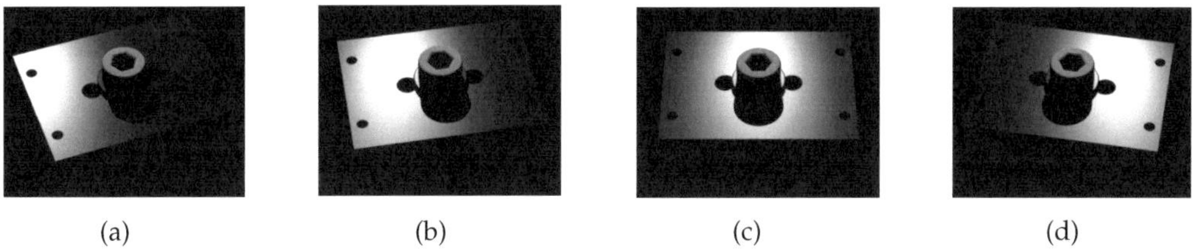

Fig. 5. Series of synthetic images of an assembly under fixed illumination and moving camera. The Phong reflection model was used with different parameters for the parts of the assembly.

3.1.3 Techniques for creation of synthetic images

Traditionally, there have been two major groups of techniques to render images in computer graphics. The aim of the first group is to create images at high framerates, required for interactive applications like CAD or computer games. They make heavy use of the increasing power of graphics processors. The aim of the second group is to create images as photorealistic as possible by implementing highly sophisticated algorithms for light propagation in a scene, e.g. Monte-Carlo sampling. We propose a dual approach as shown in Figure 6, sharing a common scene data format to switch between the two modes of rendering if required. This approach makes it also possible to use the real-time image creation part to simulate a

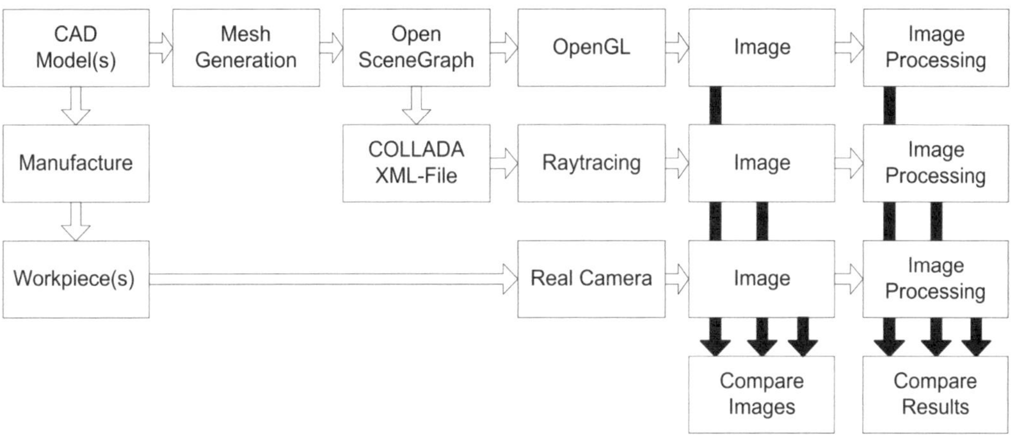

Fig. 6. Dual approach to create synthetic images including a comparison with real world images based on a common scene data format.

live virtual camera for a machine vision system as well as photorealistic rendering for an accurate prediciton of the pixel values in the real camera image. The scene data format we use is the one of the digital asset and effects exchange schema *COLLADA* (www.collada.org). This XML-based format allows us to store required data about the whole scene in a single file, including 3D models, textures, reflection model parameters, camera and lightsource parameters. It can also be used to store multiple scenes and their chronological order to be converted into frames during rendering. A core feature of *COLLADA*, which we make use of, is the possibilty to store properties of material appearance in many different ways, from reflection model parameters to shader code fragments.

Our real-time rendering application is also used as a scene composer and editor and is based on the open source scene graph framework *OpenSceneGraph* (www.openscenegraph.org). Our dual approach is designed and implemented as follows: Our simulation system expects CAD models of the workpiece(s) as a basic input of the simulation to have knowledge of the dimensions with high accuracy. To be imported by our real-time simulation application, the CAD models are converted into meshed models. This is either done by export-filters of the CAD application or by using the finite element tesselation tool *gmsh* (www.geuz.org/gmsh). The real-time application is then used to compose the desired scene consisting of camera(s), illumination and one or more models and assigning material properties to the models with immediate display of the rendering result. To simulate light propagation in the scene more accurate, especially for inner- and inter-object light ray tracing, the scene can be exported into a *COLLADA* file and then loaded in any other rendering application that supports this format. In our experiments, we used the open source software *Blender* (www.blender.org) to test this functionality.

3.1.4 Using synthetic images for algorithm parameterization

An additional usage of synthetic images in machine vision applications is the creation of large sets of samples to choose and parameterize the processing algorithms appropriated for the desired application. There are many different scenarios for this process. They range from simply supporting machine visions engineers during their system design phase through providing them a first impression on how the camera view on the scene will look like in the final application to an in-depth performance analysis of an image processing algorithm by calculating the error between the ground truth data and the processing result of the algorithm. The scope of this feature is to also create sample images, that may occure very rare in reality, but should still be covered by the processing algorithm or at least, should not lead to a false decision or system failure. Figure 7 demonstrates this idea by an example of an edge detection algorithm. The algorithm is parameterized by processing an image series of the workpiece to be detected under various lighting conditions. For this example, we defined two metrics as input for the optimization process. The first one was the total number of detected edge pixels inside the regions of interest. The second one was the distance of the edge pixels from a straight line determined by using linear regression. As parameters of the edge detector to be optimized the gray levels thresholds around an edge pixel are choosen. To aquire the intrinsic and extrinsic camera parameters of the experimental setup, a camera calibration was executed. We used the publically available *Camera Calibration Toolbox*

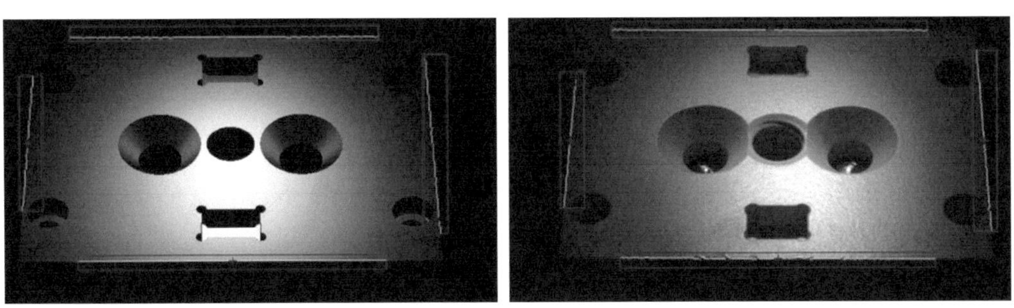

Fig. 7. Edge detection on a synthetic image (left) and a real image (right).

for MATLAB (www.vision.caltech.edu/bouguetj/calib_doc) for this calibration. The material parameters of the workpiece were aquired by using a robot-based gonioreflectometer at our institute which was also used to measure the LED spotlight parameters. Finally, the algorithm performance was then tested on a real workpiece that was manufactured after the CAD model to verify successful parameterization as shown in Figure 6.

3.2 Using reflectance features in automated visual inspection

3.2.1 Illumination series in machine vision

The choice of an appropriate illumination design is one of the most important steps in creating successful machine vision systems for automated inspection tasks. Since in image acquisition all information about a scene is encoded in its exitant light field, the incident light field provided by the illumination must be able to reveal reflectance features of the test object that are relevant to the inspection task. The relationship between incident and extant light field is described by the scene's *reflectance field* (Debevec et al. (2000)) which models light transport of a scene in a global manner.

For many inspection tasks it is difficult or even impossible to find a single optimal illumination condition and therefore, inspection images under multiple different illumination conditions have to be evaluated. In a widely used technique, inspection images are captured under directional light from different directions which yields a so-called *illumination series* of the object (see Section 2.1.2). Thus, recording an illumination series corresponds to sampling the reflectance field of the scene.

In the field of automated visual inspection, the acquisition and evaluation of illumination series have been studied and applied to solve difficult surface inspection tasks. Puente León (1997) proposed an image fusion algorithm to compute images with maximal contrast from an illumination series. To this end, an energy functional is introduced and minimized which specifies the desired requirements on the image fusion result. Lindner & Puente León (2007) proposed several methods for surface segmentation using varying illumination directions and demonstrated the superiority of illumination series over a single image. For this purpose, different reflectance features are extracted for each pixel and used for unsupervised clustering. With this approach, a wide variety of textures on structured surfaces can be segmented. Grassi et al. (2006) used illumination series to detect and classify varnish defects on wood surfaces. By constructing invariant features, good detection and classification ratios can be achieved.

In order to segment images into material types, Wang et al. (2009) used illumination series acquired with a dome providing lighting from many directions. A hemispherical harmonics model is fit to the measured reflectance values and the model coefficients are used to train a multi-class support vector machine. To account for geometric dependencies on the measured reflectance, photometric stereo is applied to estimate the surface normal at each pixel and to transform the measurements to the local surface reference frame. Jehle et al. (2010) used a random forest classifier to learn optimal illumination directions for material classification by using embedded feature selection. For illumination series acquisition, an illumination device very similar to the one presented in this chapter is used. However, our device, developed independently, differs in the wax coating of the parabolic mirror to obtain a homogeneous illumination of the test object.

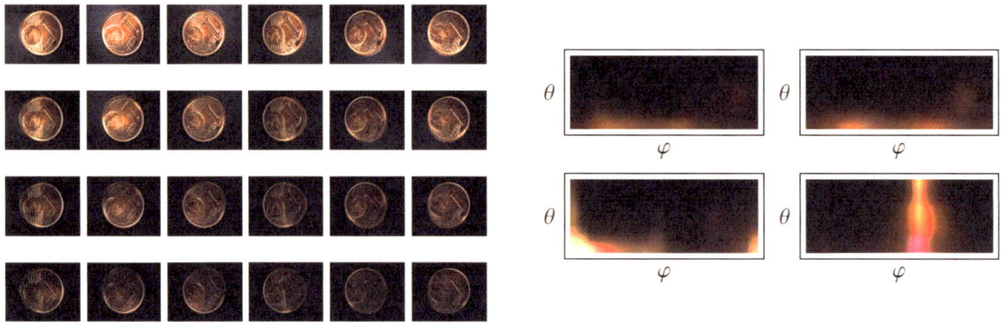

(a) (b)

Fig. 8. (a) Illumination series of a coin for various illumination directions. (b) Reflectance functions at different pixel locations. The upper two reflectance functions depict the illumination-depend appearance of two surface points with similar surface geometry. The lower reflectance functions correspond to a surface edge (left) and shadowed surface point (right), respectively.

3.2.2 Acquisition of reflectance features

The illumination device introduced in Section 2.1.2 is used to capture images of small test objects under varying illumination directions. For this purpose, the illumination pattern shown in Figure 2 is projected for different r and φ so that we get equidistant illumination directions $L(\varphi,\theta)$ along the angular domains $\varphi \in [0, 2\pi)$ and $\theta \in [\theta_{min}, \pi)$. Note that due to the position of the camera, the polar angle θ is limited to θ_{min} to prevent a direct illumination of the camera. Since the spatial extend of the objects to be illuminated is small ($\approx$ 15 mm) compared to the diameter of the parabolic reflector (600 mm), we make the approximation that the projection of the illumination pattern emulates a distant point light source. This means, for each scene point the illumination originates from the same direction with the same intensity. As consequence, a single-channel image can be written as mapping

$$g : \Omega \times \Psi \to \mathbb{R}_0^+ , \tag{3}$$

where $\Omega \subset \mathbb{Z}^2$ is the domain of pixel positions and $\Psi = [0, 2\pi) \times [\theta_{min}, \pi)$ the space of illumination parameters. Debevec et al. (2000) refer to g as the *reflectance field* of a scene, which describes the optical response of a scene illuminated by a distant light source. By varying the illumination direction and capturing n images $g(\mathbf{x}, \boldsymbol{\omega}_i)$, we obtain an illumination series

$$S := \{g(\mathbf{x}, \boldsymbol{\omega}_i), i = 0, \ldots, n-1\} , \tag{4}$$

where $\mathbf{x} \in \Omega$ denotes the pixel location and $\boldsymbol{\omega}_i \in \Psi$ describes the illumination direction incident on the object. Therefore, the illumination series S can be considered as samples of the mapping g with respect to parameter space Ψ. In Figure 8(a), a small image series of a coin for various illumination directions is shown. By considering a fixed pixel location $\mathbf{x}_0$ in Ω, the *reflectance function*

$$r_{\mathbf{x}_0}(\boldsymbol{\omega}_i) := g(\mathbf{x}_0, \boldsymbol{\omega}_i), \quad i = 0, \ldots, n-1 \tag{5}$$

can be defined, describing the illumination-dependent appearance at individual pixels. Figure 8(b) shows reflectance functions for different pixel locations in the coin image series.

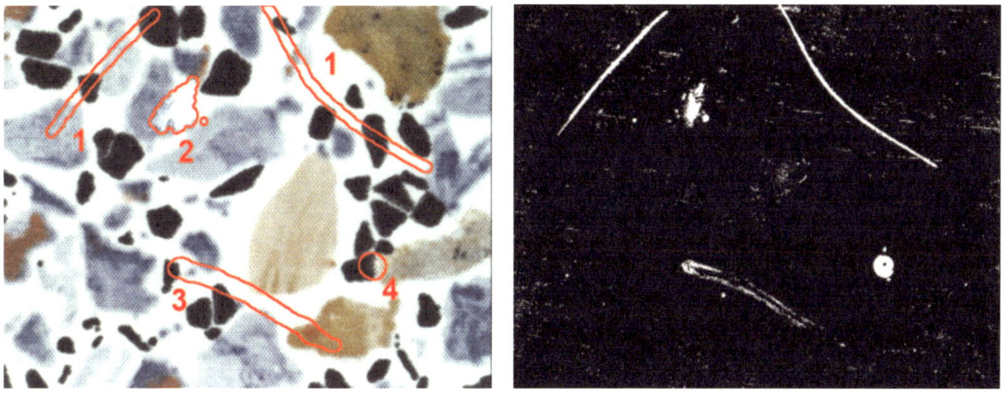

(a) (b)

Fig. 9. Visual comparison of hand-labeled ground truth and results obtained by anomaly detection. (a) Color image of a textured plastic surface under diffuse illumination. However, in our experiments we use illumination series of grayscale images without color or spectral information. Different types of surface defects are marked and labeled by hand:
(1) Scratches, (2) paint stain, (3) groove, (4) dent. (b) RX anomaly detection applied to an illumination series of the same textured plastic surface ($\gamma = 0.95$).

Note that the reflectance function is specified in an image-based coordinate frame, i.e., it includes non-local and geometry-induced illumination effects like the foreshortening term, interreflections and self-shadowing. As consequence, $r_{\mathbf{x}_0}(\boldsymbol{\omega})$ can be considered as a 2D slice of the so-called *apparent BRDF* (Wong et al. (1997)).

3.2.3 Reflectance features for unsupervised defect detection

In the following, we present an approach to unsupervised defect detection using illumination series. In automated visual inspection, collecting labeled training data is often expensive or difficult, because defects are not known a priori. However, in many cases it can be assumed that defects are rare and occur with low probability compared to the nominal state of the inspection task. This is especially true for the appearance of defects under varied illumination directions and hence for the reflectance function of defective surface regions.

In order to detect defects by their illumination-dependent appearance in an unsupervised manner, we apply the *RX anomaly detector* developed by Reed and Yu (Reed & Yu (1990)) to illumination series. The RX detector assumes a Gaussian data distribution and is widely used in hyperspectral image analysis to detect regions of interest whose spectral signature differs from the Gaussian background model without a prior knowledge. Applied to illumination series, the RX detector implements a filter specified by

$$\delta_{\mathrm{RDX}}(\mathbf{x}) = (r_{\mathbf{x}}(\boldsymbol{\omega}) - \boldsymbol{\mu})^{\mathrm{T}} \boldsymbol{C}^{-1} (r_{\mathbf{x}}(\boldsymbol{\omega}) - \boldsymbol{\mu}) \,, \tag{6}$$

where $\boldsymbol{\mu} \in \mathbb{R}^n$ is the sample mean and $\boldsymbol{C} \in \mathbb{R}^{n \times n}$ the sample covariance matrix of the reflectance functions in the image series. Therefore, the detector output $\delta_{\mathrm{RDX}}(\mathbf{x})$ is the Mahalanobis distance between a tested pixel and the mean reflectance function of the image series. Large distances correspond to low probabilities of occurrence, and hence, by displaying the detector output as grayscale image, more anomalous pixels appear brighter. In order to segment anomalous surface regions from the background, a threshold α has to be applied to the detector output. In doing so, anomalous pixels are rejected as outliers of the Gaussian background model. We determine the threshold value α by setting a confidence coefficient γ such that $P(\delta_{\mathrm{RDX}}(\mathbf{x}) < \alpha) = \gamma$.

In a practical experiment, an illumination series of $n = |\{\varphi_0, \ldots, \varphi_{23}\}| \times |\{\theta_0, \ldots, \theta_5\}| = 144$ grayscale images of a textured plastic surface with various surface defects was recorded. Figure 9(a) shows the plastic surface under diffuse illumination and the hand-labeled position of the surface defects. In Figure 9(b), the thresholded output ($\gamma = 0.95$) of the RX detector $\delta_{\mathrm{RDX}}(\mathbf{x})$ applied to the whole illumination series is shown. The result shows that nearly all surface defects become visible as anomalous pixels, demonstrating the ability of illumination series for unsupervised defect detection in textured surfaces.

3.2.4 Model-based feature extraction for material classification

Illumination series contain large amounts of information regarding the reflectance properties of the illuminated object. However, capturing a large number of images with densely sampled illumination directions results in high-dimensional reflectance features. From statistical learning theory it is known, that the complexity of any classification problem grows with the number of input features (Hastie et al. (2009)). This means, more training examples are needed to train a classifier due to the curse of dimensionality. In order to reduce the dimensionality of the feature space, methods from *feature extraction* aim to construct a reduced set of features from the original feature set without loosing discriminative information. As consequence, a less complex classifier can be applied and the reduced feature set allows a better understanding of the classification results. For an unsupervised approach to *feature selection* to reduce the dimensionality of illumination series see Gruna & Beyerer (2011).

Model-based reflectance features are extracted by fitting a parameterized reflectance model (see Section 2.2.3) to the high-dimensional reflectance measurements. The fitted reflectance model then provides a compact representation of the measurements and the estimated model parameters give a reduced set of descriptive reflectance features. Since reflectance models incorporate local surface geometry, knowledge about the scene geometry, e.g., obtained by photometric stereo (Wang et al. (2009)) or estimated from the specular reflection direction (Lindner et al. (2005)), can be utilized for feature extraction. In doing so, the extracted reflectance features become invariant to the surface normal direction.

In a practical experiment, we utilize model-based reflectance feature extraction for the material classification for printed circuit boards (PCBs). The automated visual inspection of PCBs is a challenging problem due to the mixture of different materials such as metals, varnishes, and substrates of which the PCB elements are composed (see Figure 10). Numerous approaches to PCB inspection have been described in the literature, however, most of them are based on measuring the *spectral* reflectance of the materials by multispectral imaging (Ibrahim et al. (2010)). In the presented approach, we use simple grayscale images (i.e., without color or spectral information) but evaluate *angular* resolved reflectance measurements to extract features for material classification. To this end, we record an image series with $n = 144$ grayscale images of a small PCB and utilize the Lafortune reflectance model (see Section 2.2) with one isotropic specular lobe for feature extraction. The fitting process is done

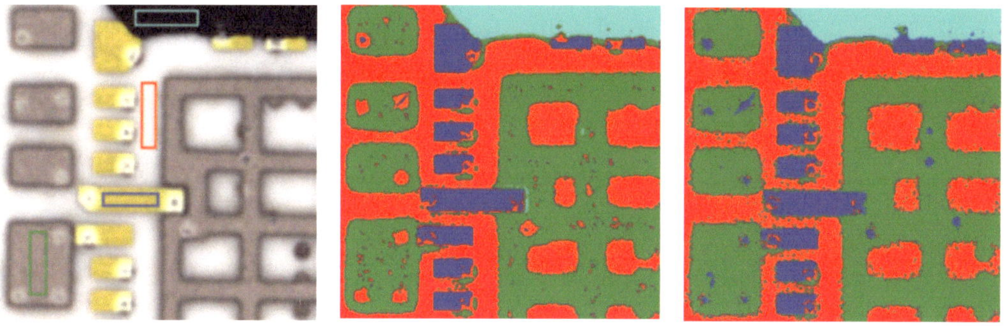

(a) (b) (c)

Fig. 10. Material classification results for a part of a PCB consisting of ground substrate (red), gold and silver conducting elements (marked blue and green, respectively) and paint (turquoise). (a) Color image of the PCB to illustrate the different material components. However, in our experiments we use illumination series of grayscale images without color or spectral information. Material samples marked with rectangles are displayed in the scatter graphs in Figure 11 according to the color encoding (b) Result of k-means clustering using measured reflectance values as feature vector directly (feature space dimension is 144). (c) Result of k-means clustering using model-based reflectance features (feature space dimension is 3).

using the Levenberg-Marquardt algorithm as described in Section 2.2.3. We assume a flat surface geometry with the surface normals aligned with the Z-axis as illustrated in Figure 2. After the reflectance model is independently fit to the reflectance function of each pixel, the estimated model parameters are extracted as feature vector $(C_z, k_d, n)^{\mathrm{T}}$, where C_z represent the inclination of the specular direction, k_d the diffuse component and n the width of the specular lobe.

In order to demonstrate the ability of the extracted reflectance feature for material classification, we compare the model-base approach to the alternative that uses the measured reflectance measures directly as high-dimensional feature vector. For unsupervised material classification we use the k-means clustering algorithm with a fixed number of $k = 4$ material classes. The classification results of a small PCB, which consists of ground substrate, gold and silver conducting elements and paint, are shown in Figure 10. Both approaches show very similar classification results, however, model-based feature extraction with the Lafortune model is able to reduce the dimensionality of feature space from 144 to a 3-element feature vector without losing relevant information. Furthermore, a closer examination of Figure 10(c) reveals that with the model-based features soldering points are identified as conducting elements and not as ground substrate as in Figure 10(b).

Due to the dimension reduction, the new feature space can be plotted and analyzed visually for relevant structures. In Figure 11, the feature space spanned by the Lafortune reflectance parameters is illustrated with scatter plot graphs. By depicting the hand-annotated material samples from Figure 10(a) in the scatter plot graphs, it can be seen that the different material samples are well separated in the model-based feature space. For a more in-depth analysis of reflectance features for material classification see Hentze (2011).

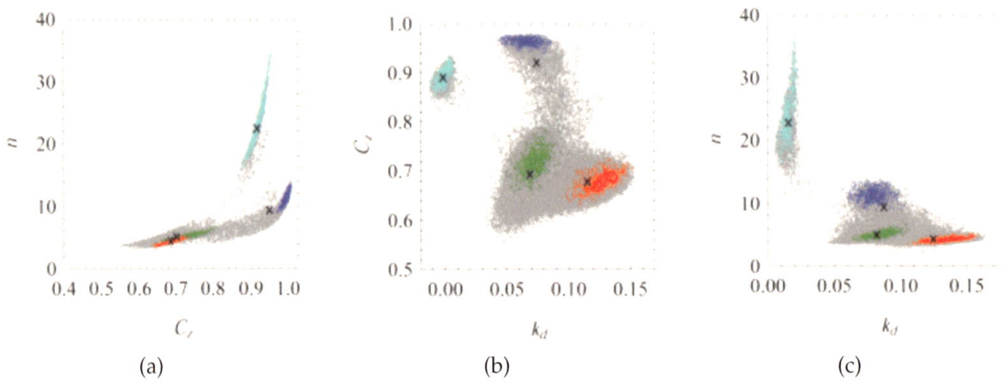

Fig. 11. Illustration of the feature space of the PCB spanned by the Lafortune reflectance model parameters. Material samples from the PCB in Figure 10(a) are marked by different colors (red: ground substrate, blue: gold conducting elements, green: silver conducting elements, turquoise: paint, gray: unlabeled data). Cluster centers found by *k*-means clustering are marked as black crosses.

4. Summary and conclusions

Machine Vision for automated visual inspection can greatly benefit from computer graphics methods for reflectance measuring and modeling. We gave an overview on different ways how to measure the reflection properties of material surfaces and how this data is either used to fit reflection models or evaluated as illumination series.

In the main part of this chapter we presented practical applications of these techniques from an image synthesis and image analysis point of view. We showed, how the reflectance models can be used to create synthetic images of scenes under varying illumination and viewing conditions. To achieve the goal of raising the degree of realism in simulating materials under varying illumination conditions, we discussed the required steps from reflection data aquisition to using it for simulating the appearance of a material in rendering applications. We drawed a bow to how this is can used to create synthetic images that can be used to support the design and development process of machine vision systems. We see two main benefits in this approach: The first one is to simplyfiy the creation of a large set of sample images for training of machine vision algorithms including the possibility to create samples with varying scene setups, e.g., simulated surface defects moving along the surface. This would close a gap in vision algorithm development where often the small set of sample images is a limiting factor. The second one is to make machine vision systems more robust against changes in scene illumination or changes in the view position, e.g. cameras mounted on the head of a moving humanoid robot.

From an image analysis point of view, we demonstrated the use of angular-resolved reflectance measurements in a machine vision application for visual inspection. By applying a density-based anomaly detection method on the high-dimensional measurements we were able to detect surface defects on a highly textured surface. Thereby, we demonstrated the potential of illumination series for unsupervised visual inspection.

In another application example, illumination series were used for inspecting printed circuit boards (PCBs). Here, we demonstrated the feasibility of model-based reflectance features for material classification. To this end, the Lafortune reflectance model was used for feature extraction and it was shown, that the dimension of original feature space can be reduced to 3 model parameters without losing relevant material reflectance information.

While the benefit of using angular-resolved reflectance measurements instead of single images has previously been reported in the literature (Lindner & Puente León (2007); Jehle et al. (2010); Wang et al. (2009); Gruna & Beyerer (2011)), using reflectance measurements in combination with modeling and simulating complex machine vision systems is a new research field and has the potential to be subject of future works.

5. Acknowledgements

We thank the reviewers for their valuable feedback on this article. Parts of the research leading to these results has received funding in the program "KMU-innovativ" from the German Federal Ministry of Education and Research under grant agreement no 01IS09036B.

6. References

Ashikhmin, M., Premože, S. & Shirley, P. (2000). A microfacet-based BRDF generator, *SIGGRAPH '00: Proceedings of the 27th annual conference on Computer graphics and interactive techniques*, ACM Press/Addison-Wesley Publishing Co., New York, NY, USA, pp. 65–74.

Ashikhmin, M. & Shirley, P. (2000). An Anisotropic Phong BRDF Model, *Journal of Graphics, GPU, & Game Tools* 5(2): 25–32.

Ben-Ezra, M., Wang, J., Wilburn, B., Xiaoyang Li & Le Ma (2008). An LED-only BRDF measurement device, *CVPR 2008: IEEE Conference on Computer Vision and Pattern Recognition*, IEEE Service Center, Piscataway, NJ.

Blinn, J. F. (1977). Models of light reflection for computer synthesized pictures, *SIGGRAPH Comput. Graph.* 11(2): 192–198.

Chessa, M., Solari, F. & Sabatini, S. (2011). Virtual reality to simulate visual tasks for robotic systems, *in* J.-J. Kim (ed.), *Virtual Reality*, InTech, pp. 71–92.

Cook, R. L. & Torrance, K. E. (1982). A Reflectance Model for Computer Graphics, *ACM Trans. Graph.* 1(1): 7–24.

Dana, K. J. (2001). BRDF/BTF measurement device, *ICCV 2001: Eighth IEEE International Conference on Computer Vision, 2001. ICCV 2001.*, Vol. 2, IEEE Computer Society, Los Alamitos, Calif, pp. 460–466.

Dana, K. J., van Ginneken, B., Nayar, S. K. & Koenderink, J. J. (1999). Reflectance and texture of real-world surfaces, *ACM Trans. Graph.* 18(1): 1–34.

Debevec, P., Hawkins, T., Tchou, C., Duiker, H.-P., Sarokin, W. & Sagar, M. (2000). Acquiring the reflectance field of a human face, *Proceedings of the 27th annual conference on Computer graphics and interactive techniques*, SIGGRAPH '00, ACM Press/Addison-Wesley Publishing Co., New York, NY, USA, pp. 145–156.

Dong, Y., Wang, J., Tong, X., Snyder, J., Lan, Y., Ben-Ezra, M. & Guo, B. (2010). Manifold bootstrapping for SVBRDF capture, *SIGGRAPH '10: ACM SIGGRAPH 2010 papers*, ACM, New York, NY, USA, pp. 1–10.

Gardner, A., Tchou, C., Hawkins, T. & Debevec, P. (2003). Linear light source reflectometry, *ACM Transactions on Graphics* 22(3): 749.

Ghosh, A., Achutha, S., Heidrich, W. & O'Toole, M. (2007). BRDF Acquisition with Basis Illumination, *ICCV 2007: IEEE 11th International Conference on Computer Vision*, IEEE Service Center, Piscataway, NJ, pp. 1–8.

Grassi, A. P., Perez, M. A. A., León, F. P. & Campos, R. M. P. (2006). Detection of circular defects on varnished or painted surfaces by image fusion, *2006 IEEE International Conference on Multisensor Fusion and Integration for Intelligent Systems*, IEEE, pp. 255–260.

Gruna, R. & Beyerer, J. (2011). Acquisition and evaluation of illumination series for unsupervised defect detection, *Proc. IEEE Instrumentation and Measurement Technology Conference*, Hangzhou, China, pp. 192–197.

Han, J. Y. & Perlin, K. (2003). Measuring bidirectional texture reflectance with a kaleidoscope, *SIGGRAPH '03: ACM SIGGRAPH 2003 Sketches & Applications*, ACM, New York, NY, USA, pp. 741–748.

Hanrahan, P. & Krueger, W. (1993). Reflection from layered surfaces due to subsurface scattering, *SIGGRAPH '93: Proceedings of the 20th annual conference on Computer graphics and interactive techniques*, ACM, New York, NY, USA, pp. 165–174.

Hastie, T., Tibshirani, R. & Friedman, J. (2009). *The Elements of Statistical Learning: Data Mining, Inference, and Prediction*, 2nd edn, Springer.

He, X. D., Torrance, K. E., Sillion, F. X. & Greenberg, D. P. (1991). A comprehensive physical model for light reflection, *SIGGRAPH Comput. Graph.* 25(4): 175–186.

Hentze, B. (2011). Feature selection to determine illumination parameters for material classification, Karlsruhe Institute of Technology. Bachelor's thesis.

Hünerhoff, D., Grusemann, U. & Höpe, A. (2006). New robot-based gonioreflectometer for measuring spectral diffuse reflection, *Metrologia* 43(2): S11.

Ibrahim, A., Tominaga, S. & Horiuchi, T. (2010). Spectral imaging method for material classification and inspection of printed circuit boards, *Optical Engineering* 49(5): 057201.

Jackett, M. J. & Frith, W. J. (2009). *Measurement of the reflection properties of road surfaces to improve the safety and sustainability of road lighting*, NZ Transport Agency, Wellington, N.Z.

Jehle, M., Sommer, C. & Jähne, B. (2010). Learning of optimal illumination for material classification, *in* M. Goesele, S. Roth, A. Kuijper, B. Schiele & K. Schindler (eds), *Pattern Recognition*, Vol. 6376 of *Lecture Notes in Computer Science*, Springer Berlin / Heidelberg, pp. 563–572.

Kajiya, J. T. (1985). Anisotropic reflection models, *SIGGRAPH '85: Proceedings of the 12th annual conference on Computer graphics and interactive techniques*, ACM, New York, NY, USA, pp. 15–21.

Kautz, J. & McCool, M. D. (1999). Interactive rendering with arbitrary brdfs using separable approximations, *IN EUROGRAPHICS RENDERING WORKSHOP*, pp. 281–292.

Khawaja, K., Maciejewski, A., Tretter, D. & Bouman, C. (1996). Camera and light placement for automated assembly inspection, *Robotics and Automation, 1996. Proceedings., 1996 IEEE International Conference on*, Vol. 4, pp. 3246–3252.

Koenderink, J. J., Van Doorn, A. J. & Stavridi, M. (1996). Bidirectional reflection distribution function expressed in terms of surface scattering modes, *Computer vision - ECCV '96*, Vol. 1065 of *Lecture notes in computer science*, Springer, Berlin, pp. 28–39.

Kurt, M., Szirmay-Kalos, L. & Křivánek, J. (2010). An anisotropic BRDF model for fitting and Monte Carlo rendering, *SIGGRAPH Comput. Graph.* 44(1): 1–15.

Lafortune, E. P. F., Foo, S.-C., Torrance, K. E. & Greenberg, D. P. (1997). Non-linear approximation of reflectance functions, *SIGGRAPH '97: Proceedings of the 24th annual conference on Computer graphics and interactive techniques*, ACM Press/Addison-Wesley Publishing Co., New York, NY, USA, pp. 117–126.

Lensch, H. P. A., Kautz, J., Goesele, M., Heidrich, W. & Seidel, H.-P. (2001). Image-based reconstruction of spatially varying materials, *Rendering Techniques 2001, Proceedings of the 12th Eurographics Workshop on Rendering*.

Lewis, R. R. (1994). Making Shaders More Physically Plausible, *In Fourth Eurographics Workshop on Rendering*, pp. 47–62.

Lindner, C., Arigita, J. & Puente León, F. (2005). Illumination-based segmentation of structured surfaces in automated visual inspection, *in* W. Osten, C. Gorecki & E. L. Novak (eds), *Optical Measurement Systems for Industrial Inspection IV*, Vol. 5856 of *Proceedings of SPIE*, SPIE, pp. 99–108.

Lindner, C. & Puente León, F. (2007). Model-Based segmentation of surfaces using illumination series, *Instrumentation and Measurement, IEEE Transactions on* 56(4): 1340–1346.

Malzbender, T., Gelb, D. & Wolters, H. (2001). Polynomial texture maps, *SIGGRAPH '01: Proceedings of the 28th annual conference on Computer graphics and interactive techniques // SIGGRAPH 2001 conference proceedings*, Vol. 2001 of *Annual conference series*, ACM and ACM Press, New York, NY, USA, pp. 519–528.

Marquardt, D. W. (1963). An algorithm for least-squares estimation of nonlinear parameters, *SIAM Journal on Applied Mathematics* 11(2): 431.

Marschner, S. R., Westin, S. H., Lafortune, E. P. F. & Torrance, K. E. (2000). Image-based bidirectional reflectance distribution function measurement, *Applied Optics* 39(16): 2592.

Matusik, W., Pfister, H., Brand, M. & McMillan, L. (2003). A data-driven reflectance model, *SIGGRAPH '03: ACM SIGGRAPH 2003 Sketches & Applications*, ACM, New York, NY, USA, pp. 759–769.

Mueller, G., Bendels, G. H. & Klein, R. (2005). Rapid Synchronous Acquisition of Geometry and BTF for Cultural Heritage Artefacts, *The 6th International Symposium on Virtual Reality, Archaeology and Cultural Heritage (VAST)*, Eurographics Association, pp. 13–20.

Mukaigawa, Y., Sumino, K. & Yagi, Y. (2007). High-Speed Measurement of BRDF using an Ellipsoidal Mirror and a Projector, *CVPR 2007: IEEE Conference on Computer Vision and Pattern Recognition, 2007*, IEEE Computer Society, Los Alamitos, Calif, pp. 1–8.

Murray-Coleman, J. F. & Smith, A. M. (1990). The Automated Measurement of BRDFs and their Application to Luminaire Modeling, *Journal of Illuminating Engineering Society* pp. 87–99.

Neumann, L., Neumann, A. & Szirmay-Kalos, L. (1999). Compact Metallic Reflectance Models, *Computer Graphics Forum* 18: 161–172.

Ngan, A., Durand, F. & Matusik, W. (2005). Experimental analysis of brdf models, *in* Kavita Bala & Philip Dutré (eds), *SR '05 Rendering Techniques*, Eurographics Association, Konstanz, Germany, pp. 117–126.
URL: *http://www.eg.org/EG/DL/WS/EGWR/EGSR05/117-126.pdf*

Oren, M. & Nayar, S. K. (1994). Generalization of Lambert's reflectance model, *SIGGRAPH '94: Proceedings of the 21st annual conference on Computer graphics and interactive techniques*, ACM, New York, NY, USA, pp. 239–246.

Peers, P., Hawkins, T. & Debevec, P. (2006). A reflective light stage, *Technical Report ICT-TR-04.2006*, University of Southern California Institute for Creative Technologies.

Phong, B. T. (1975). Illumination for computer generated pictures, *Commun. ACM* 18: 311–317.

Poulin, P., Fournier, A. & W, V. (1990). A Model for Anisotropic Reflection, *COMPUTER GRAPHICS* 24: 273–282.

Puente León, F. (1997). Enhanced imaging by fusion of illumination series, *in* O. Loffeld (ed.), *Sensors, Sensor Systems, and Sensor Data Processing*, Vol. 3100, SPIE, Munich, Germany, pp. 297–308.

Reed, I. & Yu, X. (1990). Adaptive multiple-band CFAR detection of an optical pattern with unknown spectral distribution, *Acoustics, Speech and Signal Processing, IEEE Transactions on* 38(10): 1760–1770.

Reiner, J., Laurent Mazuray, Rolf Wartmann, Wood, A., Jean-Luc Tissot & Jeffrey M. Raynor (2008). Rendering for machine vision prototyping, *Optical Design and Engineering III* 7100(1): 710009.

Ren, P., Wang, J., Snyder, J., Tong, X. & Guo, B. (2011). Pocket reflectometry, *ACM SIGGRAPH 2011 papers on - SIGGRAPH '11*, ACM Press.

Sandmeier, S. R. (2000). Acquisition of Bidirectional Reflectance Factor Data with Field Goniometers, *Remote Sensing of Environment* 73(3): 257–269.

Sattler, M., Sarlette, R. & Klein, R. (2003). Efficient and Realistic Visualization of Cloth, *Eurographics Symposium on Rendering 2003*.

Schlick, C. (1994). An Inexpensive BRDF Model for Physically-based Rendering, *Computer Graphics Forum* 13(3): 233–246.

Schröder, P. & Sweldens, W. (1995). Spherical wavelets: efficiently representing functions on the sphere, *SIGGRAPH '95: Proceedings of the 22nd annual conference on Computer graphics and interactive techniques*, ACM, New York, NY, USA, pp. 161–172.

Sillion, F. X., Arvo, J. R., Westin, S. H. & Greenberg, D. P. (1991). A global illumination solution for general reflectance distributions, *SIGGRAPH '91: Proceedings of the 18th annual conference on Computer graphics and interactive techniques*, ACM, New York, NY, USA, pp. 187–196.

Stam, J. (1999). Diffraction shaders, *SIGGRAPH '99: Proceedings of the 26th annual conference on Computer graphics and interactive techniques*, ACM Press/Addison-Wesley Publishing Co., New York, NY, USA, pp. 101–110.

Torrance, K. E. & Sparrow, E. M. (1967). Theory for Off-Specular Reflection From Roughened Surfaces, *Journal of the Optical Society of America* (57): 1105–1114.

Wang, O., Gunawardane, P., Scher, S. & Davis, J. (2009). Material classification using BRDF slices, *Computer Vision and Pattern Recognition, IEEE Computer Society Conference on*, Vol. 0, IEEE Computer Society, Los Alamitos, CA, USA, pp. 2805–2811.

Ward, G. J. (1992). Measuring and modeling anisotropic reflection, *SIGGRAPH Comput. Graph.* 26(2): 265–272.

Westin, S. H., Arvo, J. R. & Torrance, K. E. (1992). Predicting reflectance functions from complex surfaces, *SIGGRAPH '92: Proceedings of the 19th annual conference on Computer graphics and interactive techniques*, ACM, New York, NY, USA, pp. 255–264.

White, D. R., Saunders, P., Bonsey, S. J., van de Ven, J. & Edgar, H. (1998). Reflectometer for Measuring the Bidirectional Reflectance of Rough Surfaces, *Applied Optics* 37(16): 3450.

Wong, T., Heng, P., Or, S. & Ng, W. (1997). Image-based rendering with controllable illumination, *Proceedings of the Eurographics Workshop on Rendering Techniques '97* pp. 13–22. ACM ID: 731971.

12

Towards the Optimal Hardware Architecture for Computer Vision

Alejandro Nieto, David López Vilarino and Víctor Brea Sánchez
Centro de Investigación en Tecnoloxías da Información (CITIUS)
University of Santiago de Compostela
Spain

1. Introduction

Computer Vision systems are experiencing a large increase in both range of applications and market sales (BCC Research, 2010). From industry to entertainment, Computer Vision systems are becoming more and more relevant. The research community is making a big effort to develop systems able to handle complex scenes focusing on the accuracy and the robustness of the results. New algorithms provide more advanced and comprehensive analysis of the images, expanding the set of tools to implement applications (Szeliski, 2010).

Although new algorithms allow to solve new problems and to approach complex situations with a high degree of accuracy, not all the algorithms are adequate to be deployed in industrial systems. Parameters like power consumption, integration with other system modules, cost and performance limit the range of suitable platforms. In most cases, the algorithms must be adapted to achieve a trade-off solution and to take advantage of the target platform.

Conventional PC-based systems are constantly improving performance but their use is still limited to areas where portability, power consumption and integration are not critical. In case of highly complex algorithms, with an irregular execution flow, complex data representation and elaborated patterns to access to data, a significant gain is not achieved when moving to an *ad hoc* hardware design. In this case, a high-end CPU and a GPU with general-purpose capabilities (GPGPU) is a flexible and very powerful combination that will outperform other options (Castano-Díez et al., 2008).

However, when a conventional system does not meet the requirements of the application, a more aggressive planning is needed. For instance, migrating the application to a dedicated device such as a DSP, an FPGA or a custom chip (Shirvaikar & Bushnaq, 2009). At this point, the designers have to consider alternatives as to reduce operating range, accuracy and robustness of the results, or to remove expensive operations in order to simplify the hardware that will be implemented (Kolsch & Butner, 2009). PC-based systems enable a great flexibility at cost of performance so pure software-based algorithms hardly match pure hardware implementations. This is a serious limitation because it can compromise the efficiency of the application. This is the reason why the industry is making great efforts to develop novel architectures that enable greater flexibility to adapt the algorithms without compromising the quality of the results.

Computer Vision applications are often divided into three stages: low, medium and high level processing. Low level operations are quite simple and repetitive but applied over a large set of data with a very reduced data dependent program flow, so massive parallelism is essential for performance. In the medium level stage, temporal and task parallelism is key as the data set is smaller and the program flow is quite data-dependent. High level operations are performed over complex data representations but reduced data sets and with high precision as requirement. An efficient Computer Vision system must deal with all these stages. The selection of the computation model will determine the performance of the device in each one of the stages. So, which is the optimal paradigm for such applications? Increase the number of concurrent instructions? Increase the number of data elements processed simultaneously? A combination of both?

Taking into account the application requirements, a suitable platform to build the system must be selected. Besides general parameters (performance, cost and integration), there is a set of factors that restricts the computing paradigms and the devices that can be selected. For instance, a critical parameter is the way in which the data are transferred to the device because I/O operations are one of the bottlenecks in high-performance systems. Data type and representation will affect the computational units. The program flow will constrain the inner connections between these units and the storage elements.

This chapter addresses an analysis of different computing paradigms and platforms oriented to image processing. Previously, a representative set of Computer Vision algorithms covering the three levels of processing is reviewed. This study will lead us to observe the algorithms in terms of a set of common characteristics: operations, data type, program flow, etc. This is critical to design new hardware architectures in order to maximize performance. The analysis from the hardware point of view will highlight the best features of the most used computing paradigms in order to establish a relationship between the type of operation, data, programming model and hardware architecture. An efficient architecture for Computer Vision must combine all the selected features. The analysis of the characteristics of the different algorithms will lead us naturally to an optimized general-purpose hardware architecture for Computer Vision.

2. The problem of computer vision

Traditionally, Computer Vision (CV) applications include building blocks from three computation levels: low-, mid- and high-level vision computing. The type of operations, the data representation and the flow execution of programs depend deeply on the considered level of this hierarchy. Nevertheless, current CV-algorithms are composed of many different processing steps regarding the type of data and the way these are computed which makes difficult to classify them only in one subgroup. Following, a rather rough classification of widely used CV-algorithms is made, keeping in mind the data domain and the complexity of the involved operations.

2.1 Low-level vision

After image acquisition, some preprocessing steps are often required. These are intended to provide reliable input data for subsequent computing stages. Some typical operations are noise reduction, color balancing, geometrical transformation, etc. Most of these operations are based on point or near-neighborhood operations. Point operations are performed at pixel-level in such a way that the output only depends on the value of any individual pixels from one or several input images. With this type of operation it is possible to modify the pixel intensity to enhance parts of the image, by increasing contrast or brightness. Equally, simple pixel-to-pixel arithmetic and Boolean operations also enable the construction of operators as *alpha blending*, for image combination or color space conversion. Neighborhood operations take also into account the value of adjacent pixels. This operation type is the basis of filtering, binary morphology or geometric transformation. They are characterized by simple operations, typically combining weighted sums, Boolean and thresholding processing steps.

After preprocessing stages, useful information has to be extracted from the resulting images. Common operations are edge detection, feature extraction or image segmentation. Edges are usually defined as step discontinuities in the image signal so finding local maxima in the derivative of the image or zero-crossings in the second derivative are suitable to detect boundaries. Both tasks are usually performed by the convolution of the input image with spatial filtering masks that approximate a first or second derivative operator.

Feature points are widely used for subsequent computing steps in multiple CV-applications. Basically, a feature represents a point in the image which differs from its neighborhood. One of the benefits of local features is the robustness against occlusion and the ability to manage geometric deformations between images when dealing with viewpoint changes. In addition, they improve accuracy when, in the same scene, objects are at different planes, (i.e. at different scales). One of the most popular techniques is that proposed by Harris (Harris & Stephens, 1988) to detect corners. It is widely used due to its strong invariance to rotation, image noise and no large illumination changes. It uses the local auto-correlation function, which describes the gradient distribution in a local neighborhood of each image point to detect the location of the corners. Using the locally averaged moment matrix from the image gradients, corners will be located at the maximum values. Another frequently used technique is the Scale-Invariant Feature Transform (SIFT) (Lowe, 2004). SIFT localizes extrema both in space and scale. Using the *Difference of Gaussians* as scale-space function, the images are filtered with Gaussian kernels of different sizes (scales). This is performed for different image sizes (octaves). The response of each filter is subtracted from the immediately following in the same octave. The interest points are scale-space extrema so local maxima and minima are extracted by comparing the neighborhood points in the same, the previous and the posterior scales. To improve accuracy, a sub-pixel approximation step is done, interpolating the location of the feature inside the scale-space structure. The number of octaves and scales can be tuned to meet the system requirements. SIFT provides invariance against scale, orientation and affine distortion, as well as partial occlusion and illumination changes. Other algorithms were proposed to improve accuracy or performance like, the *Speeded Up Robust Features* (SURF) (Bay et al., 2006) or the *Gradient Location and Orientation Histogram* (GLOH) (Mikolajczyk & Schmid, 2005). This kind of detectors are quite complex and their performance can be low even using custom hardware. For this reason, less reliable algorithms are still in use, as Harris Corner Detector, FAST (Trajkovi & Hedley, 1998) or the Smallest Univalue Segment Assimilating Nucleus (SUSAN) (Smith & Brady, 1997) corner detectors, because of their efficiency under controlled situations and their low hardware requirements.

Segmentation refers to the process of separating the data into several sets according to certain characteristics. There are several techniques to carry out this task, either based on boundaries or regions (Pal & Pal, 1993) (Haralick & Shapiro, 1985). Nevertheless, most of them rely on near-neighborhood operations. Particular attention deserves the clustering methods like the popular k-means which partitions the data set into several clusters according to a proximity criterion defined by a distance function. These methods are not restricted to image data. N-dimensional sets of abstract data can also be partitioned. Furthermore, information about the scene or domain can be introduced (number and characteristics of the target clusters). Therefore, they might be classified either as a low or mid-level vision stage.

2.2 Mid-level vision

Mid-level CV stages usually operate on images from previous processing steps, often binary images, and produce a lower amount of data but with a higher concentration of information. Some common operations are object classification and scene reconstruction.

One of the goals of Computer Vision is to recognize objects in a scene. Based on object location, pose or 2D/3D spatial relations between the objects, the algorithms have to be able to analyze the scene and its content. This involves issues such as dealing with object models, classifiers and the ability to integrate new information in the models. In the literature, a large amount of techniques can be found, usually classified as global methods, more intended for object detection and local feature-based methods for object recognition. In all of them good image registration is essential for both accuracy and performance (Zitova & Flusser, 2003). As for the global methods, common techniques are based on *template-based matching*, which employs a convolution mask or template to measure the similarity between an object patch and the template. In this sense, *normalized cross-correlation* (NCC), *sum of squared differences* (SSD) or *sum of absolute differences* (SAD) are widely used. As for local methods, local feature descriptors play an important role. Roughly speaking, a descriptor is an abstract characterization of a feature point based on its environment. One of the most popular techniques is the proposed in second part of the SIFT algorithm based on stacked orientation histograms which associate a high dimension vector to each keypoint. In order to reduce the amount of false positives and negatives during the matching stage the search area is limited by using strategies like the *nearest neighbor search* (NNS) which attempts to find the nearest points of a given one in a vector space. An *indexing structure* allows to search for features near a given feature rapidly. This is the case of the *K-dimensional trees* , which organize points in a k-dimensional space in such a way that each node has at most two child nodes.

Scene reconstruction consists of the generation of scene models starting from their parts. There are different techniques to reconstruct one or several objects in a scene. To build a 3D model, coordinates of scene points have to be calculated from the objects. If the location of the camera is known, 3D coordinates of a scene point can be determined from its projection on image planes of different viewpoints. The whole process starts with feature extraction and matching. Using geometric consistency tests it is possible to eliminate wrong matches. There are different solutions to estimate the fundamental matrix, as the *RANdom SAmple Consensus* (RANSAC). Once the matches between images are consistent, camera pose and scene geometry is reconstructed using *Structure from Motion* methods and refined with *Bundle adjustment* techniques (Triggs et al., 2000).

2.3 High-level vision

The high-level stage often starts from an abstract representation of the information. This stage is highly application dependent but due to the variety of operations, data structures, memory access patterns and program flow characteristics are often only compatible with a general purpose processor. High-level processing is characterized by the use of a small set of data to represent knowledge about the application domain. More complex data structures are needed to store and process this information efficiently, making the operations and the memory access patterns more elaborated. This, together with the inherent complexity of decision making, makes the program flow very variable.

Robust pattern recognition, object identification, complex decision making or system adaptation are some of the benefits of integrating Artificial Intelligence methods with Computer Vision. Otherwise, the system would be limited to a predetermined set of actions. Machine learning makes computers capable of improving automatically with experience. This way it is possible to generalize the behavior from unstructured information with techniques such as neural networks, decision trees, genetic algorithms, regression models or support vector machines. Machine learning has emerged as a key component of intelligent computer vision systems, contributing to a better understanding of complex images (Viola & Jones, 2001) (Oliver et al., 2000).

Data mining is the process of analyzing data using a set of statistical techniques in order to summarize into segments of useful information. This makes possible analyze data from different dimensions or angles, categorizing and summarizing the identified relationships, making evident hidden relationships or patterns between events. On the contrary of Machine learning, data mining focuses on discovering hidden patterns instead of generalizing known patterns using the new data.

It is very difficult to establish a classification of tasks and operations for high-level processing. Some of the performed tasks fall within the scope of the measurement of application specific parameters such as size and pose of objects, fault detection and monitoring specific events such as traffic situations, for example. The algorithms and technologies are very diverse and most of them lie in statistical analysis and artificial intelligence domains.

As it was previously mentioned some tasks that initially fit into the low or medium level stages due to its context actually are more like high-level operations by the type of operations performed. This is related with the commonly used bottom-up image analysis, which starts from raw data to extend the knowledge of the scene. However, new approaches include feedback to perform top-down analysis. This way, low- and mid-level stages can be controlled with general knowledge of the image, improving the results.

3. Computing platforms

Given the wide range of algorithms and applications of Computer Vision, it is clear that it does not exist a unique computing paradigm or an optimal hardware platform. The type of operations, the complexity of data structures and especially the data access patterns greatly determine parameters such as the range of application, performance, power consumption or cost. This section presents some of the most prominent platforms in image processing, focusing on their strengths and weaknesses.

3.1 Computing paradigms

The Flynn's taxonomy (Flynn, 1972) classifies the computer architectures in four big groups according to the number of concurrent instructions and data sets processed. Image processing tasks perform more or less efficiently depending on the selected paradigm. In order to develop a Computer Vision application it is crucial to exploit the spatial (data) or temporal (task) parallelism to meet trade-offs among performance, power consumption or cost.

SISD

Single Instruction Single Data (SISD) refers to the conventional computing model. A single processing unit executes a sequence of instructions on a unique data stream. Most modern computers are placed under this category and, although only one processor and one memory element are present, those which are able to pipeline their data-path are generally classified under the SISD category as they are still *serial computers*.

This paradigm performs better when spatial and temporal parallelism are hard to exploit. As seen previously, high-level image processing fit the SISD paradigm because most tasks are sequential, with a complex program flow and strong dependences between data. As processing is done sequentially, most optimizations aims to enhance the access between the memory and the arithmetic unit. Memory and processor speed are the main constraints. Furthermore, some kind of parallelism can be exploited when pipelining the data-path. Data allocation, pre-fetching and reducing stalls in the pipeline are some of the possible optimizations.

SIMD

SIMD (*Single Instruction Multiple Data*) computers have an unique control unit and multiple processing units. This control unit sends the same instruction to all processing units, which operate over different data streams. This paradigm focuses on exploiting the spatial parallelism. It is also possible to pipeline the data-path or to employ several memories to store the data in order to increase the bandwidth. SIMD computers are commonly specific-purpose, intended to speed-up certain critical tasks.

One of the drawbacks of SIMD machines is data transference. A network is required to both supply data to each processing unit and share data among them. Its size grows with the number of connected nodes so SIMD architectures have a practical limitation. Another restriction is data alignment when gathering and scattering data into SIMD units. This results in a reduction of flexibility in practical implementations. It is needed to determine the correct memory addresses and reordering data adequately, affecting performance. In addition, as this paradigm exploits spatial parallelism, program flow is heavily limited because all units execute the same instruction. Additional operations are needed to enable at least simple flow control tasks.

Low-level image processing benefits greatly of SIMD units. As it was described previously, most operations are quite simple but repetitive over the whole set of data. In addition, certain tasks of mid and high-level processing can also take advantage of SIMD units when using in conjunction with others paradigms. The simplicity of the arithmetic units and the memory access patterns make feasible to design efficient units, which is crucial to increase the parallelism. Memory bandwidth and data distribution among the processors is also key for performance.

Two types of SIMD accelerators can be distinguished based on the number of processing units: fine-grain and coarse-grain processors, where the major difference is the number of processing units. While the first includes a large amount of very simple processors with a rigid network, the second features major flexibility although with a much lower parallelism. When using in low-level image processing, fine-grain processor arrays match with massively parallel operations such as filters or morphological operations. Using a processor-per-pixel scheme and local communications, neighborhood operations are completed in just a few instructions. On the contrary, when the parallelism level is lower a configuration as *vector processor* is usually preferable. By reducing communications and increasing core complexity, they are much more flexible and efficient not only for low-level operations but also for other processing stages.

MIMD

Multiple Instruction Multiple Data (MIMD) refers to architectures where several data streams are processed using multiple instruction streams. MIMD architectures have several processing units executing different instructions to exploit task parallelism. Processors perform independently and asynchronously. MIMD systems are classified depending on the memory architecture.

In *Shared Memory Systems*, all processors have access to an unique memory. Connection hierarchy and latencies are the same for all processors. This scheme eases data transference among processors although simultaneous access must be taken into account to avoid data hazards. Scalability is also reduced because it is hard to increase the memory bandwidth at the same rate as the number of processors.

If each processor has its own and private memory it is possible to upscale more easily as memory and processors are packet as a unit. This scheme is known as *Distributed Memory System*. In addition, local memory access is usually faster. The major disadvantage is the access to data which are located outside the private memory because dedicated buses and a *message passing system* to communicate with the processors are needed. This can result in high access times and an increase of hardware requirements.

In a *Distributed Shared Memory System*, the processors have access to a common shared memory but without a shared channel. Each processor is provided with local memory which is interconnected with other processors through a high-speed channel. All processors can access to different banks a global address space. Access to memory is done under the schemes as *Non-Uniform Memory Access* (NUMA), which takes less time to access the local memory than to access the remote memory of other processor. This way scalability is not compromised.

Very long Instruction Word (VLIW) and *superscalar* architectures are also classified within MIMD paradigm because they exploit instruction-level parallelism, executing multiple instructions in parallel. *Pipelining* also executes multiple instructions but splitting them in independent steps to keep all the units of the processor working at a time.

Mid-level image processing and some operations of the other processing levels of Computer Vision can exploit MIMD processors. Operations are relatively simple, with data-dependent program flow. Temporal and task parallelism are easier to exploit than spatial parallelism although a reduced degree is usually present in this type of algorithms. Each MIMD processing element can include SIMD units. This way it is possible to process complex tasks more efficiently, from kernel operations as when pre-processing images concurrently in multi-view vision systems to high level tasks as multiple object recognition and tracking. In general, any set of tasks with weak dependences between them to reduce internal communications can take advantage of this paradigm.

MISD

There is one more paradigm, *Multiple Instruction Simple Data* (MISD), which achieves higher parallelism than SISD executing different instructions over the same data set employing several computing units.

Systolic arrays are regular n-dimensional arrays of simple cores with nearest-neighbors interconnections. Each core operates on the input data and shares the result to its neighbor, flowing the data synchronously usually with different flow in different directions. They are employed for tasks such as image filtering or matrix multiplication. Pipelined architectures belong to this type, as they are considered an one-dimensional systolic array, but they are commonly considered an improved version of the other aforementioned paradigms.

This paradigm is rarely used for Computer Vision as the other paradigms match better and offer higher performance and flexibility when dealing with real Computer Vision problems.

Remarks

Low-level Computer Vision entails the largest processing times in most applications. Data sets are usually very large and the kind of operations simple and repetitive. However, operations are inherently massively parallel and the data access patterns are regular. It is feasible to exploit these features to design very optimized SIMD custom hardware accelerators or to migrate the algorithms to existing hardware.

It is harder to extract parallelism in the mid-level stage because operations involve more complex data-flow. In addition, the data set is smaller so the benefits of including dedicated units to speed-up the computation are lower than expected. Despite this, hybrid processors (SIMD-MIMD) able to exploit both spatial and temporal parallelism can overcome this limitation.

Finally, the amount of data involved in the high-level stage is usually small so it is rarely necessary to sacrifice precision in order to get better performance. Moreover, unlike in previous stages, the disparity in the type of data makes the use of floating-point often a requirement. Another characteristic of this stage is the program flow, far more complex, which may even consume more computation time than the arithmetic operations. The kind of computation performed at this stage is so varied that the best option is often a general purpose SISD processor.

3.2 Current devices

There are different possibilities to implement the aforementioned computation paradigms. There is not an unique and direct correspondence between a paradigm and its hardware implementation. On the one hand, it can be designed a dedicated hardware which follows the original conception. On the other hand, the paradigm can be emulated both in hardware or software.

Microprocessors

Microprocessors, SISD machines, are the most straightforward devices to develop a Computer Vision application. Their main advantage is their versatility, the ability to perform very different tasks for a low cost. They can perform any type of data processing, although its efficiency, measured in parameters such as cost, power or integration capabilities, is not always optimal because of their general-purpose condition. The large variety of available technologies, libraries, support and programs cut down the *cold start*, enabling to get the system ready for development in a short time. Developers can focus on the problem itself instead of technical issues (Bradski & Kaehler, 2008).

Basically, they are composed by a main memory and a processing unit which includes the arithmetic and the control modules. From this basic structure more optimized microprocessors can be designed. From caches to cut down the memory access times, tightly coupled high-speed memory controllers or specialized units for critical tasks, the variability of architectures is as large as the amount of fields in the market (Singhal, 2008). However, despite the evolution of the industry pure SISD microprocessors do not offer adequate performance for a large set of tasks. That is why a wide range of accelerator modules have been included, as specific-purpose arithmetic units and sets of instructions or co-processors. The inclusion of SIMD units is decisive for tasks such as video encoding and decoding, but any data-intensive algorithm can take advantage of them (Franchetti et al., 2005).

As it will be discussed later, the advances in the semiconductor industry allows to increase the integration density so it is possible to include more processing power on the same Silicon area. This has led to abandon the race for speed (to increase the working frequency) to more efficient systems where energy consumption is vital and parallelism is the way to overcome the limitations of Moore's Law (Naffziger, 2009). Most of modern processors are multi-core and the number of cores is expected to grow in the near future. Programming languages and techniques as well as image processing algorithms have to be adapted to this new reality (Chapman, 2007).

Microprocessors are employed in a wide range of applications, from developing and testing algorithms such as autonomous driving (Urmson et al., 2008) to final platforms as medical image reconstruction (Chu & Chen, 2009). Even its use in restrictive stand-alone devices is also viable such as autonomous flight (Meier et al., 2011). Microprocessors stand out in high-level tasks, as the latest stages of image retrieval (Deselaers et al., 2008) and scene understanding (Li et al., 2009), where handling image databases, storing and communicating data are fairly complex to implement them in specific purpose devices. Video surveillance tasks can take advantage of these features for image processing (Ahmed & Terada, 2010) and event control in complex distributed systems (Chen et al., 2008).

Mobile processors have become a benchmark in innovation and development after the explosion of the mobile market. As discussed below, they integrate several general purpose cores, graphics processing units and other co-processors on a single chip keeping power consumption very low. The applications they can address are increasingly complex (Taylor et al., 2009) (Wither et al., 2011) (Ren et al., 2010).

Graphics Processing Units

A *Graphics Processing Unit* (GPU) is a specialized co-processor for graphic processing to reduce the workload of the main microprocessor in PCs. They implement highly optimized graphic operations or *primitives*. Current GPUs provide a high processing power and exploit the massively spatial parallelism of these operations. Because of their specialization, they can perform operations faster than a modern microprocessor even at lower clock rates. GPUs have hundreds of independent processing units working on floating point data. Memory access is critical to avoid processing downtimes, both in bandwidth and speed.

Their high processing power makes GPUs an attractive device for non related graphic tasks. *General Purpose GPU* (GPGPU) is a technique to perform general computation not related to graphics on these devices (Owens et al., 2007) (Che et al., 2008). This makes possible to use its specialized and limited pipeline to perform complex operations over complex data types. In addition, it eases memory management and data access. Flow control, as looping or branching, is restricted as in other SIMD processors. Modern GPUs architectures as (Seiler et al., 2008) or (Lindholm et al., 2008) have added support for these operations although slightly penalizing throughput. Word size is also a limitation in GPGPU techniques. It was reduced to increase the integration density as graphic operations do not usually require high precision. However, since this was a serious limitation for scientific applications, large word-sizes support was added later (Thall, 2006).

GPUs are effective when using *stream processing*, a paradigm related to SIMD (Rixner et al., 1998). A set of operations (*kernel*) is applied to each element of a set of data (*stream*). The flexibility is reduced to increase the parallelism and to lower the communication requirements when involving hundreds of processing elements. Otherwise, providing data to hundreds of processors would be a bottleneck. Processors are usually pipelined in a way that results pass from one arithmetic unit to the next one. This way, the locality and concurrency are better exploited, reducing communication requirements because most of the data are stored on-chip.

The use of GPUs to speed-up the computing has greatly increased, specially after the optimization of libraries and functions that mask low-level technical difficulties. Most image processing kernels and algorithms were adapted to work in GPGPUs, obtaining significant improvements. Basic image processing (Podlozhnyuk, 2007), FFT transforms (Govindaraju & Manocha, 2007), feature extraction (Heymann et al., 2007) or stereo-vision (Lui & Jarvis, 2010), all of them computationally expensive, benefit greatly of the massively parallelism of GPU. They can be used also to emulate other computing paradigms frequently used in low level vision (Dolan & DeSouza, 2009). However, some authors argue that the gap between GPUs and CPUs is not as large as it seems if key optimizations are carried out (Lee et al., 2010).

Digital Signal Processors

A *Digital Signal Processor* (DSP) is a microprocessor-based system with a set of instructions and hardware optimized for intensive data applications. They are specially useful for real-time processing of analog signals but they offer a high throughput in any data intensive application. DSP market is well established and offers a large range of devices, optimized for each particular task (Schneiderman, 2010). Apart from attached processors, to assist a general purpose host microprocessor, DSPs are often used in embedded systems including all necessary elements and software.

They are able to exploit parallelism both in instruction execution and data processing. In a von Neumann architecture, instruction and data share the same memory space. However, DSP applications usually require several memory accesses to read and write data per instruction. To exploit concurrency, many DSPs are based on a Harvard architecture, with separate memories for data and instructions. Many modern devices are based on *Very Long Instruction Word* (VLIW) architectures so they are able to execute several instructions simultaneously (Lin et al., 2008). Compilers are fundamental to find the parallelism in the instructions and a large improvement can be obtained after an efficient placement of data and programs in memory. Superscalar processors and pipelined data-paths also improve the overall performance although this is done by hardware. To operate, they include specialized hardware for intense calculation such as *multiply-accumulate* operation, which is able to produce a result in one clock cycle. Although many DSPs have floating-point arithmetic units, fixed-point units fit better in battery-power devices. Formerly, floating-point units were slower and more expensive but this gap is getting smaller and smaller. They also include zero-overhead looping, rounding and saturated arithmetic or dedicated units for address management (Talavera et al., 2008; Texas Instruments, 2002; Wang et al., 2010).

DSPs are usually designed for intensive data processing so performance can be penalized in mixed tasks. However, current all-in-one devices are able to handle complete applications efficiently. In addition, DSPs are not only available as independent devices but also as part of integrated circuits such as FPGAs or SoCs.

DSPs have a large tradition on image processing tasks. As optimized versions of conventional processors, they were used to accelerate the most expensive operations. These operations were related mainly with low-level image processing (Baumgartner et al., 2009), where parallelism and data access enable a large performance increase. Stereo vision (Lin & Chiu, 2008), Fourier transform (Sun & Yu, 2009) or video matching and tracking (Shah et al., 2008) are some samples. However, higher level algorithms also suit for DSPs, specially in industrial tasks (Suzuki et al., 2007) (Neri et al., 2005). Nowadays, DSPs are able to handle large sets of operations efficiently both for co-processing (Rinnerthaler et al., 2007) or standalone (Bramberger & Rinner, 2004).

Field Programmable Gate Arrays

A *Field Programmable Gate Array* (FPGA) is a device with user-programmable hardware logic. It is made of a large set of *logic cells* connected together through a network. Both elements are programmable in such a way that the logic cells emulate combinational functions and the network permits to join them to build more complex functions. In addition, the *program* can be rewritten as many times as needed.

The main advantage of FPGAs is their high density of interconnections between cells, which provides a very high flexibility. This network has a complex hierarchy with optimizations for specific functions. It provides specialized lines to propagate clock or reset signals across all the FPGA or to build buses with high *fan-out* within acceptable time delays. With these very basic elements it is possible to build highly complex modules as arithmetic units, controllers or even embedded microprocessors. In addition to simple elements such as 1-bit *flip-flops*, FPGAs have a large set of embedded modules. Large memory elements, DSP arithmetic units, networking and memory controllers or even embedded microprocessors are available on modern FPGAs (Leong, 2008).

These devices are an excellent mechanism to build proof-of-concept prototypes. On the one hand due to their flexibility any computing paradigm can be implemented restricted mainly for the number of available cells. On the other hand, thanks to the re-programmability it is possible to debug and test on real hardware. Even more, nowadays some final products are implemented exclusively on FPGAs instead of migrating the design to custom integrated circuits. The achieved performance can be tens of times higher with lower power consumption than standard PC-based approaches (Fasang, 2009). FPGAs are widely employed as co-processors in personal computers such as GPUs or as accelerators in specific purpose devices as high-capacity network systems (Djordjevic et al., 2009) or high-performance computing (Craven & Athanas, 2007). Nowadays, it is possible to embed full systems on a single FPGA.

One of the major disadvantages of FPGAs is the set-up time, still higher than in pure-software approaches. Traditional FPGA programming is done with HDL languages, forcing to design at a very low level. High-level languages, such as C extensions, are more friendly for software engineers, although the control over the design is much lower (Coussy et al., 2010). The FPGA-based designs can be exported and distributed as *IP Cores* because theses languages are platform-independent unless very specific features of a given FPGA are employed. This way, *non-recurring engineering* costs (NRE) are cut-down. Therefore FPGAs are in an intermediate stage between software and hardware. Algorithms are programmed by software and *compiled* to a hardware architecture, so an careful hardware/software codesign is fundamental (Moreno et al., 2010). They are able to exploit both spatial and temporal parallelism very efficiently. Since logic cells are independent many arithmetic units can process concurrently, with custom routing between them. In addition, the memory subsystem can be tuned to exploit the on-chip memory banks, reducing the access to the external memories.

Regarding Computer Vision, FPGAs are widely used both in industry and research. They offer a high degree of flexibility and performance to handle many different applications. Most compute-intensive algorithms were migrated to FPGAs: stereo vision (Jin et al., 2010), geometric algebra (Franchini et al., 2009), optical flow (Martineau et al., 2007), object recognition (Meng et al., 2011) or video surveillance (Nair et al., 2005) (Salem et al., 2009) to name some examples. Low and mid-level image processing stages, based on SIMD/MIMD paradigms can be efficiently implemented. However, high-level processing, although it could be possible to implement on an FPGA, fits better on conventional processors. Nevertheless, FPGAs can use external processors connected through high-speed off-chip communications or include general-purpose processors. There are available soft-core (emulated) (Lysecky & Vahid, 2005) and hard-core (Veale et al., 2006) embedded processors. While the first offer a large degree of configuration, adapting all parameters of the design to the particular needs of the applications, the last feature better performance. This way FPGAs are able to implement all the stages of a complete application (Shi, 2010).

Application-specific integrated circuits

An *Application-Specific Integrated Circuit* (ASIC) is a device designed for a particular task instead of for general purpose functionality. In this case, designers have to work at the very bottom level of design so this process is long and error-prone. As there are elements present in almost all ICs, a set of *libraries* is usually provided making easier system-design. This way it is possible to work at different levels of abstraction. *Full Custom* designs require a greater effort because it is necessary to design both functionality and physical layout. However, it allows better optimizations and performance. *Standard Cells* allow to focus on the logic operation instead on the physical design, splitting the process into two parts. Physical design is usually done by the manufacturer, who provides from simple logic gates to more complex units such as *Flip-Flops* or adders. Apart from simple units, third party manufacturers provide more complex modules for specific functions, known as *IP Cores*. This way, they can be used as subcomponents in a large design. There is a large variety of cores to tackle all needs, as it happens with the FPGAs, and there is available from IO controllers (RAM, PCI-Express, ethernet) to arithmetic cores (signal processing, video and audio decoding) or even complete microprocessors.

The IC-level design allows to build embedded systems, *System-on-Chip*, efficiently. It is possible to include all the elements of the system on a single chip, even if there are digital, analog and mixed-signal modules. Nowadays, custom ASICs are the unique alternative for complex SoCs although FPGAs size grows with each generation and are becoming a viable alternative. IC design leads to high costs, both in design and manufacturing when production volume is low. However, some applications still need ICs because the alternatives do not match with performance, power consumption or size (Kuon & Rose, 2007).

It is true that some of the devices previously mentioned are in some way ASICs. However, the design of a custom architecture for a concrete algorithm provides the best possible results. Many custom chips were designed and built instead of mapping them in a programmable device, for specific algorithms (Kim et al., 2008), domain applications (Stein et al., 2005) or complete general purpose SoCs (Khailany et al., 2008a).

Because of this flexibility a large set of *exotic* devices can be found in the market or in the specialized literature. While some aims to address very specific tasks or novel computing paradigms, others are taking their first steps on the market after that technology has allowed its viability. *Pixel-parallel Processor Arrays* are the natural platform for low-level vision. They are massively parallel SIMD processors laid down on a 2D grid with a processor-per-pixel correspondence and local connections among neighbors. Each processor, very simple, can also include an image sensor to eliminate the IO bottleneck. Some representative examples are (Foldesy et al., 2008; Garrido et al., 2008; Lopich & Dudek, 2008). There are also approaches closer to the biological vision, as (Koyanagi et al., 2001) or (Constandinou et al., 2004). More information is available in (Zarándy, 2011). *Massively Parallel Processor Arrays* (MPPAs) provide hundreds to thousands of processors. They are encapsulated and independent and have their own program and memories. They work in MIMD mode but also include internal improvements as pipelines, superscalar capabilities or SIMD units. Some examples are (Bell et al., 2008; Butts et al., 2007; Duller et al., 2003).

3.3 Discussion

As described previously in this section, there is a wide range of hardware devices suitable for Computer Vision. Depending on the application requirements, a compromise between performance, cost, power consumption and development time is needed. Commercial applications are heavily constrained by the time-to-market so suboptimal solutions are preferable if the development cycle is shorter. This way, software-based solutions are usually better from the commercial point of view.

As discussed before, the large amount of highly optimized libraries make PCs (conventional microprocessors) the first choice both as development and production platform. Multi-core and SIMD programming are key for performance, although this can significantly increase the development time. One of the benefits of choosing a PC as platform is that a GPU is included *"at no cost"*. This is, most application requires some kind of graphical display so including a GP-capable GPU provides a much greater benefit with a very low cost/performance ratio, even using a low-cost card. The combined use of CPU-GPU has proved to be very effective, although very restricted in terms of form factor and specially in power consumption. Only if these parameters are very constrained DSP-based solutions are preferable. This is the case of mobile applications, although low-power microprocessors are taking advantage in this field. As these are extensively used, development kits, compilers and libraries are very optimized, helping to cut down time-to-market and related costs.

However, if the aforementioned devices do not provide acceptable results, it will be required to go into the hardware. As application developers, we need to search for *exotic* devices such as MPPAs or dedicated image processors. In contrast to the previous devices, these are not normally industry standards therefore a greater effort during development is needed. However, we are still under the software coverage. On the contrary, if the requirements are very strict or if the production volume is very high, a custom chip is the unique alternative to reach the desired performance or to lower the cost per unit. FPGAs are an excellent platform for testing before manufacturing the final design in a custom chip. As they are reconfigurable, different architectures can be evaluated before sending to fabric. On the other hand, some authors claim than software development is the bottleneck in the current ASIC development. The *software stage* can not start until a device where to do tests has been built. Although emulators (both functional an cycle-accurate) are used, their performance is very low, resulting in very large test cycles and poor feedback for those programmers at the hardware control layer. This is why FPGAs are widely used as proof-of-concept devices, as they enable software development cycles many months before a test chip was built.

Tab. 1 shows performance comparison of a computationally intensive algorithm as SURF (Bay et al., 2006) for different platforms. PCs offer uneven performance if heavily use multithreading programming (Zhang, 2010) or a straightforward implementation (Bouris et al., 2010). GPUs feature very large performance at the expense of high power consumption. However, FPGA-based implementation delivers the best performance in terms of speed and power consumption. Tab. 2 shows a comparison between a low-power CPU and GPU, compared to an standard laptop CPU. Authors conclude that optimization is critical and a carefully analysis of low-level operations must be performed but the achieved performance is quite close to standard conventional processors. Complex algorithms which include a higher abstraction level as Viola-Jones detector (Viola & Jones, 2001) are also candidate for mobile platforms. In (Aby et al., 2011), a Beagleboard xM board with a Texas Instruments DM3730 SoC (ARM Cortex-A8 and TMS320C64X DSP) achieves around $0.5x$ speed-up compared with a conventional Intel 2.2 GHz processor, both using openCV (Bradski & Kaehler, 2008) library. In (Arth & Bischof, 2008), a DSP-based embedded system for object recognition achieves up to 4fps, including SIFT-based feature detection and description and object recognition. Although with lower performance than other approaches, the major advantage is that the whole application fits in a single device reducing also power consumption.

Some operations, as the Fourier transform, are computationally very expensive and yet required in many applications, including low-power, low-cost or high performance devices. Optimized libraries for both CPU (Takahashi, 2007) and GPU (Ogata et al., 2008) aim to exploit SIMD units and multitasking, achieving high performance with regard to straightforward implementations. In particular, Fourier transform operation is very suitable for FPGAs an ASICs if performance and power consumption is critical. In (He & Guo, 2008), a FFT core design for FPGAs is proposed, consuming less than 1W in the worst case and lowering manufacturing cost more than $x15$ compared with a DSP implementation. In (Guan et al., 2009), a more aggressive approach is done, developing an *application-specific instruction set processor* (ASIP). With very little hardware overhead and a consumption of few tenths of a watt, outperforms standard software and DSP implementations more than $x800$ and $x5$ times respectively. However, these designs have larger development cycles, as (Pauwels et al., 2011) depicts. In this work, some low-level operations (phase-based optical flow, stereo and local image features) are compared both on FPGA and GPU. Tab. 3 summarizes some of the results of this work. As authors conclude, high-performance or low-cost implementations should be done on CPUs with GPU co-processing. GPUs overcome FPGAs in terms of absolute performance due to their memory throughput. However, if a standalone platform is needed, an FPGA board should meet the requirements or establish the basis for testing and validating an ASIC design.

Finally, general-purpose custom designs converge form factor, power consumption and performance. For instance, SCAMP processor (Dudek & Hicks, 2005) exploits the massively spatial parallelism of low-level operations, integrating processing units and sensors in a processor-per-pixel fashion. As it is an analog design, the integration density and the performance is very high, keeping power consumption under 240 mW. Current digital solutions also offer similar performance and many advantages as faster development and array scalability. ASPA processor (Lopich & Dudek, 2007) includes novel techniques to increase performance, specially on global operations, without sacrificing the other trade-offs. Hardware-oriented algorithms are also key to take advantage of custom hardware. Tab. 4 shows a comparative between different specific-purpose processors executing an active contour algorithm for retinal-vessel tree extraction. The algorithm is designed for focal-plane processing arrays. The *cycles-per-pixel* parameter illustrates how specific platforms are able to lower hardware requirements and even to increase performance. As well as other custom designs depicted in this chapter, they are not suitable to handle a whole Computer Vision application as they are intended to reduce the workload of the main processor in the more computational expensive tasks. Approaches as (Khailany et al., 2008b) or (Fijany & Hosseini, 2011) can completely embed highly complex applications without compromise its efficiency.

	Device	Performance	Power (W)
Bouris et al. (2010)	Intel Core 2 Duo 2.4 GHz	< 7 fps	N/S
Zhang (2010)	Intel Core 2 Duo P8600 2.4 GHz	33 fps	25
Terriberry et al. (2008)	nVidia GeForce 880 GTX	56 fps	200
Bouris et al. (2010)	Xilinx Virtex 5XC5VFX130T	70 fps	< 20

Table 1. Summary of different SURF (Bay et al., 2006) implementations on different platforms for images of 640 $\times$ 640 px.

Device	Un-optimized	Optimized
Intel Core 2 Duo Merom 2.4GHz	9.03 fps	16.37 fps
Intel Atom 1.6GHz	2.59 fps	5.48 fps
GMA X3100 GPU 500MHz	1.04 fps	5.75 fps

Table 2. Performance of SIFT (Lowe, 2004) implementations in low-power devices for images of 640 $\times$ 640 px. See (Murphy et al., 2009) for details.

Device	Power (W)	Cost ($)	Time-to-Market (months)
nVidia GeForce GTX 280	236	N.S.	2 (1 persons)
nVidia GeForce GTX 580	244	499	2 (1 persons)
Xilinx Virtex4 xc4vfx100	7.2	2084	15 (2 persons)
Xilinx Virtex5 xc5vlx330t	5.5	12651	12 (2 persons)

Table 3. Main GPU and FPGA costs for optical flow, stereo and local image features implementation. See (Pauwels et al., 2011) for complete details and performance results.

Device	Cores (used/aval.)	Performance (s)	Cycles-per-pixel
Intel Core i7 940 2.93GHz	4/4 (8 threads)	13.7	357993
SCAMP-3 Visual Processor	16384/16384	0.230	8950
Xilinx Spartan-3 sc3s4000-5	90/90	1.349	14070
Xilinx Virtex-6 xc6vlx240t-1	192/192	0.080	3425
Ambric Am2045 MPPA	125/360	0.0087	742

Table 4. On-chip retinal vessel-tree extraction (hardware-oriented algorithm) on different platforms. See (Nieto et al., 2011) for complete performance analysis. *Note: the algorithm was slightly adapted for SCAMP-3 and substantially lightened for Ambric Am2045. The FPGA implementation was carried out faithfully the original algorithm. Virtex-6 results are an estimation.*

Looking ahead

The progress of new technologies, marked by Moore's Law allows increasingly integration density. More hardware resources, with higher clock frequencies, are available for the designer. However, although the ultimate goal is to increase the performance, other parameters come into play. Nowadays, one of the critical trade-offs is power consumption, directly related with energy efficiency and power dissipation, some of the most decisive limitation design constraints (Kim et al., 2003).

Power consumption is driven by two sources, *dynamic* and *static*. Static consumption is a result of the leakage current and it refers when all inputs are held so the circuit is not changing state. On the contrary, the dynamic term refers to the circuit switching at a given frequency. This power is dominated in today CMOS circuits, being directly proportional to frequency. This is one of the capital reasons why the semiconductor industry moves from a *race for frequency* to a *race for parallelism*. In recent years, the industry is making a big effort to increase the parallelism of most devices to keep the performance increase rate. Apart from more arithmetic units, leading architectures integrate more systems previously contained on separated circuits, as microcontrollers or GPUs. To achieve these results, it is still necessary to scale down the transistors. In this sense, the advent of emerging technologies like CMOS-3D (Philip Garrou, 2008) will permit to integrate heterogeneous functions on the same monolithic solution more easily. A vision-oriented ASIC could integrate the image acquisition stage to an eventual processor. At the same time, more conventional solutions as PCs or FPGAs would yield large parallelization using this and other advances such as the Tri-Gate technology (Intel, 2011). However, this involves problems as the increment of leakage currents, thereby increasing static power consumption, not negligible at all nowadays (Koch, 2005) (Kim et al., 2003). In addition, new manufacturing methods are more expensive because the yield is lower and more time is needed to discount the investments. Or equivalently, it is necessary to sell more devices to continue growing at the rate set by Moore's Law.

Conventional microprocessors are in the leading edge of evolution. There is a large market which justifies large investments in R&D to meet the growing needs of consumers, especially by large increase in media consumption. This way, it is now possible to find low-cost multicore microprocessors. It is expected that the current evolution towards a greater number of cores will be maintained but increasingly including more elements previously located on external chips, reducing the bottleneck when communicating with off-chip elements (Singhal, 2008). New parallel computing techniques need to be developed to take advantage of the available multithreading capabilities.

PCs also benefit of GPU capabilities. GPU performance grows at a higher rate than microprocessors. As they are very specialized devices, although featuring general purpose computing, the technical improvements in the semiconductor industry are clearly more beneficial. As discussed previously, there are available hardware resources to increase the parallelism and enhance the datapath pipeline. Leading GPUs have more than 1000 processing units and high speed and bandwidth memories. It is also possible to combine multi-core GPUs to work together, achieving a very large throughput. Still, their major disadvantage is being the power consumption. New architectures are taking advantage of the fixed-function hardware to improve area usage and power efficiency. GPU design will focus entirely on improving GPGPU computing (Brookwood, 2010a).

DSPs are also moving to multicore architectures. As specialized microprocessors, they can take advantage of all the improvements in the consumer market, both in hardware and software improvements such as compilers or other optimization techniques. Although competitors are strong, DSP will continue to be used because they lead to compact circuit boards, lower power consumption and cost, if the appropiate device is selected based on the application requirements. In addition, the benefits of the extensive experience in DSP development, with shorter time-to-market thanks to the very optimized compilers and libraries. This is specially relevant in embedded applications, to take advantage of the multi-core capabilities of modern DSPs. This way it is possible to integrate several DSP cores, each one optimized for a specific task, on a single chip (Friedmann, 2010). Low-power devices which still keeping reasonable performance are fundamental in handled and portable devices, where traditional microprocessors are not suitable.

Microprocessors and GPUs tend to converge on a single chip. Apart from the obvious benefits of integration, reducing cost, size, power consumption, the performance will increase because the reduction of off-chip communications. In addition, architectures as AMD Fusion integrate in the same units 3D acceleration, parallel processing and other functions of GPUs (Brookwood, 2010b). On the other hand, mobile microprocessors are becoming more important. These microprocessors embed very low-power GPUs and auxiliary DSP units for co-processing in the same chip (NVIDIA, 2011).

Programmable systems, not only FPGAs, are able to get the same performance as recent past ASICs, keeping time-to-market and non-recurring engineering costs lower compared to custom ICs. As discussed previously, FPGAs are between software and hardware solutions. Modern FPGAs experienced a large increase in hardware resources, both in dedicated units and logic cells. High-level programming languages are another major reason why FPGAs are becoming increasingly competitive, specially when dealing with complex FPGAs and to maintain and keep the designs portable (Singh, 2011). Nowadays these devices can address complete SoCs, integrating memory and IO controller natively. Manufacturer roadmaps show their inclusion in a very near future and it is expected a big leap in performance and flexibility (DeHaven, 2010).

4. Summary and conclusions

The large variety of Computer Vision applications makes difficult to classify them into tight categories. As a result, it is extremely difficult to design a unique hardware architecture which handle efficiently all processing stages of any Computer Vision algorithm. In the literature there are available several studies where different platforms are tested under the same conditions (Asano et al. (2009); Baumgartner et al. (2009); Kisacanin (2005); Wnuk (2008)). They show that to tune-up is key for performance and that new parallel computing techniques are a requirement to exploit parallel devices. However, the increase of the market makes investment in new platforms that implement different algorithms a necessity.

The most accessible platform is a Personal Computer equipped with a GP-capable GPU. As test or final platform, it cuts down developing time and costs. GPUs give enough performance for most intensive tasks, while using the CPU multimedia extensions it is possible to meet the requirements in the other stages. In addition, they include all necessary elements for user IO, communication, storage and information display. The availability of models is large enough to select the adequate platform according to the application trade-offs. When CPU performance is not adequate, DSPs are a serious alternative. In addition, it becomes almost mandatory when dealing with embedded devices without compromising performance, where power consumption and form factor are very restrictive. They are widely used for prototyping custom ICs but FPGA-based applications have their own niche. Integration and high flexibility besides a large number of available IP Cores allow to drop NRE costs. Although all devices described in this chapter are ASICs, they were not conceived for an unique application. To lower costs, the manufacturer expands their range of application although it is possible to find families specialized in specific tasks. But there are available devices very specific for critical tasks, where the requirements are very tight and any other device complies with them. Flexibility is complete and there is not restriction to employ cutting-edge technologies which are not available in commercial devices until a near future.

Almost all Computer Vision applications need to face all processing stages in a lesser or a greater degree. Generally, this leads to implement efficient mechanisms to tackle massively spatial parallelism, mixed spatial and temporal parallelism and sequential processing. Each stage matches with a level of processing so all mechanisms have to be implemented in most applications. Low-level stages benefit of massively parallelism with simple data distribution systems as operations. When the data abstraction level grows, during mid-level tasks, more information about the problem is required by the algorithms increasing their complexity. This leads to complex architectures, where information distribution and sharing makes difficult to exploit spatial parallelism, although it is usually present. Task-parallel architectures are able to exploit better their possibilities. Low and mid-level processing stages can be implemented in pure hardware solutions because they often implement kernel operations. However, high-level is closer to software and designers can take advantage of this to build complex systems easier by using general purpose processors. In addition, the device which performs the image processing related tasks needs to communicate or to control other devices. This is not strictly related with the Computer Vision domain but it is clearly a requirement in the final solution. In this case, the use of a general purpose processor is beneficial because it allows easier control and it increases the flexibility of the whole system.

Although it is almost impossible to develop a system able to run all operations in an optimal way due to the rich nature of the Computer Vision applications, it is desirable to provide the capability to perform any operation. The design must be scalable to adapt it to the specific needs of each application. This way, a product ranging from low to high-end devices can be easily built. The internal architecture should be also modular, so that from a basic outline more features could be added without dramatic changes. In general, a high-end microprocessor is a requirement to manage complex operations and communications between the system and the external components of the complete system. An on-chip SIMD-MIMD hybrid co-processor would tackle the most expensive computation, reconfiguring its internal interconnections according to the current task. Embedded high-speed memory controllers are also key to reduce the data-access bottleneck. All these elements, together, are able to face efficiently most of the situations described throughout this chapter.

5. Acknowledgments

This work is funded by Xunta de Galicia under the projects 10PXIB206168PR and 10PXIB206037PR and the program Maria Barbeito.

6. References

Aby, P., Jose, A., Jose, B., Dinu, L., John, J. & Sabarinath, G. (2011). Implementation and optimization of embedded face detection system, *Signal Processing, Communication,*

Computing and Networking Technologies (ICSCCN), 2011 International Conference on, IEEE, pp. 250–253.

Ahmed, A. & Terada, K. (2010). A general framework for multi-human tracking, *Journal of Software* 5(9): 966–973.

Arth, C. & Bischof, H. (2008). Real-time object recognition using local features on a dsp-based embedded system, *Journal of Real-Time Image Processing* 3(4): 233–253.

Asano, S., Maruyama, T. & Yamaguchi, Y. (2009). Performance comparison of fpga, gpu and cpu in image processing, *Field Programmable Logic and Applications, 2009. FPL 2009. International Conference on*, IEEE, pp. 126–131.

Baumgartner, D., Roessler, P., Kubinger, W., Zinner, C. & Ambrosch, K. (2009). Benchmarks of low-level vision algorithms for dsp, fpga, and mobile pc processors, *Embedded Computer Vision* pp. 101–120.

Bay, H., Tuytelaars, T. & Van Gool, L. (2006). Surf: Speeded up robust features, *Computer Vision–ECCV 2006* pp. 404–417.

BCC Research (2010). Machine Vision: Technologies and Global Markets, *Report IAS010C*, BCC Research.

Bell, S., Edwards, B., Amann, J., Conlin, R., Joyce, K., Leung, V., MacKay, J., Reif, M., Bao, L., Brown, J. et al. (2008). Tile64-processor: A 64-core soc with mesh interconnect, *Solid-State Circuits Conference, 2008. ISSCC 2008. Digest of Technical Papers. IEEE International*, IEEE, pp. 88–598.

Bouris, D., Nikitakis, A. & Papaefstathiou, I. (2010). Fast and efficient fpga-based feature detection employing the surf algorithm, *2010 18th IEEE Annual International Symposium on Field-Programmable Custom Computing Machines*, IEEE, pp. 3–10.

Bradski, G. & Kaehler, A. (2008). *Learning OpenCV*, O'Reilly.

Bramberger, M. & Rinner, B. (2004). An embedded smart camera on a scalable heterogeneous multi-dsp system, *In Proceedings of the European DSP Education and Research Symposium (EDERS 2004*, Citeseer.

Brookwood, N. (2010a). Amd fusion family of apus – enabling a superior, immersive pc experience, *AMD white paper* .

Brookwood, N. (2010b). Amd fusion family of apus – enabling a superior, immersive pc experience, *AMD white paper* .

Butts, M., Jones, A. & Wasson, P. (2007). A structural object programming model, architecture, chip and tools for reconfigurable computing, *Field-Programmable Custom Computing Machines, 2007. FCCM 2007. 15th Annual IEEE Symposium on*, IEEE, pp. 55–64.

Castano-Díez, D., Moser, D., Schoenegger, A., Pruggnaller, S. & Frangakis, A. S. (2008). Performance evaluation of image processing algorithms on the GPU, *Journal of Structural Biology* 164(1): 153 - 160.

Chapman, B. (2007). The multicore programming challenge, *Lecture Notes in Computer Science* 4847: 3.

Che, S., Boyer, M., Meng, J., Tarjan, D., Sheaffer, J. & Skadron, K. (2008). A performance study of general-purpose applications on graphics processors using cuda, *Journal of Parallel and Distributed Computing* 68(10): 1370–1380.

Chen, W., Chen, P., Lee, W. & Huang, C. (2008). Design and implementation of a real time video surveillance system with wireless sensor networks, *Vehicular Technology Conference, 2008. VTC Spring 2008. IEEE*, IEEE, pp. 218–222.

Chu, C. & Chen, S. (2009). Parallel implementation for cone beam based 3d computed tomography (ct) medical image reconstruction on multi-core processors, *World Congress on Medical Physics and Biomedical Engineering, September 7-12, 2009, Munich, Germany*, Springer, pp. 2066–2069.

Constandinou, T., Georgiou, J. & Toumazou, C. (2004). Towards a bio-inspired mixed-signal retinal processor, *Circuits and Systems, 2004. ISCAS'04. Proceedings of the 2004 International Symposium on*, Vol. 5, IEEE, pp. V–493.

Coussy, P., Takach, A., McNamara, M. & Meredith, M. (2010). An introduction to the systemc synthesis subset standard, *Proceedings of the eighth IEEE/ACM/IFIP international conference on Hardware/software codesign and system synthesis*, ACM, pp. 183–184.

Craven, S. & Athanas, P. (2007). Examining the viability of fpga supercomputing, *EURASIP Journal on Embedded systems* 2007(1): 13–13.

DeHaven, K. (2010). Extensible processing platform ideal solution for a wide range of embedded systems, *Xilinx White Paper* .

Deselaers, T., Keysers, D. & Ney, H. (2008). Features for image retrieval: an experimental comparison, *Information Retrieval* 11(2): 77–107.

Djordjevic, I., Arabaci, M. & Minkov, L. (2009). Next generation fec for high-capacity communication in optical transport networks, *Journal of Lightwave Technology* 27(16): 3518–3530.

Dolan, R. & DeSouza, G. (2009). Gpu-based simulation of cellular neural networks for image processing, *Proceedings of the 2009 international joint conference on Neural Networks*, IEEE Press, pp. 2712–2717.

Dudek, P. & Hicks, P. (2005). A general-purpose processor-per-pixel analog simd vision chip, *Circuits and Systems I: Regular Papers, IEEE Transactions on* 52(1): 13–20.

Duller, A., Panesar, G. & Towner, D. (2003). Parallel processing-the picochip way, *Communicating Processing Architectures* pp. 125–138.

Fasang, P. (2009). Prototyping for industrial applications [industry forum], *Industrial Electronics Magazine, IEEE* 3(1): 4–7.

Fijany, A. & Hosseini, F. (2011). Image processing applications on a low power highly parallel simd architecture, *Aerospace Conference, 2011 IEEE*, IEEE, pp. 1–12.

Flynn, M. (1972). Some computer organizations and their effectiveness, *Computers, IEEE Transactions on* 100(9): 948–960.

Foldesy, P., Zarándy, Á. & Rekeczky, C. (2008). Configurable 3d-integrated focal-plane cellular sensor–processor array architecture, *International Journal of Circuit Theory and Applications* 36(5-6): 573–588.

Franchetti, F., Kral, S., Lorenz, J. & Ueberhuber, C. (2005). Efficient utilization of simd extensions, *Proceedings of the IEEE* 93(2): 409–425.

Franchini, S., Gentile, A., Sorbello, F., Vassallo, G. & Vitabile, S. (2009). An embedded, fpga-based computer graphics coprocessor with native geometric algebra support, *Integration, the VLSI Journal* 42(3): 346–355.

Friedmann, A. (2010). Enabling small cells with tiâ€™s new multicore soc, *Texas Instruments White Paper* .

Garrido, S., Listán, J., Alba, L., Utrera, C., Rodríguez-Vázquez, S., Domínguez-Castro, R., Jiménez-Espejo, F. & Romay, R. (2008). The Eye-RIS CMOS Vision System, *Analog circuit design: sensors, actuators and power drivers; integrated power amplifiers from wireline to RF; very high frequency front ends* pp. 15–32.

Govindaraju, N. & Manocha, D. (2007). Cache-efficient numerical algorithms using graphics hardware, *Parallel Computing* 33(10-11): 663–684.

Guan, X., Lin, H. & Fei, Y. (2009). Design of an application-specific instruction set processor for high-throughput and scalable fft, *Proceedings of the Conference on Design, Automation and Test in Europe*, pp. 1302–1307.

Haralick, R. M. & Shapiro, L. G. (1985). Image segmentation techniques, *Computer Vision, Graphics, and Image Processing* 29(1): 100 – 132.

Harris, C. & Stephens, M. (1988). A combined corner and edge detector, *Alvey vision conference*, Vol. 15, Manchester, UK, p. 50.

He, H. & Guo, H. (2008). The realization of fft algorithm based on fpga co-processor, *Second International Symposium on Intelligent Information Technology Application*, IEEE, pp. 239–243.

Heymann, S., Maller, K., Smolic, A., Froehlich, B. & Wiegand, T. (2007). Sift implementation and optimization for general-purpose gpu, *Proceedings of the International Conference in Central Europe on Computer Graphics, Visualization and Computer Vision*, Citeseer.

Intel (2011). Intel 22nm 3-d tri-gate transistor technology, *Intel Documentation* .

Jin, S., Cho, J., Dai Pham, X., Lee, K., Park, S., Kim, M. & Jeon, J. (2010). Fpga design and implementation of a real-time stereo vision system, *Circuits and Systems for Video Technology, IEEE Transactions on* 20(1): 15–26.

Khailany, B., Williams, T., Lin, J., Long, E., Rygh, M., Tovey, D. & Dally, W. (2008a). A programmable 512 gops stream processor for signal, image, and video processing, *Solid-State Circuits, IEEE Journal of* 43(1): 202 –213.

Khailany, B., Williams, T., Lin, J., Long, E., Rygh, M., Tovey, D. & Dally, W. (2008b). A programmable 512 gops stream processor for signal, image, and video processing, *Solid-State Circuits, IEEE Journal of* 43(1): 202–213.

Kim, K., Kim, J.-Y., Lee, S., Kim, M. & Yoo, H.-J. (2008). A 211 gops/w dual-mode real-time object recognition processor with network-on-chip, *Solid-State Circuits Conference, 2008. ESSCIRC 2008. 34th European*, pp. 462 –465.

Kim, N., Austin, T., Baauw, D., Mudge, T., Flautner, K., Hu, J., Irwin, M., Kandemir, M. & Narayanan, V. (2003). Leakage current: Moore's law meets static power, *Computer* 36(12): 68–75.

Kisacanin, B. (2005). Examples of low-level computer vision on media processors, *Computer Vision and Pattern Recognition-Workshops, 2005. CVPR Workshops. IEEE Computer Society Conference on*, IEEE, pp. 135–135.

Koch, G. (2005). Discovering multi-core: extending the benefits of mooreâ€™s law, *Technology* 1.

Kolsch, M. & Butner, S. (2009). Hardware Considerations for Embedded Vision Systems, *Embedded Computer Vision, Springer* pp. 3–26.

Koyanagi, M., Nakagawa, Y., Lee, K., Nakamura, T., Yamada, Y., Inamura, K., Park, K. & Kurino, H. (2001). Neuromorphic vision chip fabricated using three-dimensional integration technology, *Solid-State Circuits Conference, 2001. Digest of Technical Papers. ISSCC. 2001 IEEE International*, IEEE, pp. 270–271.

Kuon, I. & Rose, J. (2007). Measuring the gap between fpgas and asics, *Computer-Aided Design of Integrated Circuits and Systems, IEEE Transactions on* 26(2): 203 –215.

Lee, V., Kim, C., Chhugani, J., Deisher, M., Kim, D., Nguyen, A., Satish, N., Smelyanskiy, M., Chennupaty, S., Hammarlund, P. et al. (2010). Debunking the 100x gpu vs. cpu myth:

an evaluation of throughput computing on cpu and gpu, *ACM SIGARCH Computer Architecture News*, Vol. 38, ACM, pp. 451–460.

Leong, P. (2008). Recent trends in fpga architectures and applications, *Electronic Design, Test and Applications, 2008. DELTA 2008. 4th IEEE International Symposium on*, IEEE, pp. 137–141.

Li, L., Socher, R. & Fei-Fei, L. (2009). Towards total scene understanding: Classification, annotation and segmentation in an automatic framework, *Computer Vision and Pattern Recognition, 2009. CVPR 2009. IEEE Conference on*, IEEE, pp. 2036–2043.

Lin, C. & Chiu, Y. (2008). The dsp based catcher robot system with stereo vision, *Advanced Intelligent Mechatronics, 2008. AIM 2008. IEEE/ASME International Conference on*, IEEE, pp. 897–903.

Lin, T., Liu, C., Tseng, S., Chu, Y. & Wu, A. (2008). Overview of itri pac project-from vliw dsp processor to multicore computing platform, *VLSI Design, Automation and Test, 2008. VLSI-DAT 2008. IEEE International Symposium on*, IEEE, pp. 188–191.

Lindholm, E., Nickolls, J., Oberman, S. & Montrym, J. (2008). Nvidia tesla: A unified graphics and computing architecture, *Micro, IEEE* 28(2): 39–55.

Lopich, A. & Dudek, P. (2007). Global operations in simd cellular processor arrays employing functional asynchronism, *Computer Architecture for Machine Perception and Sensing, 2006. CAMP 2006. International Workshop on*, IEEE, pp. 18–23.

Lopich, A. & Dudek, P. (2008). Aspa: Focal plane digital processor array with asynchronous processing capabilities, *Circuits and Systems, 2008. ISCAS 2008. IEEE International Symposium on*, IEEE, pp. 1592–1595.

Lowe, D. (2004). Distinctive image features from scale-invariant keypoints, *International journal of computer vision* 60(2): 91–110.

Lui, W. & Jarvis, R. (2010). Eye-full tower: A gpu-based variable multibaseline omnidirectional stereovision system with automatic baseline selection for outdoor mobile robot navigation, *Robotics and Autonomous Systems* 58(6): 747–761.

Lysecky, R. & Vahid, F. (2005). A study of the speedups and competitiveness of fpga soft processor cores using dynamic hardware/software partitioning, *Proceedings of the conference on Design, Automation and Test in Europe-Volume 1*, IEEE Computer Society, pp. 18–23.

Martineau, M., Wei, Z., Lee, D. & Martineau, M. (2007). A fast and accurate tensor-based optical flow algorithm implemented in fpga, *Applications of Computer Vision, 2007. WACV'07. IEEE Workshop on*, IEEE, pp. 18–18.

Meier, L., Tanskanen, P., Fraundorfer, F. & Pollefeys, M. (2011). Pixhawk: A system for autonomous flight using onboard computer vision, *Robotics and Automation (ICRA), 2011 IEEE International Conference on*, IEEE, pp. 2992–2997.

Meng, H., Appiah, K., Hunter, A. & Dickinson, P. (2011). Fpga implementation of naive bayes classifier for visual object recognition, IEEE Computer Vision and Pattern Recognition.

Mikolajczyk, K. & Schmid, C. (2005). A performance evaluation of local descriptors, *IEEE transactions on pattern analysis and machine intelligence* pp. 1615–1630.

Moreno, F., Lopez, I. & Sanz, R. (2010). A design process for hardware–software system co-design and its application to designing a reconfigurable fpga, *Digital System Design - Architectures, Methods and Tools (DSD), 2010 13th Euromicro Conference on*, IEEE, pp. 556–562.

Murphy, M., Keutzer, K. & Wang, H. (2009). Image feature extraction for mobile processors, *Workload Characterization, 2009. IISWC 2009. IEEE International Symposium on*, IEEE, pp. 138–147.

Naffziger, S. (2009). Microprocessors of the future: Commodity or engine growth?, *Solid-State Circuits Magazine, IEEE* 1(1): 76 –82.

Nair, V., Laprise, P. & Clark, J. (2005). An fpga-based people detection system, *EURASIP journal on applied signal processing* 2005: 1047–1061.

Neri, C., Baccarelli, G., Bertazzoni, S., Pollastrone, F. & Salmeri, M. (2005). Parallel hardware implementation of radar electronics equipment for a laser inspection system, *Nuclear Science, IEEE Transactions on* 52(6): 2741 –2748.

Nieto, A., Brea, V., Vilarino, D. & Osorio, R. (2011). Performance analysis of massively parallel embedded hardware architectures for retinal image processing, *EURASIP Journal on Image and Video Processing* 2011: 10.

NVIDIA (2011). Variable smp – a multi-core cpu architecture for low power and high performance, *NVIDIA Corporation white paper* .

Ogata, Y., Endo, T., Maruyama, N. & Matsuoka, S. (2008). An efficient, model-based cpu-gpu heterogeneous fft library, *Parallel and Distributed Processing, 2008. IPDPS 2008. IEEE International Symposium on*, IEEE, pp. 1–10.

Oliver, N., Rosario, B. & Pentland, A. (2000). A bayesian computer vision system for modeling human interactions, *Pattern Analysis and Machine Intelligence, IEEE Transactions on* 22(8): 831 –843.

Owens, J. D., Luebke, D., Govindaraju, N., Harris, M., KrALger, J., Lefohn, A. E. & Purcell, T. J. (2007). A survey of general-purpose computation on graphics hardware, *Computer Graphics Forum* 26(1): 80–113.

Pal, N. R. & Pal, S. K. (1993). A review on image segmentation techniques., *Pattern Recognition* pp. 1277–1294.

Pauwels, K., Tomasi, M., Alonso, J., Ros, E. & Van Hulle, M. (2011). A comparison of fpga and gpu for real-time phase-based optical flow, stereo, and local image features, *IEEE Transactions on Computers* .

Philip Garrou, Chistopher Bower, P. R. (2008). *Handbook of 3D Integration*, Wiley-VCH.

Podlozhnyuk, V. (2007). Image convolution with cuda, *NVIDIA Corporation white paper, June* 2097(3).

Ren, F., Huang, J., Terauchi, M., Jiang, R. & Klette, R. (2010). Lane detection on the iphone, *Arts and Technology* pp. 198–205.

Rinnerthaler, F., Kubinger, W., Langer, J., Humenberger, M. & Borbély, S. (2007). Boosting the performance of embedded vision systems using a dsp/fpga co-processor system, *Systems, Man and Cybernetics, 2007. ISIC. IEEE International Conference on*, IEEE, pp. 1141–1146.

Rixner, S., Dally, W., Kapasi, U., Khailany, B., López-Lagunas, A., Mattson, P. & Owens, J. (1998). A bandwidth-efficient architecture for media processing, *Proceedings of the 31st annual ACM/IEEE international symposium on Microarchitecture*, IEEE Computer Society Press, pp. 3–13.

Salem, M., Klaus, K., Winkler, F. & Meffert, B. (2009). Resolution mosaic-based smart camera for video surveillance, *Distributed Smart Cameras, 2009. ICDSC 2009. Third ACM/IEEE International Conference on*, IEEE, pp. 1–7.

Schneiderman, R. (2010). Dsps evolving in consumer electronics applications [special reports], *Signal Processing Magazine, IEEE* 27(3): 6–10.

Seiler, L., Carmean, D., Sprangle, E., Forsyth, T., Abrash, M., Dubey, P., Junkins, S., Lake, A., Sugerman, J., Cavin, R. et al. (2008). Larrabee: a many-core x86 architecture for visual computing, *ACM Transactions on Graphics (TOG)* 27(3): 1–15.

Shah, S., Khattak, T., Farooq, M., Khawaja, Y., Bais, A., Anees, A. & Khan, M. (2008). Real time object tracking in a video sequence using a fixed point dsp, *Advances in Visual Computing* pp. 879–888.

Shi, Y. (2010). Smart cameras for machine vision, *Smart Cameras* pp. 283–303.

Shirvaikar, M. & Bushnaq, T. (2009). A comparison between DSP and FPGA platforms for real-time imaging applications, Vol. 7244, SPIE, p. 724406.

Singh, D. (2011). Higher level programming abstractions for fpgas using opencl, *ALTERA* .

Singhal, R. (2008). Inside intel® next generation nehalem microarchitecture, *Hot Chips*, Vol. 20.

Smith, S. M. & Brady, J. M. (1997). Susanâ€š a new approach to low level image processing, *International Journal of Computer Vision* 23: 45–78.

Stein, G., Rushinek, E., Hayun, G. & Shashua, A. (2005). A computer vision system on a chip: a case study from the automotive domain, *Computer Vision and Pattern Recognition - Workshops, 2005. CVPR Workshops. IEEE Computer Society Conference on*, p. 130.

Sun, T. & Yu, Y. (2009). Memory usage reduction method for fft implementations on dsp based embedded system, *Consumer Electronics, 2009. ISCE'09. IEEE 13th International Symposium on*, IEEE, pp. 812–815.

Suzuki, K., Ikeda, H., Ishimaru, K., Suzuki, J., Adachi, F. & Wang, X. (2007). New image retrieval system utilizing image directory on gigabit network for distributing industrial product information, *Industry Applications, IEEE Transactions on* 43(4): 1099 –1107.

Szeliski, R. (2010). *Computer vision: Algorithms and applications*, Springer-Verlag New York Inc.

Takahashi, D. (2007). Implementation and evaluation of parallel fft using simd instructions on multi-core processors, *Innovative architecture for future generation high-performance processors and systems, 2007. iwia 2007. international workshop on*, IEEE, pp. 53–59.

Talavera, G., Jayapala, M., Carrabina, J. & Catthoor, F. (2008). Address generation optimization for embedded high-performance processors: A survey, *Journal of Signal Processing Systems* 53(3): 271–284.

Taylor, S., Rosten, E. & Drummond, T. (2009). Robust feature matching in 2.3μs, *IEEE CVPR Workshop on Feature Detectors and Descriptors: The State Of The Art and Beyond*.

Terriberry, T., French, L. & Helmsen, J. (2008). Gpu accelerating speeded-up robust features, *Proc. Int. Symp. on 3D Data Processing, Visualization and Transmission (3DPVT)*, Citeseer, pp. 355–362.

Texas Instruments (2002). Tms320c6000 programmer's guide, *White Paper* .

Thall, A. (2006). Extended-precision floating-point numbers for gpu computation, *ACM SIGGRAPH 2006 Research posters*, ACM, pp. 52–es.

Trajkovi, M. & Hedley, M. (1998). Fast corner detection, *Image and Vision Computing* 16(2): 75–87.

Triggs, B., McLauchlan, P., Hartley, R. & Fitzgibbon, A. (2000). Bundle adjustment: A modern synthesis, *in* B. Triggs, A. Zisserman & R. Szeliski (eds), *Vision Algorithms: Theory and Practice*, Vol. 1883 of *Lecture Notes in Computer Science*, Springer Berlin / Heidelberg, pp. 153–177.

Urmson, C., Anhalt, J., Bagnell, D., Baker, C., Bittner, R., Clark, M., Dolan, J., Duggins, D., Galatali, T., Geyer, C. et al. (2008). Autonomous driving in urban environments: Boss and the urban challenge, *Journal of Field Robotics* 25(8): 425–466.

Veale, B., Antonio, J., Tull, M. & Jones, S. (2006). Selection of instruction set extensions for an fpga embedded processor core, *Parallel and Distributed Processing Symposium, 2006. IPDPS 2006. 20th International*, p. 8 pp.

Viola, P. & Jones, M. (2001). Rapid object detection using a boosted cascade of simple features, *Computer Vision and Pattern Recognition, IEEE Computer Society Conference on* 1: 511.

Wang, M., Wang, Y., Liu, D., Qin, Z. & Shao, Z. (2010). Compiler-assisted leakage-aware loop scheduling for embedded vliw dsp processors, *Journal of Systems and Software* 83(5): 772–785.

Wither, J., Tsai, Y. & Azuma, R. (2011). Mobile augmented reality: Indirect augmented reality, *Computers and Graphics* 35(4): 810–822.

Wnuk, M. (2008). Remarks on hardware implementation of image processing algorithms, *International Journal of Applied Mathematics and Computer Science* 18(1): 105–110.

Zarándy, Ã. (2011). *Focal-plane sensor-processor chips*, Springer Verlag.

Zhang, N. (2010). Computing optimised parallel speeded-up robust features (p-surf) on multi-core processors, *International Journal of Parallel Programming* 38(2): 138–158.

Zitova, B. & Flusser, J. (2003). Image registration methods: a survey, *Image and Vision Computing* 21(11): 977 – 1000.